World Soils Book Series

Series Editor

Alfred E. Hartemink
Department of Soil Science, FD Hole Soils Laboratory
University of Wisconsin–Madison
Madison, WI
USA

The World Soils Book Series publishes peer-reviewed books on the soils of a particular country. They include sections on soil research history, climate, geology, geomorphology, major soil types, soil maps, soil properties, soil classification, soil fertility, land use and vegetation, soil management, soils and humans, soils and industry, future soil issues. The books summarize what is known about the soils in a particular country in a concise and highly reader-friendly way. The series contains both single and multi-authored books as well as edited volumes. There is additional scope for regional studies within the series, particularly when covering large land masses (for example, The Soils of Texas, The Soils of California), however, these will be assessed on an individual basis.

Delaney Johnson · Mike Lilly · Birl Lowery ·
James G. Bockheim

The Soils of Mississippi

Springer

Delaney Johnson
Vicksburg, MS, USA

Mike Lilly
Brandon, MS, USA

Birl Lowery
University of Wisconsin–Madison
Madison, WI, USA

James G. Bockheim
Department of Soil Science
University of Wisconsin–Madison
Madison, WI, USA

ISSN 2211-1255 ISSN 2211-1263 (electronic)
World Soils Book Series
ISBN 978-3-031-36234-7 ISBN 978-3-031-36235-4 (eBook)
https://doi.org/10.1007/978-3-031-36235-4

This Springer imprint is published by the registered company Springer Nature Switzerland AG
The registered company address is: Gewerbestrasse 11, 6330 Cham, Switzerland

Paper in this product is recyclable.

This book is dedicated to the professional soil scientists from the Natural Resources Conservation Service and US Forest Service (US Department of Agriculture), soil and water conservations districts, and Mississippi State University Agricultural Experiment Station, who contributed to the mapping of soils in Mississippi. We acknowledge the support of these agencies and other land management agencies that have contributed to soil surveys. This book could not have been written without the support of NRCS database managers.

The NRCS served as the lead agency in mapping the soils of Mississippi. This organization began in 1899 as the Division of Soils, became the Bureau of Soils in 1901, the Soil Conservation Service in 1935, and the Natural Resources Conservation Service in 1994. The report that follows draws primarily on information gathered from more than 125 archived soil surveys with PDF online in Mississippi, as well as more recent data contained in the Web Soil Survey.

Preface

This book originated from data collected by the USDA Natural Resources Conservation Service, including the Soils Data Mart and Web Soil Survey, and by more than 100 research reports dealing with the soils, geology, and vegetation of Mississippi. The interpretations were made solely by the authors. This book should be of interest to individuals in federal, state, county governments, and nongovernment organizations who are interested in and responsible for safeguarding Mississippi's natural resources. The book will also be of interest to students in soil science and allied disciplines.

Vicksburg, USA	Delaney Johnson
Brandon USA	Mike Lilly
Madison, USA	Birl Lowery
Madison, USA	James G. Bockheim

Foreword

When one hears the word Mississippi, many images come to mind from the mighty river that bisects the United States to beaches along the Gulf Coast of the US to cotton fields to Delta Blues. For me, Mississippi represents soils. My first job after attaining my degree was spent at the USDA-Agricultural Research Service, National Sedimentation Laboratory in Oxford Mississippi. It was during those three years that I learned to appreciate and marvel at the diversity of the soils, landscape, and land use of Mississippi. This book provides the details to ensure the reader will gain the appreciation of Mississippi that it took me time to attain.

This book sets the stage for enlightenment by providing the reader with a background of the state. These are the facts and figures starting with the origin of the name Mississippi though its geography which is critical to understanding the soils of the state. For a state with only about 250 m (820 ft) of elevation change, it has a diverse landscape from the nearly flat and highly agriculturally productive Mississippi Delta (aka Mississippi Flood Plain) to the adjacent parallel region of the Loess Bluffs or Loess Hills, to the Blackland prairies and Pontotoc Ridge within the Coastal Plain area leading to the low-lying Gulf Coast land areas on the southern border of the state.

The history of the state, its demographics, and culture are discussed with an eye toward how the soils, landscape, and climate informed each other. This information leads the reader to discussion of first the general soil systems or regions within the state and then to details about the specific soils groups and taxonomic units. Included in this discussion are details about the benchmark soils, rare soils, endemic, and endangered soils in Mississippi.

The soil diversity in Mississippi is large. There are 8 soil orders, 17 suborders, 42 great groups, 98 subgroups, 187 families, and 232 soil series. These range for older Ultisols over much of the state to young Entisols along the coast and river/stream systems. Many of the soils are highly productive with nearly 45% being consider a prime farmland (Land Capability Class I and II). Not all this land is being used for agriculture as the state is over 60% forested. By describing the diversity of land use, this book provides a complete view of the state and its soils.

As I started my job with the National Sedimentation Laboratory, I was presented with E. W. Hilgard's 1860 Geology of Mississippi. This report established a valued and long history of soil investigations within the state. This rich history followed Hilgard to California where his work was formative for Hans Jenny who considered Hilgard as "the father of modern soil science in the United States." Hilgard's publication led to numerous soil surveys starting in 1901 by the former Bureau of Soils in Mississippi with over 100 archived soil surveys have been conducted. Had I also had access to this book my understanding of this soils and landscapes of Mississippi would have been even more enhanced. For all those who are interested in the natural history of the State of Mississippi or of the Gulf Coast in general, this book is a great addition to your library.

Raleigh, USA

David L Lindbo, Ph.D.
Director, Soil and Plant Science Division, USDA-NRCS; Former
President Soil Science of America; Emeritus Professor Soil Science
North Carolina State University; Certified Professional Soil Scientist

Contents

1 Introduction 1
1.1 Etymology 1
1.2 Geography 1
1.3 Demographics 1
1.4 Brief History 1
1.5 Economy 2
1.6 Culture 3
1.7 Education 4
1.8 Recreation 4
1.9 The Subsequent Chapters 5
1.10 Summary 5
References 5

2 History of Soil Studies in Mississippi 7
2.1 Introduction 7
2.2 Early Soil Books 7
2.3 Definition of Soil 7
2.4 Soil Surveys 8
2.5 Soil Series 10
2.6 Soil Classification 11
2.7 Soil Taxonomy 11
2.8 General Soil Maps 14
2.9 Soil Research 17
2.10 The State Soil 17
2.11 Summary 17
References 20

3 Soil-Forming Factors 21
3.1 Introduction 21
3.2 Climate 21
3.2.1 Current Climate 21
3.2.2 Past Climates 22
3.2.3 Recent Climate Change 22
3.3 Organisms 22
3.3.1 Loblolly Pine-Shortleaf Pine 23
3.3.2 Oak-Hickory 24
3.3.3 Oak-Gum-Cypress 24
3.3.4 Oak-Pine 24
3.3.5 Longleaf Pine-Slash Pine 25
3.3.6 Elm-Ash-Cottonwood 25
3.3.7 Other Vegetation Types 26
3.3.8 Ecoregions of Mississippi 26

3.4 Relief 29
3.5 Physiographic Provinces 29
3.6 Geologic Structure 29
3.7 Surficial Geology 31
3.8 Time 32
3.9 Humans 32
3.10 Summary 32
References 37

4 General Soil Regions of Mississippi 39
4.1 Introduction 39
4.2 Gulf Coastal Plain (MLRA 133C) 39
4.3 Southern Mississippi Valley Loess (MLRA 134) 40
4.4 Southern Mississippi Valley Alluvium (MLRA 131A) 47
4.5 Alabama and Mississippi Blackland Prairie (MLRA 135A) 51
4.6 Eastern Gulf Coast Flatwoods (MLRA 152A) 53
4.7 Gulf Coast Marsh (MLRA 151) 56
4.8 Summary 56
References 57

5 Diagnostic Horizons and Taxonomic Structure of Mississippi Soils 59
5.1 Introduction 59
5.2 Diagnostic Horizons 59
5.3 Orders 59
5.4 Suborders 59
5.5 Great Groups 60
5.6 Subgroups 60
5.7 Families 61
5.8 Soil Series 63
5.9 Summary 65
Reference 65

6 Taxonomic Soil Regions of Mississippi 67
6.1 Introduction 67
6.2 Fragiudalfs (Soil Region 1) 67
6.3 Paleudults (Soil Region 2) 67
6.4 Hapludults (Soil Region 3) 70
6.5 Hapludalfs (Soil Region 4) 72
6.6 Fragiudults (Soil Region 5) 73
6.7 Endoaquepts (Soil Region 6) 75
6.8 Endoaqualfs (Soil Region 7) 78
6.9 Paleudalfs (Soil Region 8) 79
6.10 Dystrudepts (Soil Region 9) 82
6.11 Fluvaquents (Soil Region 10) 83
6.12 Epiaquerts (Soil Region 11) 85
6.13 Dystraquerts (Soil Region 12) 87
6.14 Udifluvents (Soil Region 13) 88
6.15 Eutrudepts (Soil Region 14) 90
6.16 Dystruderts (Soil Region 15) 90
6.17 Epiaquepts (Soil Region 16) 91
6.18 Paleaquults (Soil Region 17) 91
6.19 Fraglossudalfs (Soil Region 18) 94
6.20 Hapluderts (Soil Region 19) 97
6.21 Summary 97

7 Ultisols ... 109
7.1 Distribution ... 109
7.2 Properties and Processes ... 109
7.3 Use and Management ... 110
7.4 Summary ... 110
References ... 112

8 Alfisols ... 113
8.1 Distribution ... 113
8.2 Properties and Processes ... 113
8.3 Use and Management ... 115
8.4 Summary ... 115
References ... 119

9 Inceptisols ... 121
9.1 Distribution ... 121
9.2 Properties and Processes ... 121
9.3 Use and Management ... 121
9.4 Summary ... 121

10 Vertisols ... 125
10.1 Distribution ... 125
10.2 Properties and Processes ... 125
10.3 Use and Management ... 126
10.4 Conclusions ... 126
References ... 127

11 Entisols ... 129
11.1 Distribution ... 129
11.2 Properties and Processes ... 129
11.3 Use and Management ... 129
11.4 Summary ... 131

12 Histosols, Mollisols, and Spodosol ... 133
12.1 Distribution ... 133
12.2 Properties and Processes ... 133
12.3 Use and Management ... 133
12.4 Summary ... 134
Reference ... 135

13 Soil-Forming Processes in Mississippi ... 137
13.1 Introduction ... 137
13.2 Argilluviation ... 137
13.3 Gleization ... 137
13.4 Base Cycling ... 137
13.5 Cambisolization ... 138
13.6 Vertization ... 138
13.7 Glossification ... 139
13.8 Fragification ... 139
13.9 Ferralitization ... 139
13.10 Humification ... 139
13.11 Solodization ... 139
13.12 Paludization ... 139
13.13 Sulfurization ... 139
13.14 Podzolization ... 139
13.15 Soils with Minimal Soil-Forming Processes ... 140

13.16 Summary . . . 140
References . . . 140

14 Benchmark, Endemic, Rare, and Endangered Soils in Mississippi . . . 141
14.1 Introduction . . . 141
14.2 Benchmark, Endemic, Rare, and Endangered Soils . . . 141
14.3 Highly Represented Soil Taxa . . . 141
14.4 Summary . . . 141
References . . . 142

15 Land Use in Mississippi . . . 143
15.1 Introduction . . . 143
15.2 Forest Land . . . 143
15.3 Cropland . . . 143
15.4 Pasture Land . . . 146
15.5 Developed Land and Wildlife, Watershed, and Recreation Lands . . . 146
15.6 Key Environmental Issues . . . 146
15.6.1 Soil Erosion and Water Quality . . . 147
15.6.2 Climate Change . . . 148
15.6.3 Loss of Biodiversity . . . 148
15.7 Summary . . . 148
References . . . 148

16 Yield Potential of Mississippi Soils . . . 151
16.1 Introduction . . . 151
16.2 Crop Yield Potential . . . 151
16.3 Grazing Land Quality . . . 152
16.4 Forest Site Quality . . . 153
16.5 Summary . . . 165
References . . . 165

17 Summary . . . 167

Appendix A: List of Soil Surveys . . . 169

Appendix B: Soil-Forming Factors for Soil Series in Mississippi . . . 173

Appendix C: Thicknesses of Diagnostic Horizons in Mississippi Soil Series . . . 181

Appendix D: Classification and reas of Soil Series Recognized in Mississippi . . . 187

Appendix E: Benchmark, Endemic, Rare, and Endangered Soils in Mississippi . . . 195

Appendix F: Land Use of Mississippi Soil Series . . . 199

Index . . . 207

About the Authors

Delaney Johnson received his BS degree in Soil Science from Alabama A&M University in 1983 and has later received his MS degree from Alcorn State University in Agriculture with a concentration in Agronomy in 2022. Prior to serving as State Soil Scientist, he has classified and mapped soils in four states with location included in many Major Land Resource Areas and physiographic regions. Since retiring Delaney stays active with the agency by serving as a contractor to provide training, assisting with soil health efforts and working with conservation partners. He enjoys the outdoors, house repairs, and new technology.

Mike Lilly earned his BS degree in Plant and Soil Science from Southern Illinois University in 1977. He was employed with the USDA, Natural Resources Conservation Service until his retirement in 2007. He served as Soil Survey Project Leader, Resource Soil Scientist, Assistant State Soil Scientist, and finally State Soil Scientist in Mississippi. Additionally, he served on the National Technical Committee for Hydric Soils. Since retiring, he volunteers with the Natural Resources Conservation Service. He enjoys traveling and listening to Blues music.

Birl Lowery obtained his BS degree in agricultural education from Alcorn State University, a MA degree in agricultural engineering technology from Mississippi State University, and a Ph.D. degree in soil physics from Oregon State University. He was a professor of soil science in the Department of Soil Science at the University of Wisconsin-Madison for 35 years prior to his retirement in 2014. He was Chair of the Department for five years and Senior Associate Dean in the UW College of Agriculture and Life Sciences. He enjoys skiing, hiking, biking, motorcycling, camping, and traveling.

James G. Bockheim earned his Ph.D. in forest soils at the University of Washington. He was Professor of Soil Science at the University of Wisconsin-Madison from 1975 until his retirement in 2015. He has conducted soil investigations in many parts of the US and the world. Since retiring, he has been working with retired and active soil scientists from the Natural Resources Conservation Service and other agencies to write about the soils of their state. He enjoys writing, reading, biking, photography, and traveling.

Abbreviations

CEC	Cation-exchange capacity
CLORPT	Climate, organisms, relief, parent material, and time (the five soil-forming factors)
COLE	Coefficient of linear extensibility
EPA	Environmental Protection Agency
LCC	Land capability classification
MAST	Mean annual soil temperature
MLRA	Major Land Resource Area
MS	Mississippi
MSU	Mississippi State University
NRCS	Natural Resources Conservation Service
NWR	National Wildlife Refuge
SMR	Soil moisture regime
SOC	Soil organic carbon
SSURGO	Soil survey geographic database
ST	Soil Taxonomy
STR	Soil temperature regime
USDA	United States Department of Agriculture
USFS	United States Forest Service

Authors' Note

This book provides information on the nature, properties, genesis, classification, distribution, use, and yield potential of the soils of Mississippi. It is the first summary of the soils of Mississippi since Vanderford's *Soils of Mississippi*, published in 1962. The book also includes maps of dominant great groups based on *Soil Taxonomy*. Reconnaissance soil surveys were conducted in Mississippi beginning in 1901 by the Bureau of Soils, a precursor to the Soil Conservation Service and eventually the more encompassing Natural Resources Conservation Service (NRCS). The entire state of Mississippi has received an order two or three soil survey (scales 1:15,840 to 1:24,000). Mississippi has a variety of physiographic provinces that have led to the mapping of more than 200 soil series in the state. The compilation of this book was made easier by an abundance of published natural resource maps (vegetation, geology, etc.) and other technical information. James Curtis, Mississippi State Soil Scientist, arranged for preparation of great group maps. Rachel Stout-Evans and Joxelle Velázquez-Garcia provided digital images of soil profiles.

1 Introduction

1.1 Etymology

The origin of the word "Mississippi" likely was derived from the Ojibwe word "misiziibi," which means "Great River." Mississippi is referred to as "the Magnolia state" or "the hospitality state."

1.2 Geography

Mississippi is bordered by Tennessee to the north, Alabama to the east, the Gulf of Mexico to the south, and the Mississippi River, Louisiana, and Arkansas to the west. At 125,443 km^2, Mississippi is the 32nd largest state in the United States. The state is 545 km north to south and 275 km east to west. Elevation varies from sea level in the southern coastal plain to 245 m (806 ft) on Woodall Mountain in the northeastern part of the state. For a small state, Mississippi has unusually diverse terrain including the highly agriculturally productive Mississippi Delta, a north–south-trending belt of Loess Hills, hills within the Coastal Plain that are interrupted by blackland prairies and the Pontotoc Ridge, and the Gulf Coast land areas.

Major water bodies are the Mississippi River to the west and the Gulf of Mexico to the south. The state features an array of some 95 rivers, many of which retain their Choctaw and Chickasaw names. These include the Tallahatchie, Tallahaga, Tangipahoa, Tickfaw, Tippah, Tuscumbia, Yalobusha, Yocona, Tombigbee, Yockanookany, Hushpuckena, Noxubee, Yazoo, Chickasawhay, and Okatibbee. Pascagoula, Homochitto, Bogue Chitto, Sucarnoochee, Escatawpa, and Tchoutacabouffa (Fig. 1.1). There are no natural lakes in Mississippi but damming of rivers to mitigate flooding, particularly in the Delta region, has created nine major reservoirs, including Pickwick Lake on the Tennessee River, Arkabutla Lake on the Coldwater River, Big Springs Lake and Columbus Lake on the Tombigbee River, the Ross Barnett Reservoir on the Pearl River, Sardis Lake on the Tallahatchie River, Enid Lake on the Yocona River, Grenada Lake on the Yalobusha River, and Okatibbee Lake on the Okatibbee River.

1.3 Demographics

With a population of 3 million, Mississippi is the 35th most populous US state. Major cities and 2020 populations include Jackson (167,000 persons; 592,000 in Greater Jackson Area; the state capital), Gulfport (72,000), Southaven (54,000), Hattiesburg (46,000), Biloxi (46,000), and Tupelo (38,000). Mississippi is divided into 82 counties of nearly equivalent areas (Fig. 1.2).

1.4 Brief History

The history of Mississippi was detailed by Barnwell (1997), Busbee (2015), and the Mississippi Encyclopedia. Mississippi was originally settled by indigenous peoples approximately 10,000 years ago and saw considerable development during the Woodland and Mississippi Cultures. Major tribal nations were the Chickasaw, Choctaw, Natchez, Yazoo, and Biloxi.

The European exploration period extended from 1542 to 1775. Hernando de Soto and Ponce de Leon, Spanish explorers, entered Mississippi in 1540. During the Colonial Period, French colonists were present in Mississippi in 1699 until 1783. The Mississippi Territory was created in 1775. In 1817, Mississippi was recognized as the 20th state in the USA. Mississippi was one of seven Confederate states that succeeded from the Union from 1861 to 1870 during the Civil War. In 1860, 55% of the state's population was composed of slaves. The period 1916–1970 was known as "The Great Migration," when 6 million African Americans moved from rural areas of the southern states to urban areas in the northern states. In 2019, 38% of

D. Johnson et al., *The Soils of Mississippi*, World Soils Book Series,
https://doi.org/10.1007/978-3-031-36235-4_1

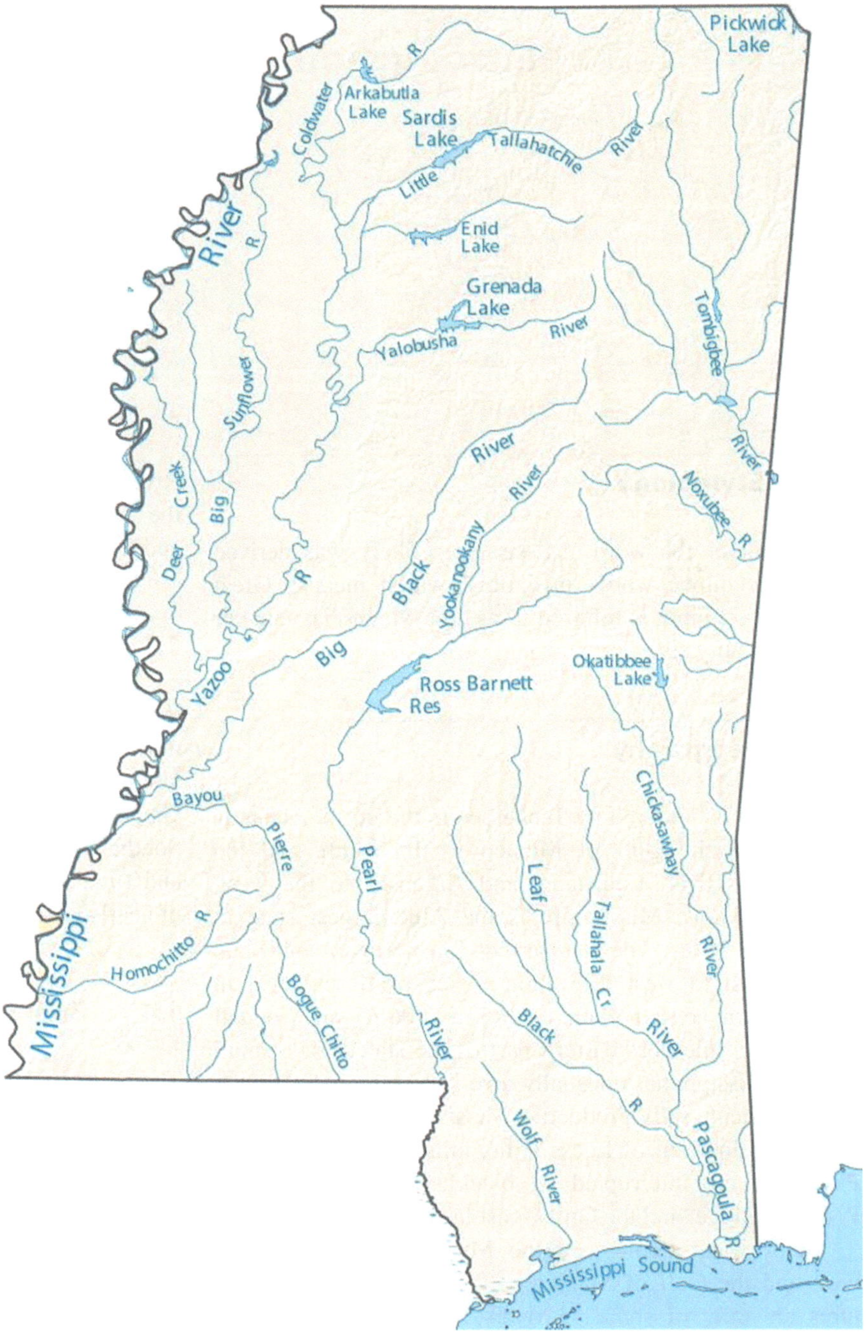

Fig. 1.1 Rivers and reservoirs of Mississippi. *Source* GISGeography.com

Mississippi's population was African American, the highest percentage of any US state. Mississippi was the site of many prominent events during the civil rights movement, including the lynching of Emmet Till in 1955, the 1963 assassination of Medgar Evers, and the 1964 Freedom Summer murders of three civil rights workers that led to the passing of the Civil Rights Act of 1964 and the Voting Rights Act of 1965.

1.5 Economy

According to the US Bureau of Economic Analysis, Mississippi's gross domestic product in 2021 was $104 billion (37th in the USA). Agriculture is Mississippi's number one industry, employing 17% of the state's workforce and accounting for 8.5% of the state's gross domestic product. There are approximately 34,700 farms in the state, with

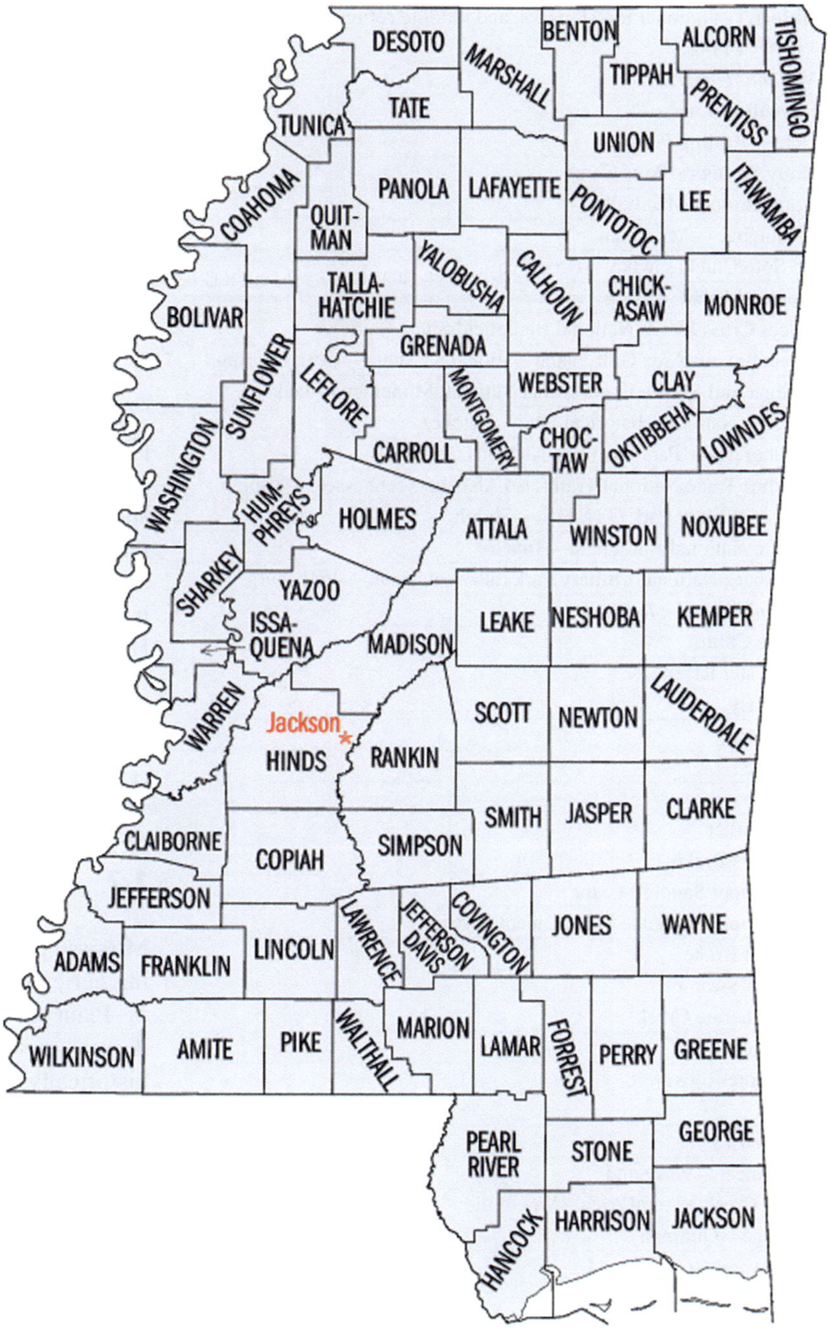

Fig. 1.2 Mississippi's 82 counties. *Source* Wikipedia.com

an average size of 120 ha (300 acres. The six major agricultural commodities are poultry/eggs ($2.42 billion), soybeans ($1.49 billion), forestry ($1.3 billion), corn ($748.3 million), cotton ($558 million), and cattle/calves ($282 million). Other important contributors to the state's economy are catfish production, entertainment and tourism, manufacturing and services, retail and wholesale trade, real estate, and health and social services.

1.6 Culture

Mississippi is especially known for its literature and music, both of which are strongly linked to the soil (Eubanks, 2021a, b). Famous Mississippi authors who "write or wrote of the soil" include William Faulkner, Eudora Welty, Jesmyn Ward, Donna Tartt, Tennessee Williams, Richard Ford, and Richard Wright, and Margaret Walker Alexander.

Table 1.1 National forest, parks, and wildlife refuges and state parks in Mississippi

National Forests
Bienville—Lake
Delta—Rolling Fork
Holly Springs—Potts Camp
Homochitto—Meadville
Tombigbee—Ackerman
De Soto/Chickasawhay—Brooklyn
National Parks
Brices Cross Roads National Battlefield Site—Baldwyn
Davis Bayous Area Gulf Island National Seashore—Ocean Springs
Medgar and Myrlie Evers Home National Monument—Jackson
Natchez National Historical Park—Natchez
Natchez Trace Parkway (also Alabama, Tennessee)
Natchez Trace National Trail (also Alabama, Tennessee)—Tupelo
Shiloh Military Park (TN, MS)—Shiloh
Tupelo National Battlefield—Tupelo
Vicksburg National Military Park (also Louisiana)—Vicksburg
National Wildlife Refuges
Bogue Chitto
Coldwater River
Dahomey
Grand Bay
Noxubee
Hillside
Holt Collier
Mathews Brake
Mississippi Sandhill Crane
Mississippi Wetlands Management District
Morgan Brake
Panther Swamp
St. Catherine Creek
Tallahatchie
Theodore Roosevelt
Yazoo
State Parks
Buccaneer—Waveland
Clark Creek Natural Area—Woodville
Clarkco—Quitman
Florewood—Greenwood
George P. Cossar—Oakland
Golden Memorial—Walnut Grove
Great River Road—Rosedale
Holmes County—Durant
Hugh White—Grenada
John W. Kyle—Sardis
J. P. Coleman—Iuka
Lake Lincoln—Wesson
Lake Lowndes—Columbus
LeFleur's Bluff—Jackson
Legion—Louisville
Leroy Percy—Hollandale
Natchez—Natchez
Paul B. Johnson—Hattiesburg
Percy Quin—McComb
Roosevelt—Morton
Shepard—Gautier
Tishomingo—Tishomingo
Tombigbee—Tupelo
Trace—Pontotoc
Wall Doxey—Holly Springs

"To understand the world, you must first understand a place like Mississippi" (W. Faulkner). "To a first-time visitor, Mississippi's rural landscape brings to mind solitude and loneliness, a place from which one escapes rather than returns" (W. Ralph Eubanks).

Famous blues musicians who hail from Mississippi include Robert Johnson, BB King, Muddy Waters, WC Handy, Charley Patten, Mississippi John Hurt, and many others. Many of these musicians wrote and sang of the soil. "I remember breaking ground, when the earth was bone dry; Digging and sweating underneath that blazing delta sky" (James Cotton, *Mississippi Mud*).

1.7 Education

Mississippi State University in Mississippi State is the leading agricultural university in the state, with the Department of Plant and Soil Sciences offering BS, MS, and PhD degrees in soil science. Alcorn State University is a public, historically Black land-grant university in Lorman, where the Department of Agriculture offers BS and MS degrees in the agricultural and environmental sciences. The University of Mississippi ("Ole Miss") in Oxford offers programs in Environmental Science and Geoscience.

1.8 Recreation

There are six national forests, nine national parks, and 16 national wildlife refuges in Mississippi (Table 1.1). In addition, there are 25 state parks spread throughout the state. The Natchez Trace is a 740-km (444 mi) parkway that extends from Nashville, Tennessee (via Tishomingo, MS) to Natchez, Mississippi. The trace is an historical travel corridor that has been used for 10,000 years by American Indians, "Kaintucks," European settlers, slave traders, and soldiers. The Natchez Trace linked eastern US ports along the Cumberland and Tennessee Rivers to those along the Mississippi River.

1.9 The Subsequent Chapters

In the following chapters, we will consider the history of soil studies in Mississippi (Chap. 2), soil-forming factors (Chap. 3), general soil regions of Mississippi (Chap. 4), diagnostic horizons and taxonomic structure of Mississippi's soils (Chap. 5), taxonomic soil regions (Chap. 6), the eight soil orders (Chaps. 7 through 12), soil-forming processes (Chap. 13), benchmark, endemic, rare and endangered soils in Mississippi (Chap. 14), land-use in Mississippi (Chap. 15), and the yield potential of Mississippi's soils (Chap. 16).

1.10 Summary

The origin of the word "Mississippi" likely was derived from the Ojibwe word "misiziibi," which means "Great River." Mississippi is the 32nd largest US state, the 35th most populated state, and is ranked 37th in gross domestic product. Agriculture is the leading industry, particularly poultry, soybeans, forestry, corn, cotton, and cattle and calves. For a small state, Mississippi has unusually diverse terrain including the highly agriculturally productive Mississippi Delta, a north–south-trending belt of Loess Hills, hills within the Coastal Plain that are interrupted by blackland prairies and the Pontotoc Ridge, and the Gulf Coast. Mississippi has been influenced by Native American tribes that extend back 10,000 years, Spanish and French colonists, and Blacks who were conscripted from Africa and forced to work on plantations. Mississippi is the heart of the civil rights movement. Mississippi is especially known for its literature and music, both of which are strongly linked to the soil.

References

Barnwell M (1997) A place called Mississippi: collected narratives. Univ Mississippi Press

Busbee WF Jr (2015) Mississippi: a History

Eubanks WR (2021a) A place like Mississippi: a journey through a real and imagined literary landscape. Timber Press

Eubanks WR (2021b) Why has Mississippi inspired so many great writers? Amer Mag 30 Sept

Mississippi Encyclopedia. http://mississippiencyclopedia.org

2 History of Soil Studies in Mississippi

2.1 Introduction

Mississippi has a rich and long history of soils investigations that began in 1860 with a report on the geology and agriculture of the state of Mississippi by E. W. Hilgard, considered by some (Jenny 1961) as "the father of modern soil science in the United States." His geological map of Mississippi is reproduced here as Fig. 2.1. The legend includes, from top to bottom, post-Tertiary (Mississippi Bottom and Bluff Formation), Tertiary Coast Pliocene, Grand Gulf, Vicksburg, Jackson, Calcareous Claiborne, Siliceous Claiborne, and Northern Lignite), Cretaceous (Ripley, Rotten Limestone, Tombigbee Sand, and Eutaw), and Carboniferous (Mountain Limestone) deposits.

2.2 Early Soil Books

In 1898, T. O. Mabry published a paper on the "brown or yellow loams" of the northern Mississippi loess region and defined the terms "buckshot" and "crawfish" lands; these terms have persisted for many years. The first book summarizing the distribution and characteristics of Mississippi soils was *Soils of Mississippi*, published in 1910 by W. L. Hutchinson of Mississippi Agricultural College's Mississippi Agricultural Experiment Station (MAES). This was followed in 1916 by W. N. Logan's *The Soils of Mississippi*, also published by the MAES. A third version of the *Soils of Mississippi*, published in 1962 by H. B. Vanderford of the MAES, included keys to soils and described soil associations, classification of soils (zonal system of Baldwin et al. 1938), main uses, and soil management problems for each of eight land resource areas in Mississippi. Pettiet (1974) provided an interpretive evaluation of soils in the Yazoo-Mississippi Delta area for crop production. Vanderford's study was updated in abbreviated form by Pettry (1977). Pettry and others (1995) studied water table fluctuations of 23 soil series from five soil orders in Mississippi, noting the difficulty in distinguishing among drainage classes from water table depth. They noted that many of the reductimorphic features in soils with restricted drainage were relict. Kushla and Londo (2014) provided an overview of soil-forming factors, soil properties, ecoregions, and management of Mississippi soils.

As of 2012, all 82 counties in Mississippi had been mapped generally at scales of 1:15,000 to 1:24,000. Although the mapping was done by the Soil Conservation Service (later the Natural Resources Conservation Service), the Mississippi State University Agricultural Experiment Station (later the Mississippi State University Agricultural and Forest Experiment Station), the US Forest Service (USFS), and county and local organizations assisted in the mapping. Research on Mississippi soils has been conducted by Mississippi universities, the Natural Resources Conservation Service (NRCS), and USFS personnel for more than 160 years.

2.3 Definition of Soil

There are many definitions of soil ranging from the utilitarian to a description that focuses on material. Soil has been recognized as (i) a natural body, (ii) a medium for plant growth, (iii) an ecosystem component, (iv) a vegetated water-transmitting mantle, and (iv) an archive of past climate and processes. In this book, we follow the definition given in the *Keys to Soil Taxonomy* (Soil Survey Staff 2014, p. 1) that the soil "is a natural body comprised of solids (minerals and organic matter), liquid, and gases that occurs on the land surface, occupies space, and is characterized by one or both of the following: horizons, or layers, that are distinguishable from the initial material as a result of additions, losses, transfers, and transformations of energy and matter or the ability to support rooted plants in a natural environment."

D. Johnson et al., *The Soils of Mississippi*, World Soils Book Series,
https://doi.org/10.1007/978-3-031-36235-4_2

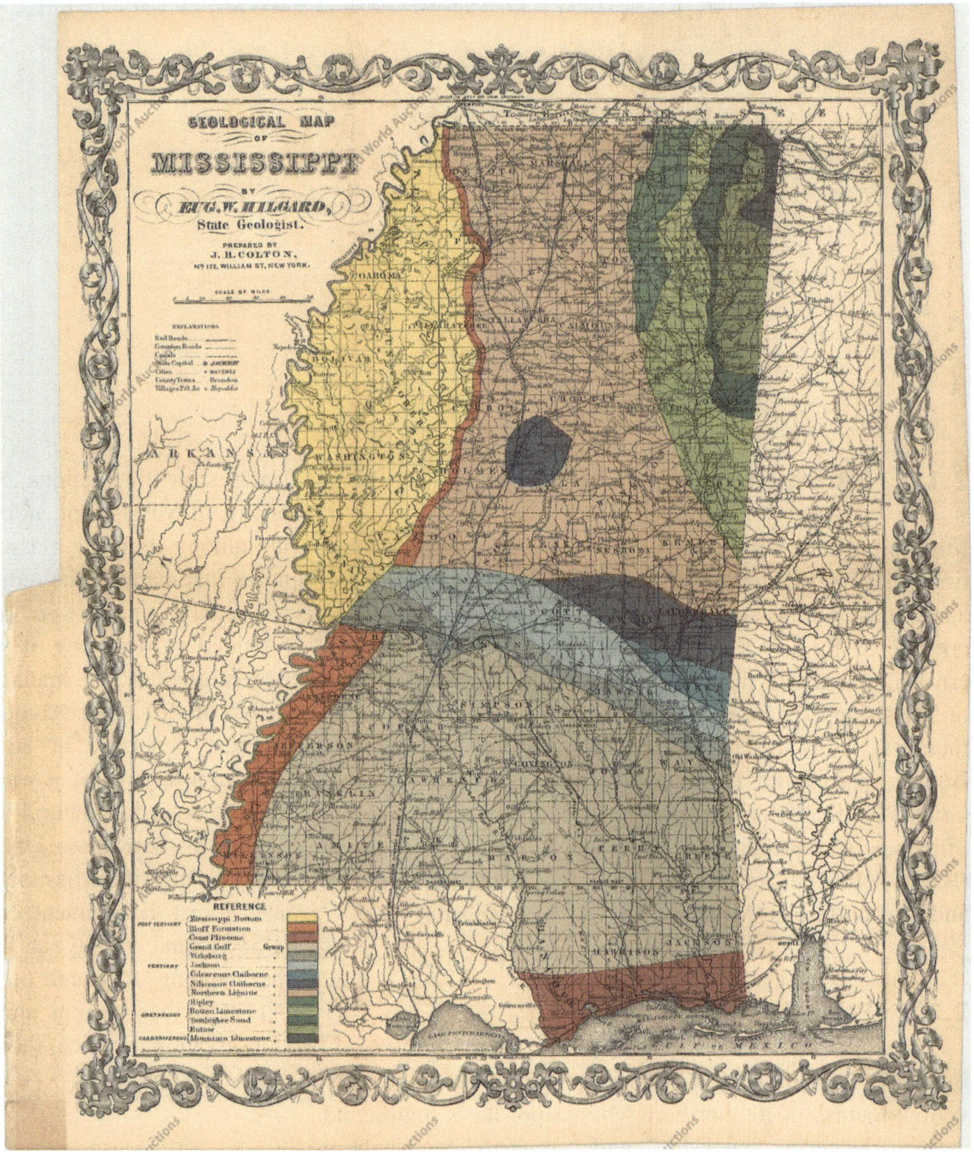

Fig. 2.1 E. W. Hilgard's geologic map of Mississippi, published in 1860. The legend includes, from top to bottom, post-Tertiary (Mississippi Bottom and Bluff Formation), Tertiary Coast Pliocene, Grand Gulf, Vicksburg, Jackson, Calcareous Claiborne, Siliceous Claiborne, and Northern Lignite), Cretaceous (Ripley, Rotten Limestone, Tombigbee Sand, and Eutaw), and Carboniferous (Mountain Limestone) deposits. *Source* Mississippi State Geological Survey

2.4 Soil Surveys

Under the auspices of Milton Whitney, the Bureau of Soils conducted 17 reconnaissance soil surveys in Mississippi between 1900 and 1910, including some of the earliest soil surveys in the US. The first survey was conducted by JA Bonsteel and party in 1901 in the Yazoo Area (Fig. 2.2). The map, published at a scale of 1:63,360, shows four soil series—the Yazoo, Memphis, Sharkey, and Lintonia—, seven soil types, and one land type. Many of the early maps of soil series were truly works of art.

Soil mapping in Mississippi increased linearly from 1900 to 2010, with more than 112 soil surveys being produced (Fig. 2.3). Many counties have been surveyed several times. Jackson County, which is in the southeastern part of the state along the Gulf Coast, was surveyed in 1904, 1927, 1964, and 2006 (publication dates), with more recent changes recorded in Web Soil Survey. Initial mapping by the Bureau of Soils in 1903 in the Biloxi area recorded the Norfolk and Orangeburg soil series and a Meadow land type. By 1927, 18 soil series and 29 soil types were recognized in the region, with the Norfolk (24% of map area) and Plummer soil series (13%) being dominant. By 1964, eight soil associations, 21 soil series, seven land types, and 56 soil map units were recognized in Jackson County. The Goldsboro (12% of map area), Rains (12%), Lynchburg (6.2%), Norfolk (6.1%), and Klej (5.1%) were the dominant soil series—none of which are recognized today. By 2006, 12 soil associations, 46 soil series, and 60 soil map units had been identified in Jackson County. The most extensive soils

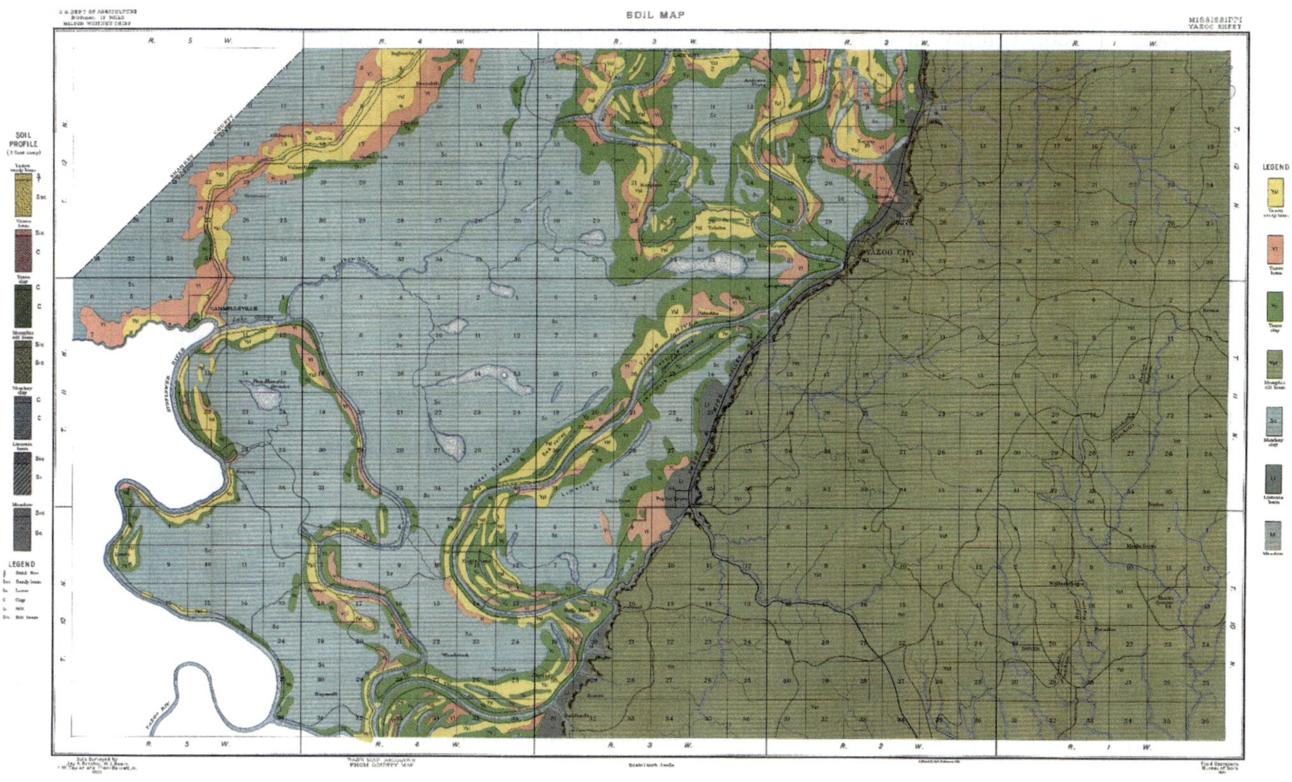

Fig. 2.2 Soil map of the Yazoo Area, Mississippi, from 1901. *Source* Bonsteel et al. (1901)

Fig. 2.3 Cumulative number of soil surveys in Mississippi *Source* by J. Bockheim

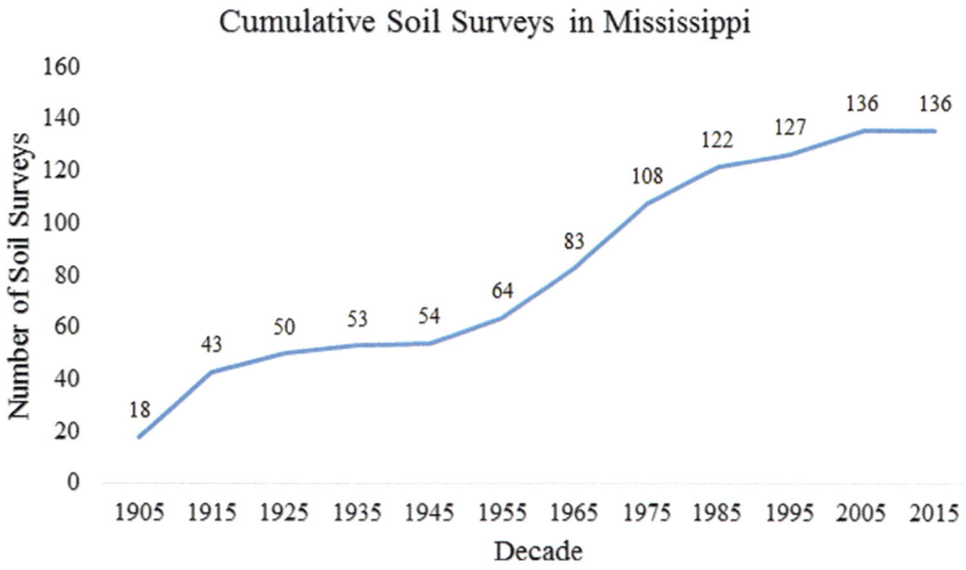

in the county in 2006 were the Benndale (8.5%), Vancleave (5.8%), Smithton (3.7%), and Malbis (3.5%) soil series. The Norfolk soil series is common in the Southern Coastal Plain, but does not occur in Mississippi. It is of interest that the earliest map (1904) recognized the Norfolk and Orangeburg soil series, which are now classified as Kandiudults; and the 1964 map recognized the Goldsboro, Rains, and Lynchburg, soil series, which, although no longer are mapped in Mississippi, are classified as Paleudults and Paleaquults. The Benndale and Malbis, Vancleave, and Smithton soil series that are recognized in Jackson County today are classified as Paleudults, Fragiudults, and Paleaquults, respectively.

There are archived PDF files available online for detailed soil surveys of each county in Mississippi except Greene,

Holmes, and Wilkinson. The last archived soil survey in Mississippi was of Scott County in 2010. At the present time mapping is reported digitally on Web Soil Survey (WSS), and 100% of the state has received detailed soil mapping.

Do Book Covers Matter?

Of the more than 100 archived soil survey reports in pdf form online for Mississippi, 52 have covers that highlight a key aspect of the soil survey area. Nearly one-third (29%) of the soil survey covers feature beef cattle (Fig. 2.4). This is not surprising in view that cattle/calves are the 6th top agricultural industry in the state. Dams and recreation areas account for one-quarter (25%) of the covers, reflecting the importance of dams in controlling flooding in the many areas of the state, particularly in the Delta. About 19% of the covers feature an agricultural crop, including soybeans, corn, and cotton, which are ranked 2nd, 4th, and 5th in importance to agriculture in the state. Forestry, the 3rd most important agricultural commodity in Mississippi, is recognized on 15% of the covers. Ten percent of the covers show buildings, either historical structures or county courthouses. The 1967 Tate County soil survey features a highway interchange on the cover, not surprising given that it borders the Memphis Metropolitan area which is the 5th most populous (1.3 million persons) area in the southeastern USA. These findings suggest that book covers do matter!

Appendix A provides a list of all of Mississippi's archived published soil surveys and their publication date. In about 2005, the USDA phased out printing soil survey reports and made the Web Soil Survey the official source for soil survey information. In most cases, published soil survey reports are available for reference and information at local NRCS offices and public libraries. However, copies for distribution are no longer available. Official soil survey information for all of the Mississippi soil survey areas is available online using Web Soil Survey at https://websoilsurvey.sc.egov.usda.gov. To start using Web Soil Survey click on one of the links under Browse by Subject or on the large Green Button that says start WSS. Pettry (2008) provided a list and the history of Mississippi soil surveys.

2.5 Soil Series

The early concept of the soil series is that all occurrences of a series would have the same, or very nearly the same, chemical composition because they were derived from the same rocks. Maps completed between 1900 and 1910 showed fewer than ten soil series but up to 20 soil types (textural subdivisions of soil series). By 1960 individual counties in Mississippi had around 20 soil series. In 2006, Jackson County had 46 soil series.

Eleven of the original 21 soil series identified in Mississippi between 1901 and 1907 are still recognized in the state, including the Catalpa, Houston, Lintonia, Memphis, Myatt, Ochlockonee, Oktibbeha, Sharkey,

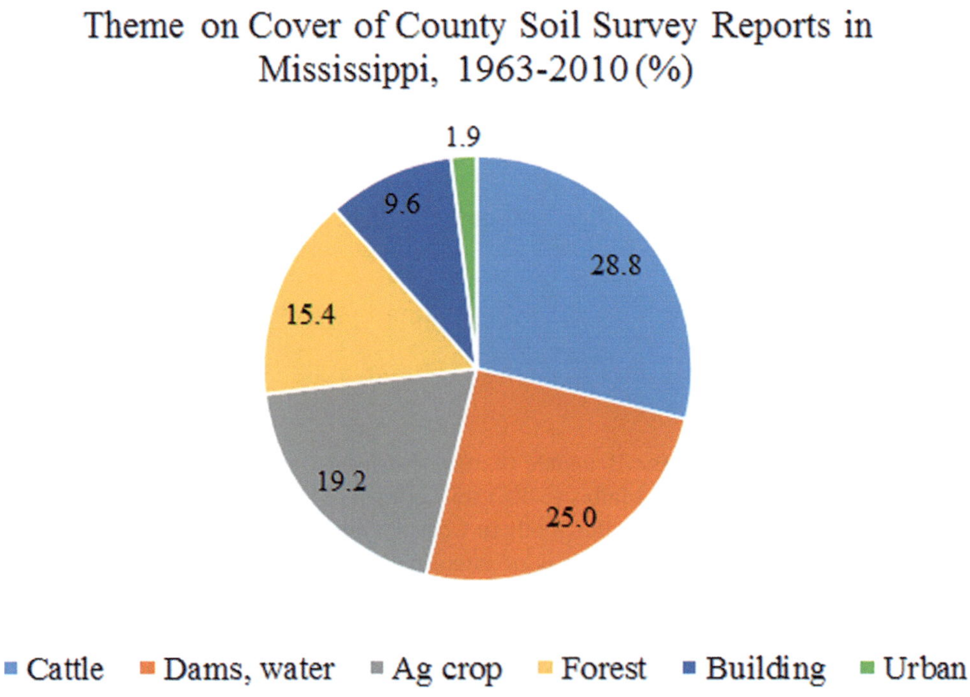

Fig. 2.4 Frequency of agricultural and other themes on the covers of soil survey reports of Mississippi archived in PDF form online *Source* by J. Bockheim

Susquehanna, Waverly, and Yazoo. By 2023, 232 soil series had been identified in Mississippi (Fig. 2.5), 10% of which occur strictly in Mississippi. The trend in cumulative soil series in Mississippi differs from the national trend in that there is a "stagnation" in the number of soil series after 1975. This is because most of the mapping had been completed by then, and the national trends reflect accelerated mapping on public lands in the western USA (Thorson et al. 2022).

2.6 Soil Classification

Early soil classification schemes developed in the United States were not used in soil survey reports. The 1938 zonal soil classification system developed by Baldwin and others was used in Mississippi from 1956 (Tunica County) to 1969. After more than a decade of development, in 1960 the *Seventh Approximation*, a new classification scheme and a precursor to *Soil Taxonomy* (Soil Survey Staff 1999) was published. *Soil Taxonomy* was first used in Mississippi in 1967 in Grenada County and has been used exclusively thereafter in the state.

2.7 Soil Taxonomy

All soils in Mississippi are now classified using *Soil Taxonomy*, which is used throughout this book. The *Keys to Soil Taxonomy* (Soil Survey Staff 2014) is an abridged companion document that incorporates all the amendments that have been approved to the system since the publication of the second edition of *Soil Taxonomy* in 1999, in a form that can be used easily in a field setting. *Soil Taxonomy* is a hierarchical classification system that classifies soils based on the properties of diagnostic surface and subsurface horizons. The *Illustrated Guide to Soil Taxonomy* (2015) is a useful guide to the use of *Soil Taxonomy*. All of these publications are available online, and readers are directed to those publications for detailed explanations of concepts.

For classification purposes, the upper limit of the soil is defined as the boundary between the soil (including organic horizons) and the air above it. The lower limit is arbitrarily set at a maximum limit of 200 cm. The definition of the classes (taxa) is quantitative and uses well-described methods of analysis for the diagnostic properties. The assumed genesis of the soil is not used in the system and the soil is classified "as it is" using morphometric observations in the field coupled with laboratory analysis and other data. The nomenclature in *Soil Taxonomy* is mostly derived from Greek and Latin sources, as is done for the classification of plants and animals.

Soil Taxonomy classifies soils, from broadest to narrowest levels, into orders, suborders, great groups, subgroups, and families. *Soil Taxonomy* is similar to the taxonomic classification of living organisms, from broadest to narrowest levels, into kingdom, phylum, class, order, family, genus, and species. Casual readers may be unfamiliar with soil taxonomic terms, which often employ abbreviations based on Latin or Greek words. Those readers will benefit from a glossary of terms, such as the online Soil Formation and Classification provided by the NRCS (2022) or Chap. 7, Nomenclature, of *Soil Taxonomy* (1999). Introductory soil science textbooks, such as Weil and Brady's *The Nature and Properties of Soils* (2016), also explain *Soil Taxonomy*, as well as other aspects of soil science. In any case, the soil taxonomic system is logical and consistent and reveals a wealth of information about soils in a concise manner.

Soil taxonomic names are constructed in an order that progresses from narrow to broad, unlike biological

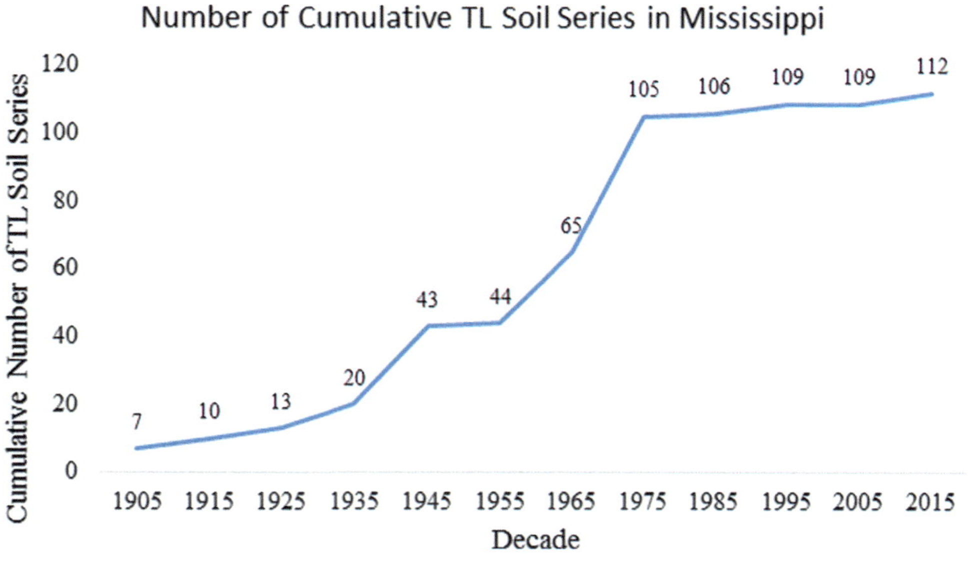

Fig. 2.5 Cumulative number of soil series in which Mississippi is the lead state (TL) over time *Source* by J. Bockheim

classifications, which progress from broad to narrow. For example, the taxonomic classification of loblolly pine progresses from kingdom (broad) to species (narrow) as follows: Plantae (kingdom)-Tracheophytes-Gymnosperms-Pinophyta-Pinopsida-Pinales-Pinaceae-*Pinus*-*Pinus taeda* (species). In contrast, a soil taxonomic classification for the Natchez soil series, a coarse-silty, mixed, superactive, thermic Typic Eutrudepts, from the family (narrow) to the order (broad) level is:

- "coarse-silty, mixed, superactive, thermic" family classification (narrow classification),
- "Typic" describes a typical soil subgroup,
- "Eutr" describes a soil great group with a high base saturation,
- "ud" describes a soil suborder without a pronounced dry season, and
- "ept" describes a soil in the order of Inceptisols (broad classification).

Using this system, taxonomic names may be applied at various levels of detail. For example, "coarse-silty," "mixed," "superactive," and "thermic" describe the family, "Typic" describes a subgroup, "Eutrudept" describes a great group, "Udept" describes a suborder, and "Inceptisol" is an order. Soil associations, soil complexes (composed of two or three major soil series and miscellaneous area components), and consociations (comprising a single major component) constitute the primary soil map units, which are shown on soil maps. For example, Loring-Smithdale-Providence association, is a soil map unit.

Eight diagnostic surface horizons (epipedons) are defined in *Soil Taxonomy* and only four of them occur in Mississippi: histic, mollic, umbric, and ochric (Table 2.1). The histic epipedon contains primarily organic materials and is saturated for prolonged periods during the year. The mollic and umbric epipedons occur in mineral soils and are thick, dark-colored, and enriched in organic matter. The mollic epipedon is enriched in base cations, such as calcium, magnesium, and potassium, while the umbric epipedon contains low amounts of these cations. The ochric epipedon is thin, commonly light-colored, and often low in organic matter content.

Eight of the 20 diagnostic subsurface horizons identified in *Soil Taxonomy* are present in the soils of Mississippi (Table 2.1). The albic horizon is composed of materials from which clay and/or free iron oxides have been removed by eluviation to a degree that primary sand and silt particles impart a light color to the horizon. The argillic horizon is enriched in clay that has moved down the profile from percolating water. The cambic horizon shows minimal development other than soil structure and color. A fragipan is a subsoil layer 15 cm or more thick that shows evidence of pedogenesis, has a high bulk density, is brittle when moist, firm or firmer when dry, and slakes when immersed in water. A glossic horizon is a subsurface horizon that is 5 cm or more thick and usually occurs between an overlying albic horizon and an underlying argillic, kandic, or natric horizon, or fragipan. Albic materials constitute 15–85% of the horizon.

The natric horizon is a type of argillic horizon, which shows evidence of clay illuviation that has been accelerated by the dispersive properties of sodium. The spodic horizon is an illuvial layer containing at least 85% spodic material and is at least 2.5 cm thick. Spodic materials contain illuvial amorphous materials of organic matter and aluminum (Al), with or without iron (Fe).

The kandic horizon is a subsurface horizon with a texture finer than that of an overlying horizon that is 30-cm or more thick and composed mainly of low-activity clays so that when the cation-exchange capacity (CEC), a measure of the soil's ability to hold and exchange cations, extracted with ammonium acetate at pH 7 divided by the percent clay present the apparent CEC_7 of the clay is 16 cmol/kg clay or less.

There are other diagnostic soil characteristics that are important in classifying soils of Mississippi, including plinthite and sulfidic materials. Plinthite is a humus-poor and iron-rich soil material that hardens irreversibly if exposed to repeated wetting and drying in place. Sulfidic materials is a mineral or organic horizon 15 cm or more thick that has a pH value of 3.5 or less due to the presence of sulfuric acid, as evidenced by the presence of yellow jarosite concretions or more than 0.05% water-soluble sulfate. Photographs of subsurface horizons are given in chapters describing soils in each of the major great groups and the orders represented in Mississippi.

Soil orders are defined primarily on the basis of diagnostic soil characteristics and diagnostic surface and subsurface horizons. Eight of the 12 orders in *Soil Taxonomy* occur in Mississippi: Mollisols, Inceptisols, Ultisols, Alfisols, Entisols, Vertisols, Histosols, and a single Spodosol (Table 2.2). Mollisols are dark-colored, base-enriched grassland soils. Inceptisols are juvenile soils that contain an epipedon and either a cambic horizon, a salic horizon, or a high exchangeable sodium percentage. Ultisols have an argillic horizon and a base saturation less than 35% at a depth of 180 cm. Alfisols are base-enriched forest soils with an argillic horizon. Entisols are very poorly developed, recent soils that may have only an anthropic or ochric epipedon. Vertisols are derived from abundant swelling clays that lead to cracks and slickensides. Histosols are organic soils. Spodosols have a spodic

Table 2.1 Definitions of diagnostic horizons present in Mississippi soils[a]

Diagnostic surface horizons (epipedons)	
Histic	Greater than 20 cm thick; organic matter content 16% or more; saturated for more than 30 days in normal years unless artificially drained
Mollic	At least 18 cm thick; dark-colored; organic C 0.6% or more; base saturation 50% or more
Ochric	An altered horizon that fails to meet the requirements of other epipedons; lacks rock structure or finely stratified fresh sediments; includes underlying eluvial horizons such as albic
Umbric	At least 18 cm thick; dark-colored; organic C 0.6% or more; base saturation less than 50%
Diagnostic subsurface horizons	
Albic	A light-colored eluvial horizon 1-cm or more in thickness; composed of albic materials
Argillic	An illuvial horizon that gives evidence of translocation of clay, based on the ratio of that in the clay-enriched horizon to an overlying eluvial horizon, the presence of clay films (argillans)
Cambic	An altered horizon that shows color and/or structure development, is at least 15 cm thick, and has a texture of very-fine sand, loamy very-fine sand, or finer
Fragipan	A subsoil layer 15 cm or more thick that shows evidence of pedogenesis, has a high bulk density, is brittle when moist, firm or firmer when moist, and slakes when immersed in water
Glossic	A subsurface horizon that is 5 cm or more thick and usually occurs between an overlying albic horizon and an underlying argillic, kandic, or natric horizon, or fragipan. Albic materials constitute 15–85% of the horizon
Kandic	A subsurface horizon with a texture finer than that of an overlying horizon; it is 30 cm or more thick, composed mainly of low-activity clays so that when the CEC of the soil by ammonium acetate at pH 7 is divided by the percent clay present the apparent CEC_7 of the clay is 16 cmol/kg clay or less
Natric	Meets the requirements of an argillic horizon but also has prismatic, columnar, or blocky structure, an exchangeable sodium percentage of 15 or more, or a sodium adsorption ratio of 13 or more
Spodic	An illuvial layer containing at least 85% spodic materials; 2.5 cm or more thick; spodic materials must have a pH value in 1:1 water of 5.9 or less; 0.6% organic C or more; an optical density of oxalate extract (ODOE) of 0.25 or more and that value that is at least twice that of an overlying eluvial horizon; an $Al_o + ½\ Fe_o$ percentage of 0.50 or more and that value is at least twice that of an overlying eluvial horizon
Other diagnostic soil characteristics	
Plinthite	A humus-poor and iron-rich soil material that hardens irreversibly if exposed to repeated wetting and drying in place
Sulfidic	A mineral or organic horizon 15 cm or more thick that has a pH value of 3.5 or less due to the presence of sulfuric acid, as evidenced by the presence of yellow jarosite concretions or more than 0.05% water-soluble sulfate

[a] Revised from Buol et al. (2011)

Table 2.2 Simplified key to soil orders in Mississippi[a]

Histosols	Soils that do not have andic soil properties in 60% or more of the upper 60 cm and have organic soil materials in two-thirds or more of the total thickness
Spodosols	Other soils with a spodic horizon within a depth of 200 cm
Vertisols	Other soils with a layer 25 cm or more thick containing either slickensides or wedge-shaped peds, have more than 30% clay in all horizons between depths of 18 and 50 cm or a root-limiting layer if shallower, and have cracks that open and close periodically
Ultisols	Other soils with an argillic horizon and a base saturated percentage at pH 8.2 less than 35 at a depth of 180 cm
Mollisols	Other soils with a mollic epipedon and a base saturation (by ammonium acetate at pH 7) of 50% or more in all depths above 180 cm
Alfisols	Other soils with an argillic or natric horizon
Inceptisols	Other soils with an umbric or mollic epipedon, or a cambic horizon, or a salic horizon, or a high exchangeable sodium percentage which decreases with increasing depth accompanied by groundwater within 100 cm of the soil surface
Entisols	Other soils

[a] Revised from: Buol et al. (2011)

horizon within 200 cm of the surface. This horizon contains an abundance of extractable aluminum and organic C.

Suborders are distinguished on the basis of soil climate for five of the eight orders occurring in Mississippi: the Alfisols, Inceptisols, Mollisols, Ultisols, and Vertisols. Soil parent materials are used to differentiate suborders of Vertisols and Entisols. Histosols are the organic soils. There are 17 suborders of soils in Mississippi. Great groups are distinguished from a variety of soil characteristics; there are 42 great groups of soils in Mississippi.

2.8 General Soil Maps

We have been able to locate three general soil maps of Mississippi. The first map, published in 1924 and prepared by W. N. Logan, showed ten soil regions, including (from top to bottom in the legend) the Northeast Highland, the Northeast Prairie, the Pontotoc Ridge, Flatwoods, Shortleaf Pine, Central Prairie, Brown Loam and Loess, Yazoo Basin, Longleaf Pine, and Gulf Coastal (Fig. 2.6). The second map, prepared in 1942 by the Mississippi State Department

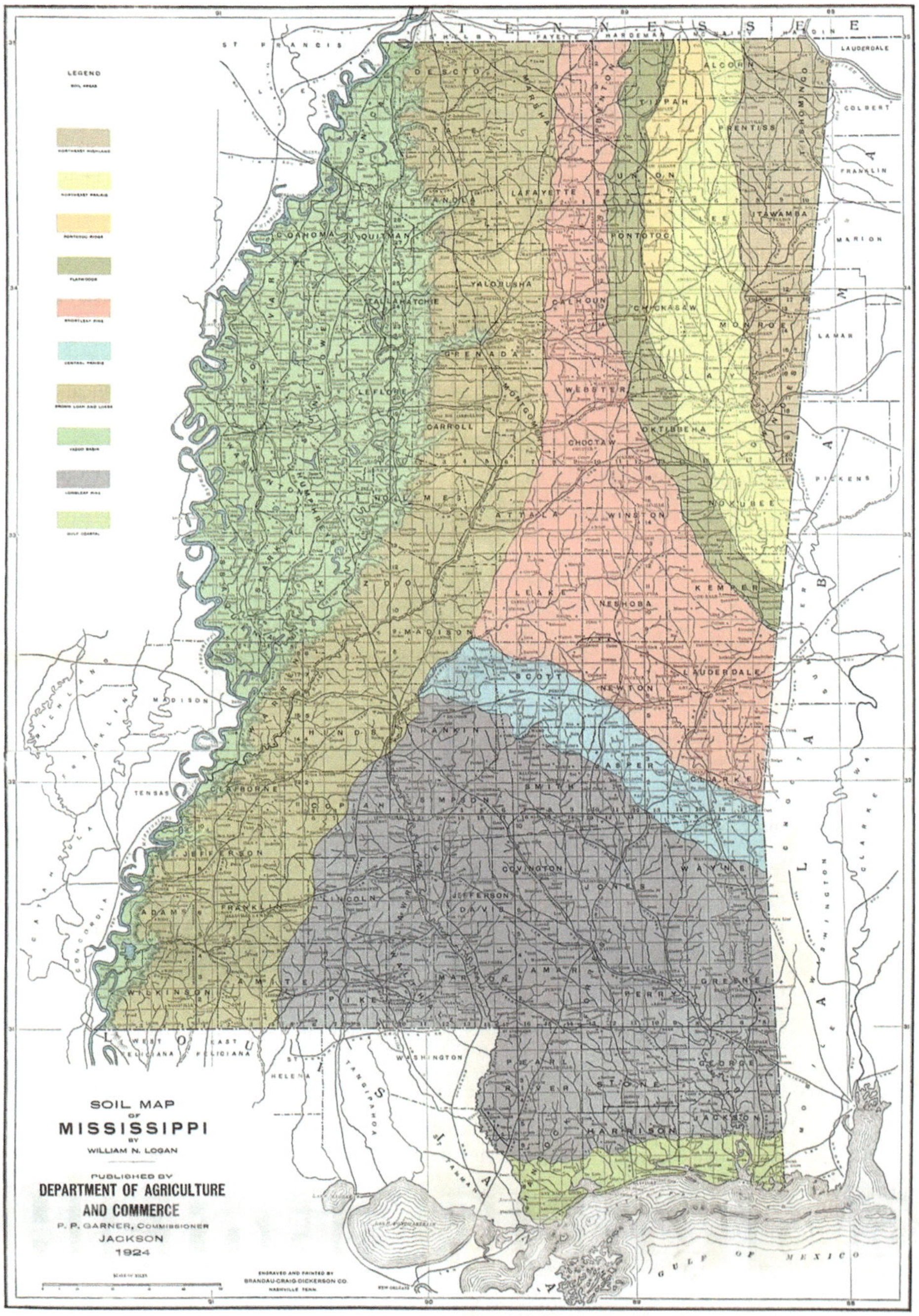

Fig. 2.6 General soil map of Mississippi prepared in 1924 by W. N. Logan. From top to bottom, the legend is the Northeast Highland, the Northeast Prairie, the Pontotoc Ridge, Flatwoods, Shortleaf Pine, Central Prairie, Brown Loam and Loess, Yazoo Basin, Longleaf Pine, and Gulf Coastal. *Source* Mississippi Department of Agriculture and Commerce

of Agriculture, also showed ten soil regions, including (from top to bottom in the legend) the Northeast Highland, North Central Prairie, Pontotoc Ridge, Flatwoods, Sand-Clay Hills (former Shortleaf Pine region), Mixed Brown Loam and Coastal Plain (former Central Prairie), Delta (former Yazoo region), Longleaf Pine, and Coastal Terrace (Fig. 2.7). The third map, published in 1974 by the former Soil Conservation Service, the Mississippi Agricultural and Forestry Experiment Station (Mississippi State University), and the Mississippi State Department of Agriculture and Commerce, is at a scale of 1:750,000 and shows six soil orders containing 29 soil associations (Fig. 2.8). A soil association consists of two or more dissimilar major components occurring in a regular and repeating pattern on the landscape.

A general map of soil orders in Mississippi, created by the Natural Resources Conservation Service in 1998, shows six soil orders, including Alfisols, Entisols, Histosols, Inceptisols,

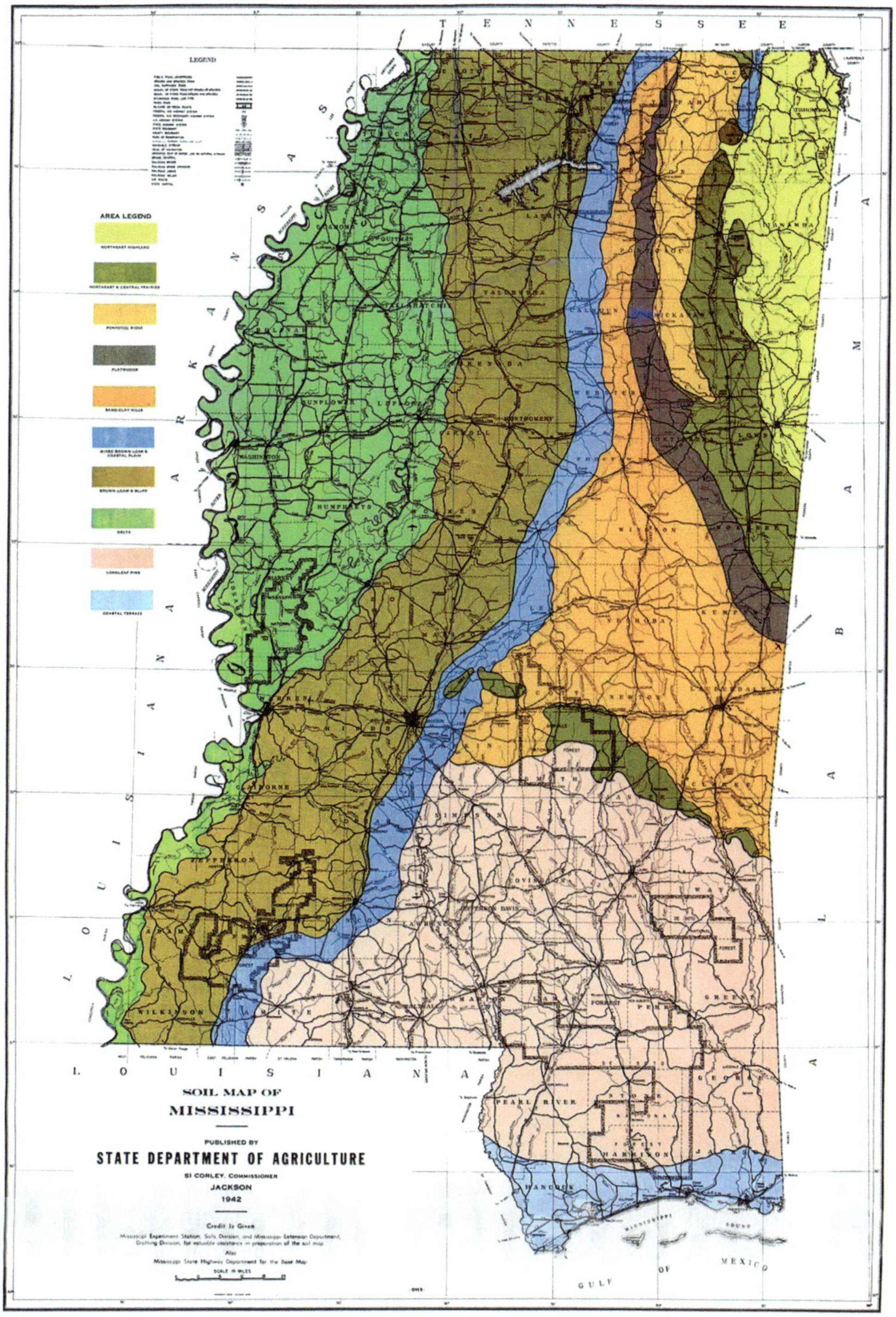

Fig. 2.7 General soil map of Mississippi prepared in 1942 by the Mississippi State Department of Agriculture. From top to bottom, the legend is the Northeast Highland, North Central Prairie, Pontotoc Ridge, Flatwoods, Sand-Clay Hills, Mixed Brown Loam and Coastal Plain, Delta, Longleaf Pine, and Coastal Terrace. *Source* Mississippi State Department of Agriculture

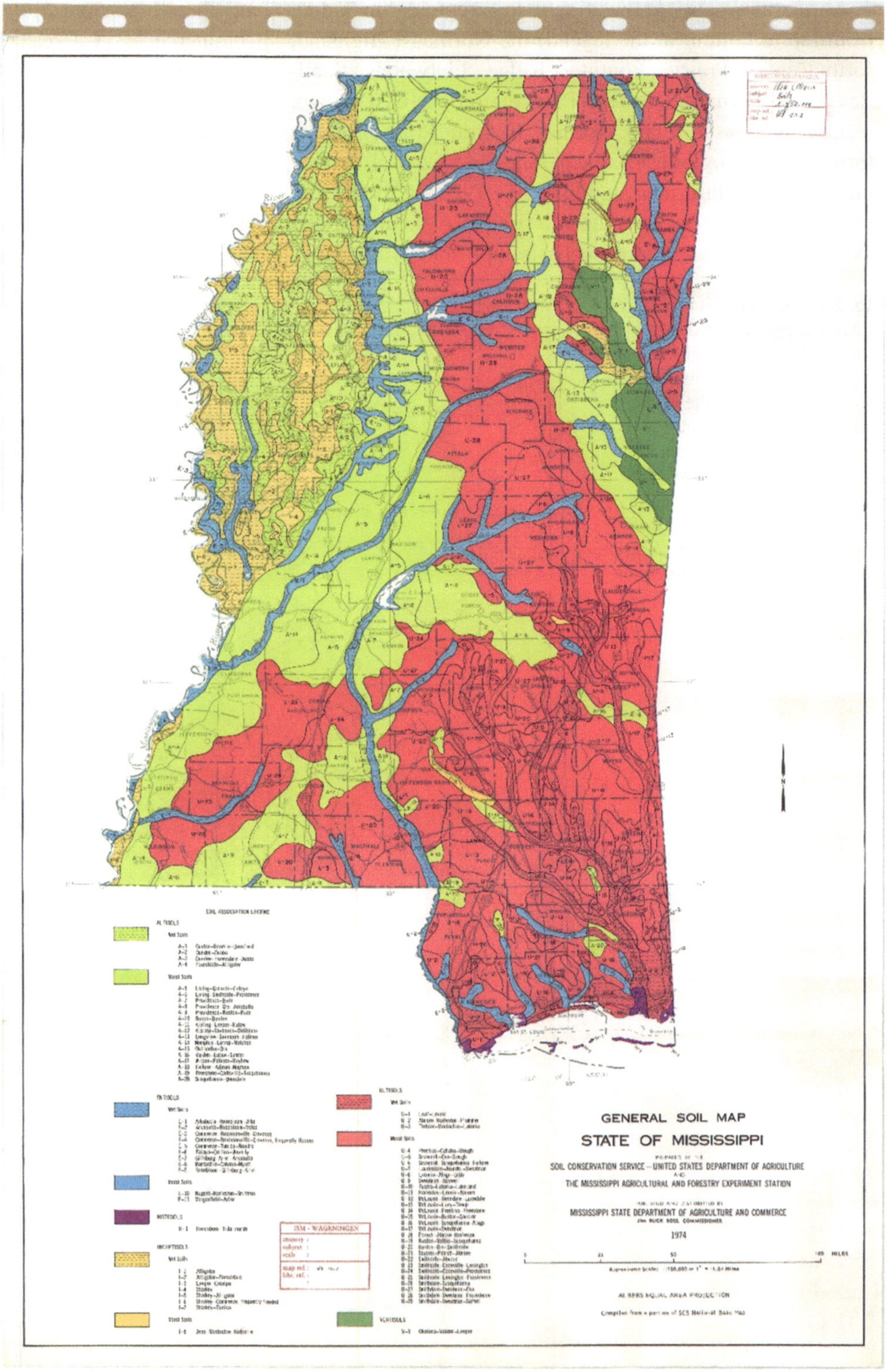

Fig. 2.8 General soil map of Mississippi from 1974. The legend is Alfisols (yellow), Entisols (blue), Histosols (purple), Inceptisols (orange), Ultisols (red), and Vertisols (green). Each order contains from one to 29 soil associations composed of from one to three soil series. *Source* USDA, Soil Conservation Service

Ultisols, and Vertisols (Fig. 2.9). Mollisols and Spodosols accounted for too small of an area to show on the map. The map differences markedly from the 1974 general soil map in that several extensive Alfisols and Inceptisols were reclassified as Vertisols in the Mississippi Delta; Histosols replaced many of the Entisols in river valleys; and Vertisols are more common than Alfisols in the Black Prairies.

In 2015, a "general soil map" was published that shows the ecoregions of Mississippi (to be discussed in Chap. 3) (Fig. 2.10). The detailed description of each ecoregion

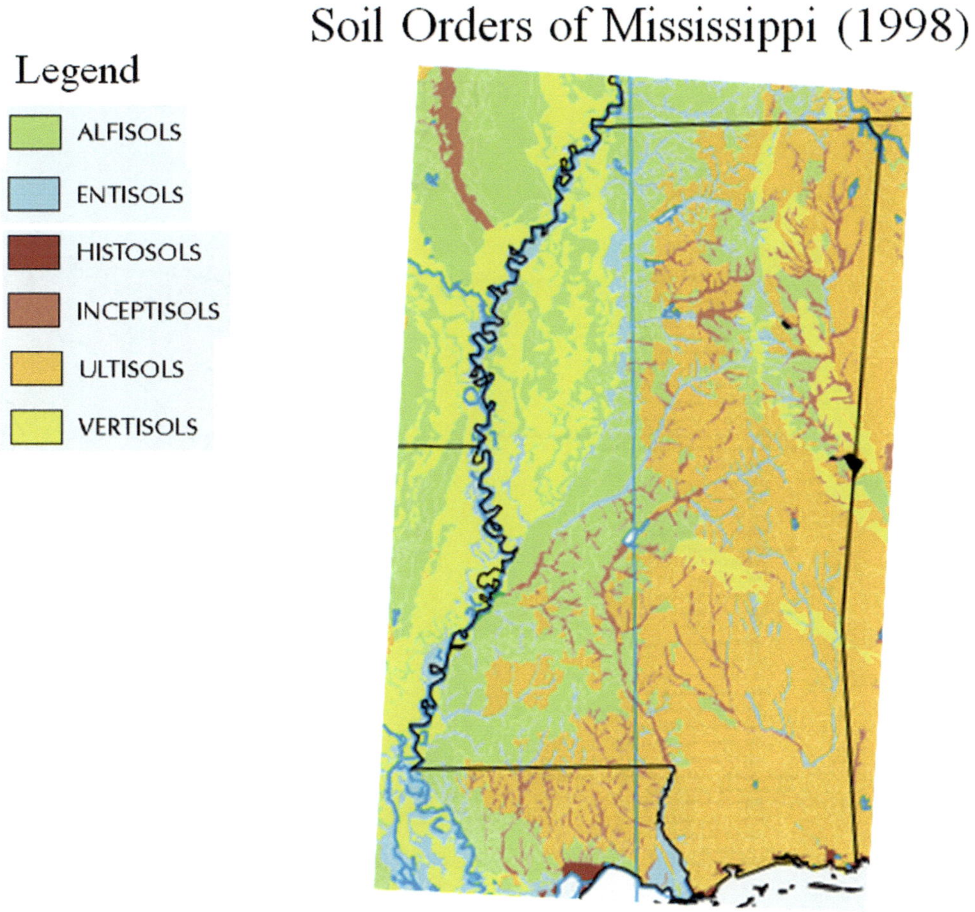

Fig. 2.9 General soil map of Mississippi from 1998. *Source* Natural Resources Conservation Service

contains a list of soil orders/great groups and common soil series. The backside of this map lists 110 major soil series representing 32 great groups. Figure 2.11 is a description of MLRA 131A, the Mississippi Alluvial Plain.

2.9 Soil Research

Mississippi has benefited from considerable soil research by university, NRCS, and USFS investigators over the past 160 years or so. The work of Hilgard (1860), Mabry (1898), and Logan (1916) has already been mentioned. During the 1950s and 1960s, W. A. Raney, W. M. Broadfoot, H. B. Vanderford, and R. C. Glenn published papers on the soils of Mississippi. More recent work has been done by D. E. Pettry, L. Oldham, and J. D. Kushla.

2.10 The State Soil

The Natchez soil series, a coarse-silty, mixed, superactive, thermic Typic Eutrudepts, was approved as the official representative soil of Mississippi in 2003 (Fig. 2.12). The Natchez soil occurs on over 900 km^2 in the Loess Hills, and the native vegetation is mixed hardwoods and loblolly pine. The soil is important for forestry in the state. A small portion of the soil has been cleared and used for pasture. The Natchez soil is formed in thick loess on strongly sloping to very steep hillsides in the dissected bluff hills section of the Southern Mississippi Silty Uplands that border the alluvial plains of the Mississippi River and its tributaries.

2.11 Summary

Mississippi has a rich and long history of soils investigations that began in 1860 with a report on the geology and agriculture of the state of Mississippi by E. W. Hilgard, considered by some as "the father of modern soil science in the United States." Earlier versions of the "Soils of Mississippi" were published by Logan (1916) and Vanderford (1962). Under the auspices of Milton Whitney, the Bureau of Soils conducted the first soil surveys in Mississippi between 1900 and 1910. The first survey was conducted by JA Bonsteel and party in 1901 in the Yazoo Area. As of 2012, all 82 counties in Mississippi had been mapped generally at a scale of 1:24,000. The concept of the soil series has been refined, and 232 soil series have been

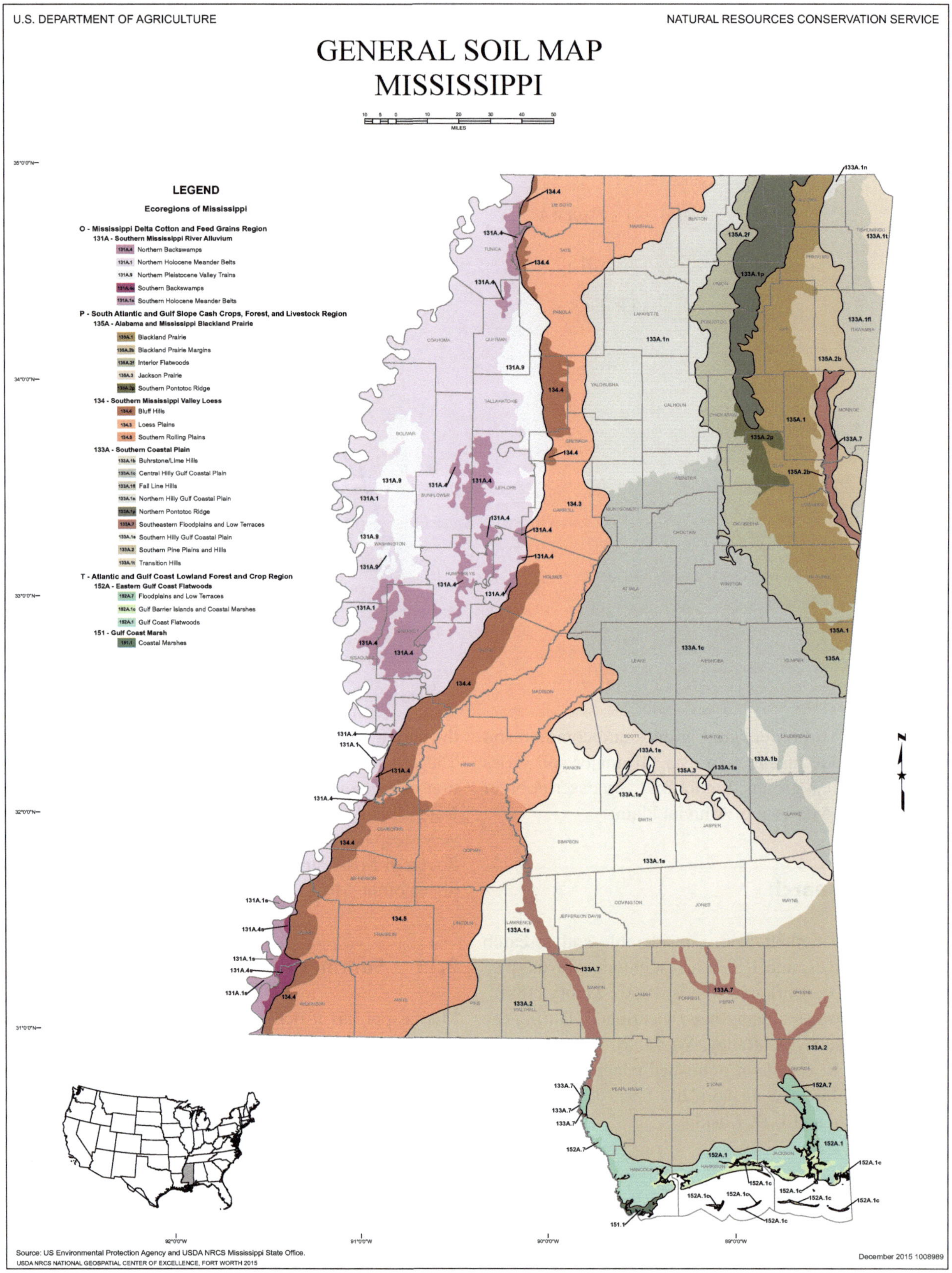

Fig. 2.10 General soil map of Mississippi based on ecoregions (see Fig. 2.11 for example of Mississippi Delta shown in purple). *Please Note* The Mississippi portion of MLRA 133A has been changed to 133C, the Gulf Coastal Plain. *Source* USDA Natural Resources Conservation Service

73. MISSISSIPPI ALLUVIAL PLAIN													
Level IV Ecoregions		Physiography		Geology	Soils			Climate			Potential Natural Vegetation	Land Use and Land Cover
	Area (square miles)		Elevation/ Local Relief (feet)	Surficial and Bedrock	Order (Great Group)	Common Soil Series	Temperature/ Moisture Regimes	Precipitation Mean annual (inches)	Frost Free Mean annual (days)	Mean Temperature January min/max; July min/max (°F)		
73a. Northern Holocene Meander Belts	5428	Flat plains and river meander belt with levees, point bars, oxbows, and abandoned channels. Large rivers and some smaller low-gradient streams, often channelized.	30-214 / 5-40	Quaternary (Holocene) alluvial sand, silt, clay, and gravel.	Inceptisols (Endoaquepts, Dystrudepts), Entisols (Udifluvents, Udipsamments), Alfisols (Hapludalfs, Endoaqualfs), Vertisols (Epiaquerts)	Commerce, Crevasse, Convent, Robinsonville, Dubbs, Beulah, Dundee, Sharkey	Thermic / Udic, Aquic	52-57	210-230	32/51 72/92	Bottomland hardwood forests. Mostly overcup oak, water hickory, Nuttall oak, willow oak, red maple, green ash, elm, and sweetgum. Riverfront forests include black willow, cottonwood, river birch, and sycamore. Bald cypress and water tupelo in wettest areas. Some forested canebrakes with open, mixed deciduous trees and giant cane.	Cropland with mostly cotton and soybeans, also corn, wheat, rice, and sorghum; catfish production, turfgrass and sod production, some deciduous forest and forested wetlands.
73b. Northern Pleistocene Valley Trains	1101	Flat plains with relict patterns of branching channels, irregular braid bars and interfluve areas; low gradient streams with silty substrates, often channelized.	100-190 / 5-25	Quaternary (Holocene and Pleistocene) alluvial sand, silt, clay, and gravel.	Alfisols (Endoaqualfs, Hapludalfs), Vertisols (Epiaquerts), Inceptisols (Epiaquepts, Endoaquepts, Eutrudepts), Entisols (Udifluvents)	Dundee, Askew, Forestdale, Dubbs, Sharkey, Tunica; on floodplains Waverly, Adler, Bruno.	Thermic / Udic, Aquic	54-56	220-230	33/52 72/92	Bottomland hardwood forests with Nuttall oak, willow oak, swamp chestnut oak, water hickory, sweetgum, sugarberry, American elm, and green ash. Some bald cypress and water tupelo in wettest areas. Some forested canebrakes with open, mixed deciduous trees and giant cane.	Cropland with mostly soybeans and some cotton, also corn, wheat, rice, and sorghum; catfish production, some deciduous forest and forested wetlands.
73d. Northern Backswamps	991	Flat plains with depressions containing ponded wetlands, swamps and lakes; some low-gradient streams with silty substrates.	70-200 / 5-25	Quaternary (Holocene) alluvial clay and silty clay, some layers of peat and sand.	Vertisols (Dystraquerts, Epiaquerts), Inceptisols (Endoaquepts, Epiaquepts), Mollisols (Hapludolls)	Alligator, Dowling, Sharkey, Tunica, Bowdre	Thermic / Aquic, some Udic	52-56	210-230	32/51 72/92	In wettest areas, cypress-gum swamps (bald cypress, water tupelo); on less flooded zones, overcup oak, Nuttall oak, willow oak, water hickory, elm, green ash, and sweetgum. Some forested canebrakes with open, mixed deciduous trees and giant cane.	Deciduous forest, forested wetlands, cropland with cotton, soybeans, corn, wheat, rice, and sorghum, catfish production, some public land (Delta National Forest).
73k. Southern Holocene Meander Belts	145	Flat plains and river meander belt with levees, point bars, oxbows, and abandoned channels. Large rivers and some smaller low-gradient streams, often channelized.	25-100 / 5-40	Quaternary (Holocene) alluvial sand, silt, clay, and gravel.	Inceptisols (Endoaquepts, Dystrudepts, Epiaquepts), Entisols (Udifluvents, Udipsamments), Alfisols (Hapludalfs, Endoaqualfs), Vertisols (Epiaquerts)	Commerce, Crevasse, Convent, Robinsonville, Dubbs, Beulah, Dundee, Sharkey, Newellton	Thermic/ Udic, Aquic	58-62	230-250	36/58 72/92	Bottomland hardwood forests. Mostly overcup oak, water hickory, Nuttall oak, willow oak, red maple, green ash, elm, and sweetgum. Riverfront forests include black willow, cottonwood, river birch, and sycamore. Bald cypress and water tupelo in wettest areas. Some forested canebrakes with open, mixed deciduous trees and giant cane.	Forested wetlands, deciduous forest, cropland with cotton, soybeans, corn, and wheat.
73m. Southern Backswamps	91	Flat plains with depressions containing ponded wetlands, swamps and lakes; some low-gradient streams with silty substrates.	35-100 / 5-25	Quaternary (Holocene) alluvial clay and silty clay, some layers of peat and sand.	Vertisols (Epiaquerts), Inceptisols (Endoaquepts, Epiaquepts, Eutrudepts), Mollisols (Hapludolls)	Sharkey, Tunica, Bowdre, Adler, Bruin, Dowling	Thermic / Aquic, some Udic	58-62	230-250	36/58 72/92	In wettest areas, cypress-gum swamps (bald cypress, water tupelo); on less flooded zones, overcup oak, water hickory, willow oak, Nuttall oak, sweetgum, elm, and green ash. Some forested canebrakes with open, mixed deciduous trees and giant cane.	Forested wetlands, deciduous forest, cropland with soybeans, corn, and wheat.

Fig. 2.11 Details for Mississippi Alluvial Plain (MLRA 131A) for Fig. 2.10 General Soil Map of Mississippi *Source* by J. Bockheim

Fig. 2.12 Natchez soil series, a coarse-silty, mixed, superactive, thermic Typic Eutrudepts, is the state soil of Mississippi. *Source* forces.si.edu/soils/interactive/statesoils/html

mapped in Mississippi. Three general soil maps have been produced for Mississippi, the most recent being a 1974 map that showed 29 soil associations within six orders. A 2015 map based on ecoregions of Mississippi lists 110 major soil series representing 32 great groups. The Natchez soil series, a coarse-silty, mixed, superactive, thermic Typic Eutrudepts, was approved as the official representative soil of Mississippi in 2003.

References

Baldwin M, Kellogg CE, Thorp J (1938) Soil classification. Soils and men. U.S. Dep. Agric. Yearbook. U.S. Govt. Print, Washington, DC, pp 979–1001

Bonsteel JA and party (1901) Soil survey of the Yazoo Area, Mississippi. Field Oper Bureau Soils 359–388

Buol SW, Southard, SJ, Graham RC, McDaniel PA (2011) Soil genesis and classification. Wiley-Blackwell 543 pp

Hilgard EW (1860) Report on the geology and agriculture of the state of Mississippi. E. Barksdale, state printer, Jackson, MS. 393 pp

Hutchinson WL (1910) Soils of Mississippi. Miss Agr Exp Stn

Jenny H (1961) E. W. Hilgard and the birth of modern soil science. Pisa, Italy, 144 pp

Kushla JD, Londo AJ (2014) Mississippi soils, Chapter 3. In: Rohnke AT, Cummins J (eds) Fish and wildlife management: a handbook for Mississippi landowners. University Press of Mississippi, Jackson, MS

Logan WN (1916) The soils of Mississippi. Miss Agr Exp Stn, Tech Bull No. 7

Mabry TO (1898) The brown or yellow loam of north Mississippi, and its relation to the northern drift. J Geol 6:273–302

Pettiet JV (1974) An interpretive evaluation of soils in the Yazoo-Mississippi Delta area for crop production. Miss Agr For Exp Bull 808

Pettry DE (1977) Soil resource areas of Mississippi. Miss State Univ Miss Agr For Exp Stn, Infor Sheet 1278, 4 pp

Pettry DE (2008) Mississippi soil surveys. Miss State Univ Miss Agric For Exp Stn, 5 pp

Pettry DE, Switzer RE, Hinton RB (1995) Temporal water table levels and characteristics of representative Mississippi soils. Miss Agr For Exp Stn Bull, p 1027

Soil Survey Staff (1999) Soil taxonomy: a basic system of soil classification for making and interpreting soil surveys. 2nd edit. USDA, NRCS, Agric Handb No. 436, 871 pp

Soil Survey Staff (2014) Keys to soil taxonomy, 12th edit. USDA, NRCS

Thorson T, McGrath C, Moberg D, Fillmore M, Campbell S, Lammers D, Bockheim J (2022) The soils of Oregon. World Soils Book Series. Springer Nature, Switzerland, 545 pp

Vanderford HB (1962) Soils of Mississippi. State Univ Miss Agr Exp Stn State College, MS, Miss

Weil RR, Brady NC (2016) The nature and properties of soils, 15th edit. Pearson, 1104 pp

3 Soil-Forming Factors

3.1 Introduction

The expression of a soil results from five factors operating collectively: climate, organisms, relief, parent material, and time. The factors interact and cause a range of soil processes (e.g., argilluviation) that result in a diversity of soil properties (e.g., high clay content in the subsoil). Human activities cause soil changes and are often considered a sixth factor. Following the "Russian school of soil science," Hans Jenny published *Factors of Soil Formation* in 1941, in which he described soil (s) as the result of climate (cl), organisms (o), topography (r), parent material (p) and time (t); the acronym CLORPT known to most pedologists today. The following is a review of the role of soil-forming factors in the development of soils in Mississippi. Soil-forming factor data for soil series in Mississippi are given in Appendix B.

3.2 Climate

3.2.1 Current Climate

The climate of Mississippi is influenced by a latitudinal difference of 4.5° along a distance of 495 km, proximity to the Gulf of Mexico, and a range in elevations from 0 to 246 m.

According to the Köppen climate system, Mississippi has a Humid Subtropical (Cfa) climate type characterized by temperate winters and long, hot summers (Fig. 3.1). The mean annual air temperature ranges from 16 °C in northern Mississippi to 20.5 °C along the Gulf Coast. The coldest area in Mississippi is the Holly Springs area where the average monthly temperature of the coldest month is 7.8 °C and the warmest is the Picayune area where the average temperature of the warmest month is 26 °C.

The mean annual precipitation varies from less than 1320 mm in the northern Mississippi River valley to more than 1725 mm along the Gulf Coast and is fairly evenly distributed through the year (Fig. 3.2). Prevailing northerly winds from the Gulf Coast create a strong gradient in moisture from the coast to about Jackson and enables locally violent and destructive thunderstorms, hurricanes, and tornadoes particularly from May through September. Heavy snowfall is uncommon in Mississippi, particularly in the south. Droughts and flooding are common in the state.

Soil temperature regimes (STRs) are determined by the mean soil temperature at a depth of 50 cm. Mississippi has a thermic soil temperature regime throughout the state, in which the mean annual soil temperature (MAST) is above 15 °C but below 22 °C. However, the Picayune area in southern Hancock County has a hyperthermic STR, where the MAST is above 22 °C.

Soil moisture regimes (SMRs) are classified according to the presence of water in the soil moisture control section, which corresponds to the rooting depths of many crops. Technically, the upper boundary of the soil moisture control section is the depth to which 2.5 cm of water will moisten a dry soil within 24 h, and the lower boundary is the depth to which 7.5 cm of water will moisten the soil within 48 h (Soil Survey Staff 1999). Mississippi has two SMRs, including an aquic SMR, in which reducing conditions occur from saturation by water, and an udic SMR in which the soil moisture control section is not dry in any part for 90 or more cumulative days and dry less than 45 consecutive days in normal years (Fig. 3.3).

Differences in atmospheric and soil temperature regimes are reflected in the plant hardiness zone map of Mississippi (Fig. 3.4). The most favorable growing conditions (the longest growing season) occur in the southern portions of the state. There are 210 growing days in northern Mississippi and 270 in the south.

D. Johnson et al., *The Soils of Mississippi*, World Soils Book Series,
https://doi.org/10.1007/978-3-031-36235-4_3

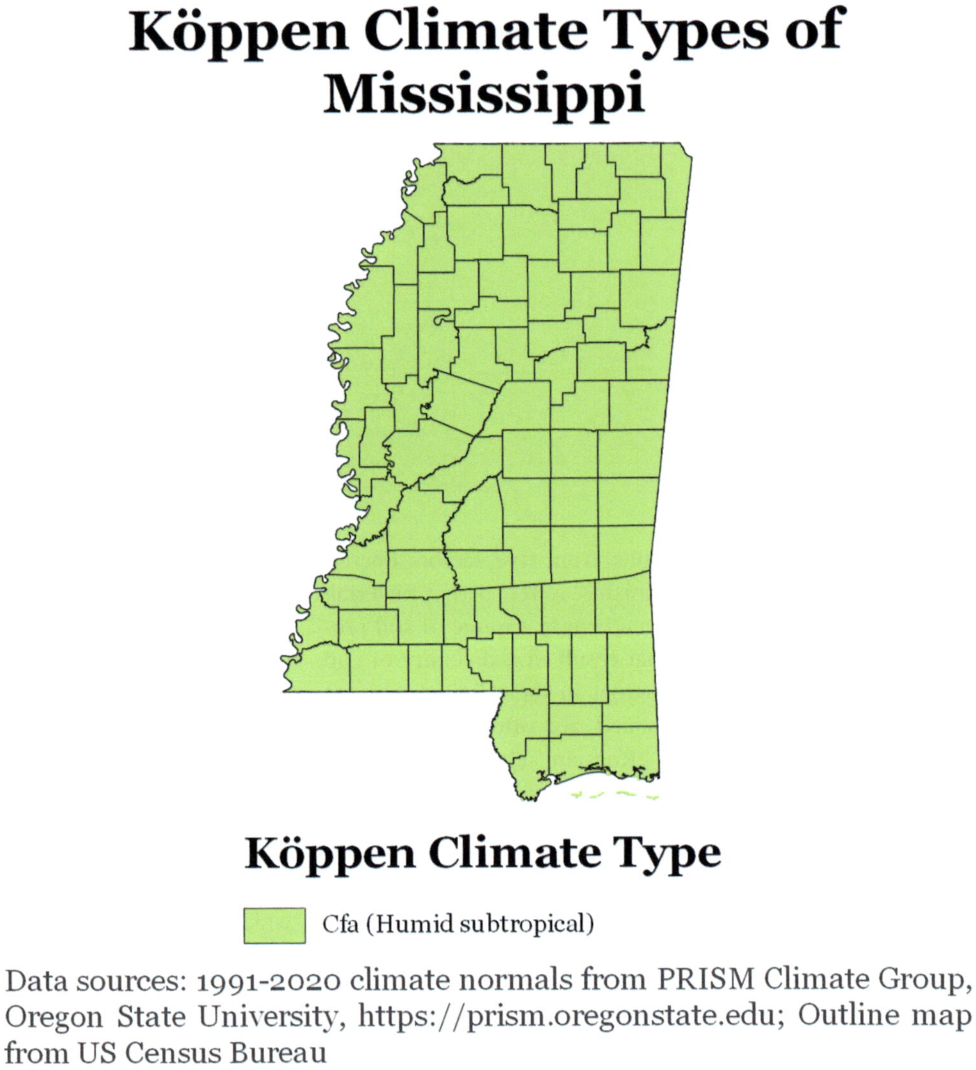

Fig. 3.1 Koppen climate type for Mississippi *Source* by J. Bockheim

3.2.2 Past Climates

The climate of Mississippi has varied considerably over geologic time, particularly over the past 1.8 million years with the onset of the Pleistocene glaciers, particularly with regards to the Mississippi River and loess deposition during glacial periods (Russell et al. 2021). The Mississippi Delta contains evidence for climate/sea-level connections during the late Holocene (González and Törnqvist 2009).

3.2.3 Recent Climate Change

As of 2016, Mississippi has not warmed substantially during the last 50–100 years (EPA 2016). However, Hattiesburg has experienced 3.1 more days with temperatures above 35 °C in the past 70 years (http://statesatrisk.org). Global warming is associated with a 35-cm rise in sea level along the Gulf Coast during the past century. Additional impacts of climate change of the natural resources of Mississippi are considered in Chap. 16.

3.3 Organisms

The soil-forming factor—organisms—includes flora and fauna. In Mississippi, two examples of fauna influencing soil development include fire ants and crayfish. Mounds created by fire ants (*Solenopsis* spp.) contained higher amounts of clay, phosphorus, and potassium and lower amounts of soil organic matter than in areas not influenced by these organisms (Green et al. 1998). Crayfishes occur throughout Mississippi, not only in swamps, marshes, vernal pools, and floodplains, but also in savannas with little or

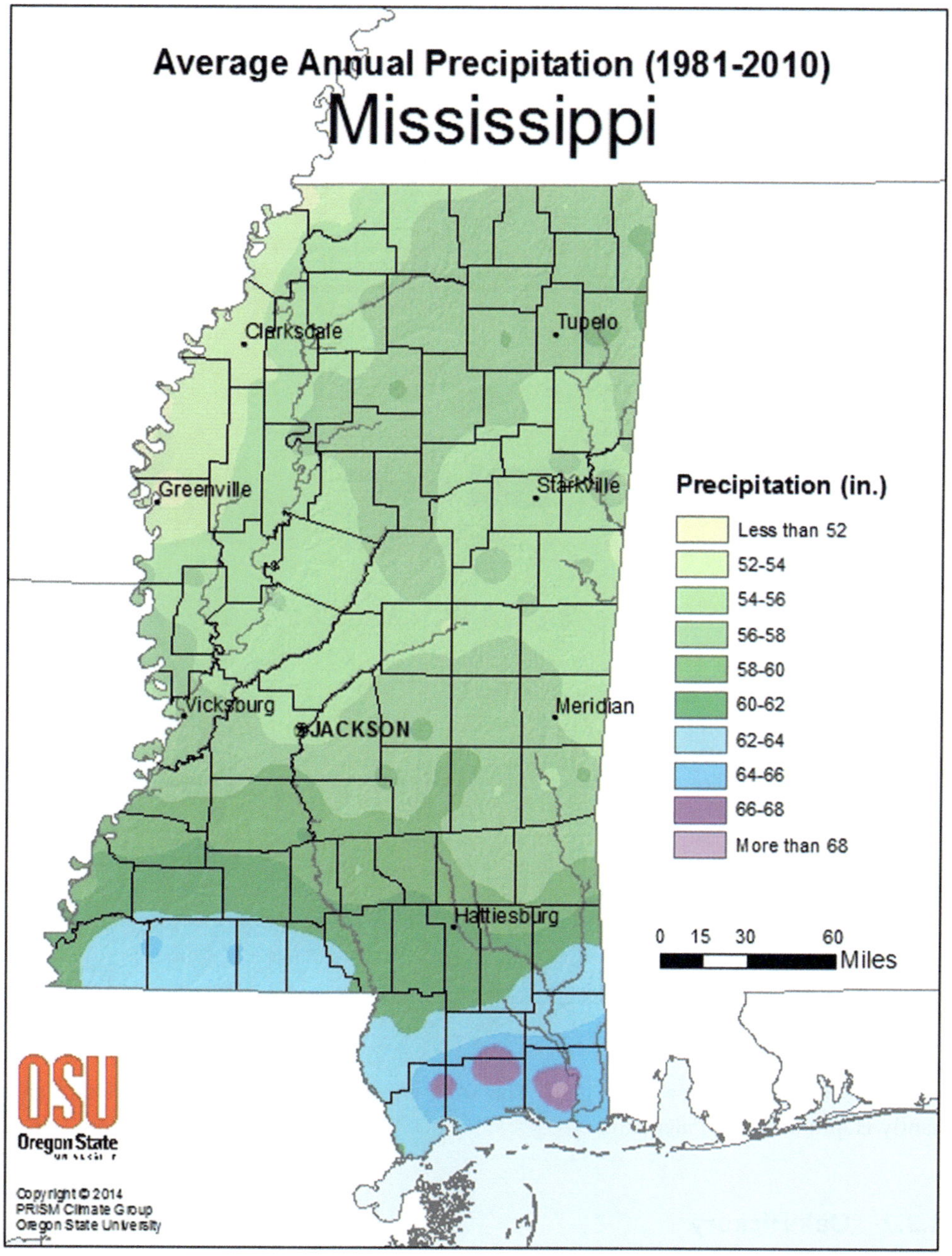

Fig. 3.2 Distribution of mean annual precipitation in Mississippi. *Source* PRISM, Oregon State Univ

no water. An examination of burrows of Camp Shelby crayfish (*Fallicambarus gordoni*) showed an annual soil disturbance of 82 T/ha/yr (81 t/a/yr) (Welch et al. 2008).

The role of flora on soil development has received considerably more attention nationally than in Mississippi. Historically, about 90% of Mississippi was covered by forest. Today, forest covers from 60 to 65% of the state, with the remainder in agricultural, grazing, or developed land (Fig. 3.5). The Mississippi Forestry Commission recognizes 13 forest communities in the state. These are shown in Fig. 3.5 primarily in green. The US Forest Service has identified six dominant forest types in Mississippi, which are described below. Table 3.1 compares the two systems, along with forest cover types recognized by the Society of American Foresters in Mississippi.

3.3.1 Loblolly Pine-Shortleaf Pine

Loblolly pine-shortleaf pine is the most extensive forest cover type in Mississippi, occupying 38% of the state. Although it is widespread, it is most common in the North Central Hills physiographic province. The indicator species are loblolly pine and shortleaf pine (Fig. 3.6a). Common

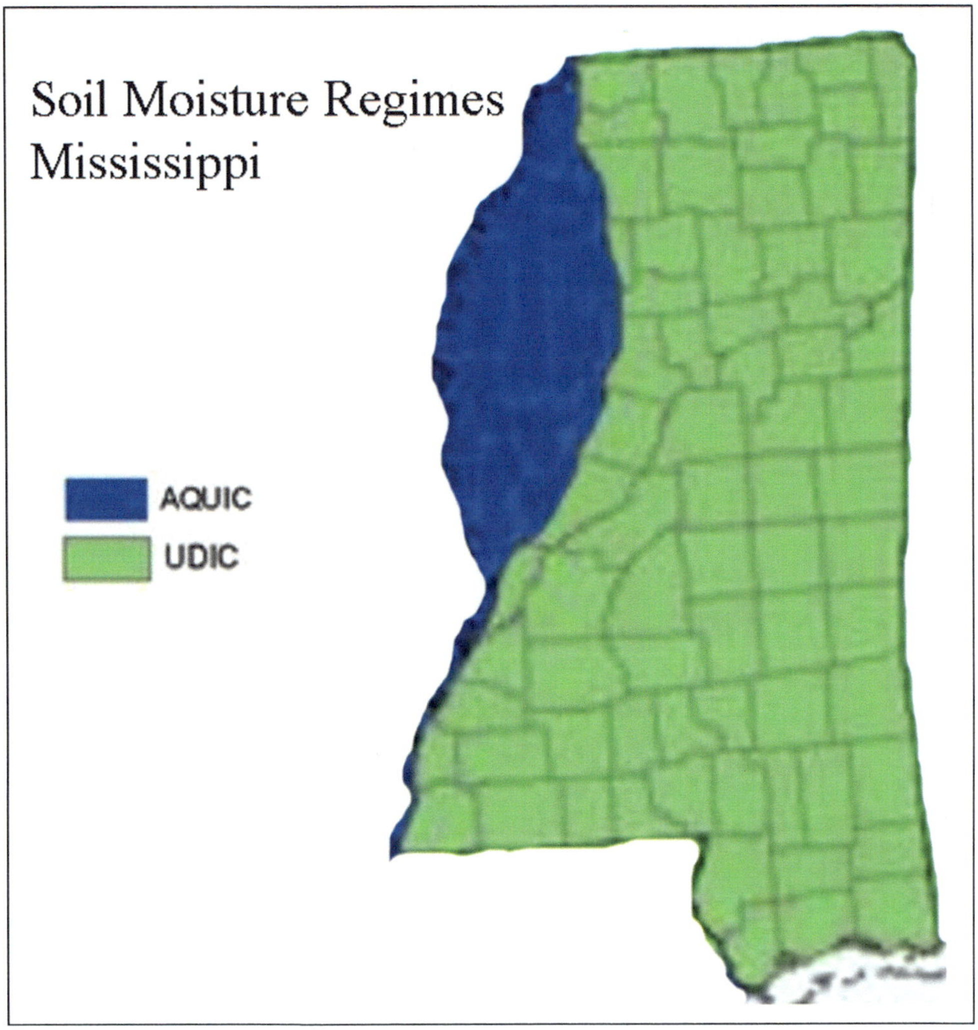

Fig. 3.3 Soil moisture regimes for Mississippi. *Source* USDA, Natural Resources Conservation Service

associates are oak, hickory, and gum. The soils are dominantly Hapludults and Paleudults.

3.3.2 Oak-Hickory

Oak-Hickory is the second most common forest type in Mississippi, occupying 26% of the land area, mainly in the Loess Hills physiographic province. This forest type contains mainly oak and hickory but common associates are yellow poplar, winged elms, red maple, and black walnut (Fig. 3.6b). The dominant soils are Hapludalfs, Paleudalfs, Fraglossudalfs, Fragiudalfs, and Fragiudults.

3.3.3 Oak-Gum-Cypress

Oak-Gum-Cypress is the third most common forest type in Mississippi, composing 12% of the state, primarily in the Mississippi Delta and in the bottomlands of the major streams of the state. The indicator species are water tupelo, black tupelo, sweetgum, oak, and baldcypress, but common associates are cottonwood, willow, ash, elm, hackberry, and maple (Fig. 3.6c). Common soils are Endoaquepts, Epiaquepts, Dystrudepts, Haplosaprists, and Udifluvents.

3.3.4 Oak-Pine

Oak-Pine is the fourth most common forest type in Mississippi, accounting for 10% of the state area and occurring primarily in the upper Southern Coastal Plain, Blackland Prairies, and Interior Flatwoods. This forest type contains a variety of oaks and pines, as well as eastern red cedar (Fig. 3.6d). Common associates are sweetgum, black tupelo, hickory, and yellow poplar. Common soils are Hapludults, Paleudults, Hapludalfs, Paleudalfs, and Fragiudults.

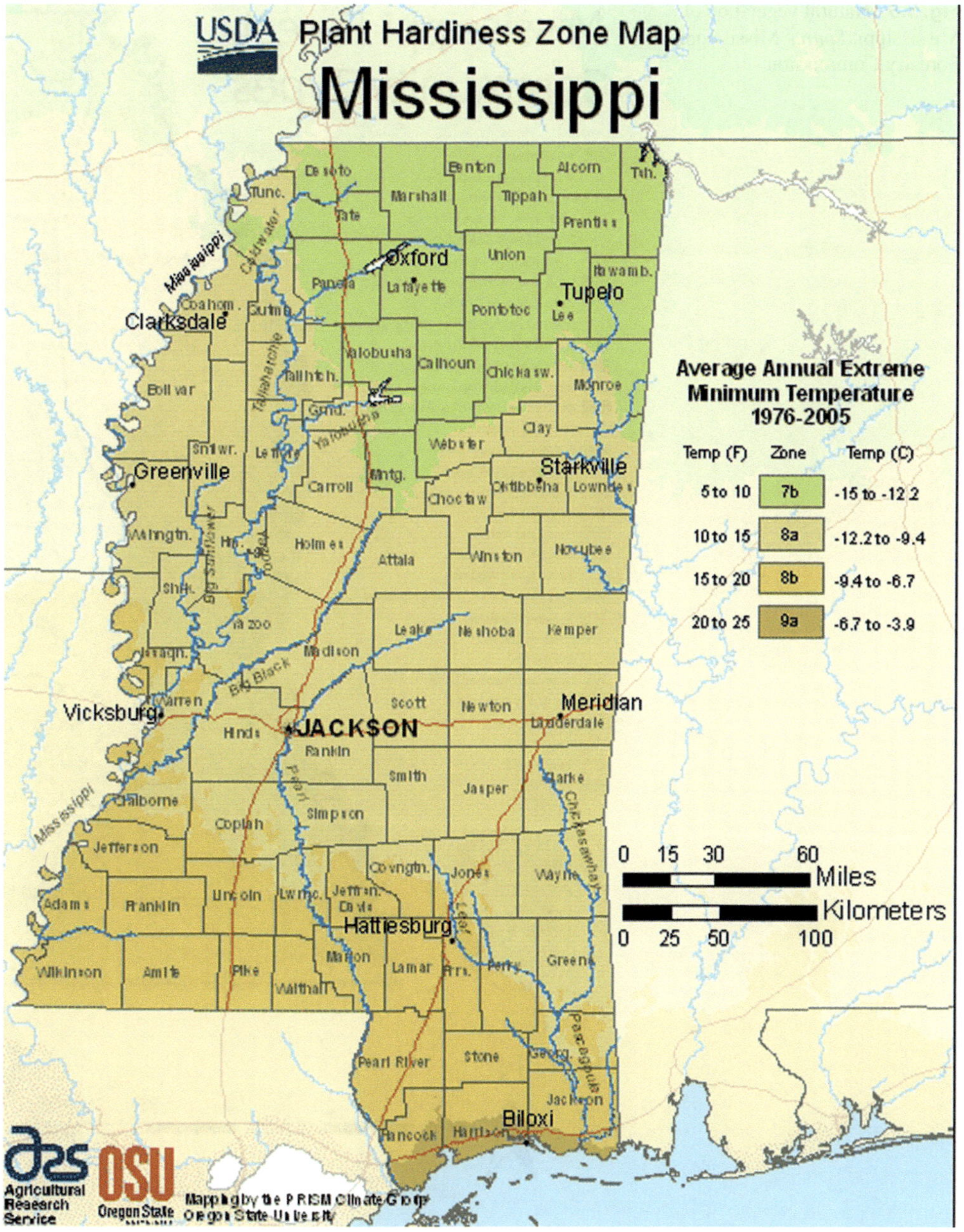

Fig. 3.4 Plant hardiness zones for Mississippi. *Source* PRISM, Oregon State Univ

3.3.5 Longleaf Pine-Slash Pine

The longleaf pine-slash pine forest cover type comprises 4.4% of the area of Mississippi, primarily in the lower Coastal Plain or Southern Pine Hills physiographic province. Whereas longleaf pine and slash pine are the indicator species, common associates are other pines, oak, and gum (Fig. 3.6e). Dominant soil great groups are Paleudults, Paleaquults, and Hapludults.

3.3.6 Elm-Ash-Cottonwood

Elm-Ash-Cottonwood comprises 4.2% of the state and occurs mainly along the bottoms of major streams. Elm, ash, and cottonwood are the indicator species, but common associates are willow, sycamore, beech, and maple. Common soils are in the Eutrudepts and Udifluvents great groups.

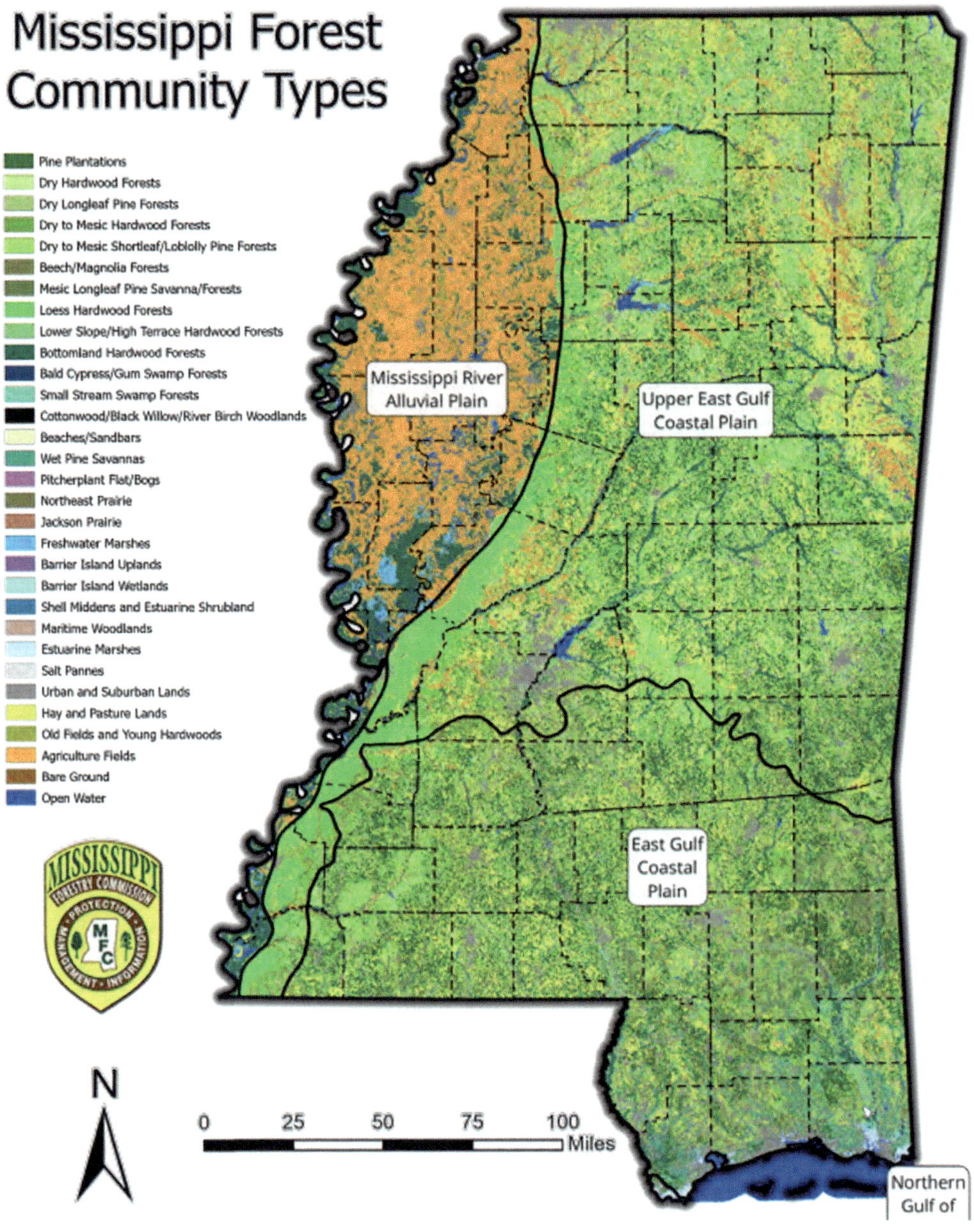

Fig. 3.5 Natural vegetation of Mississippi. *Source* Mississippi Forestry Commission

3.3.7 Other Vegetation Types

Other less common vegetation types in Mississippi include prairies, marshes.

3.3.8 Ecoregions of Mississippi

The US Environmental Protection Agency has identified four level III and 21 level IV ecoregions in Mississippi (Fig. 3.7). The ecoregion map is at a scale of 1:1 million. Ecoregions denote areas of general similarity in ecosystems and in the type, quality, and quantity of environmental resources, including geology, physiography, vegetation, climate, soils, land use, wildlife, and hydrology. The level IV ecoregions is comparable with the 2015 general soil map of Mississippi (Fig. 2.7). The three main differences are (i) the Southern Plains (ecoregion 65) includes MLRA 135A, the Alabama and Mississippi Blackland Prairie, and 133C, the Gulf Coastal Plain; (ii) the Northern Pontotoc Ridge (135A.1) and Southern Pontotoc Ridge (135A.2p) are contained within the Blackland Prairie (ecoregion 65a); and (iii) the Gulf Barrier Islands and Coastal Marshes (ecoregion 75k) is separated into the Gulf Barrier Islands and Coastal Marshes (MLRA 152A) and Coastal Marshes (MLRA 151) (Table 3.2).

Table 3.1 Forest cover type classification schemes for Mississippi

US forest service MS	Mississippi forestry commission	Society of American Foresters MS
Loblolly pine-shortleaf pine	Loblolly pine-shortleaf pine	Loblolly pine-shortleaf pine
	Loblolly-shortleaf-mixed hardwood with scattered prairies	
	Shortleaf/longleaf pine-Loblolly pine-upland hardwood	
	Shortleaf/Longleaf pine-upland hardwood-loblolly pine	
Oak-hickory	Oak-hickory-magnolia-poplar	
Oak-gum-cypress	Bottomland hardwood (oak-gum-cottonwood-cypress)	Baldcypress-tupelo
Oak-pine	Upland hardwood-shortleaf pine-loblolly pine	Loblolly pine-hardwood
	Loblolly/shortleaf pines-oak	Shortleaf pine-oak
		Loblolly pine-hardwood
	Post oak-blackjack oak-shortleaf pine	
	Red oak-hickory-shortleaf pine	
	Shortleaf-loblolly pine-Post/blackjack oak	
Longleaf pine-slash pine	Longleaf pine with loblolly pine-slash pine	Longleaf pine-slash pine
	Slash pine with longleaf pine-red bay-savannas	
Elm-ash-cottonwood		Cottonwood
		Sugarberry-American elm-Green ash
	Intermittent prairies with blackjack oak-post oak-hawthorn	Post oak-blackjack oak

(a)

Fig. 3.6 **a** Loblolly pine and shortleaf pine forest in Mississippi. *Source* Miss. State Univ. Ext. Serv. **b** Oak-hickory forest along the Natchez Trace. Photo by US National Park Service. **c** Bottomland hardwoods near Sledge, Mississippi. Photo by J. Bockheim. **d** Oak-pine forest near Noxubee National Wildlife Refuge. Photo by Dr. JoVonn G. Hill, Department of Biochemistry, Molecular Biology, Entomology and Plant Pathology, Mississippi State University. **e** Longleaf and slash pine near Gautier, Mississippi. Photo by J. Bockheim

(continued)

(continued)

(e)

Fickle (2001) provided an interesting history of how people have used Mississippi's forests since settlement.

3.4 Relief

Relief is a measure of surface roughness, or quantitatively, the measurement of elevation change in a landscape. Relatively smooth terrain may be described as having low relief, whereas areas deeply dissected and with steep slopes have high relief (Fig. 3.8). Relief gives rise to slope morphometry (e.g., length, gradient, aspect, position, and shape), which in turn influence erosional, gravitational, and depositional processes. Relief is an important soil-forming factor that results in catenas, topographic or hydromorphic sequences of soils. These are illustrated in block diagrams in Chap. 5. Switzer and Pettry (1992) reported that fragipan development, total phosphorus, and maximum clay content in loessial Fragiudalfs decreased with an increase in slope.

3.5 Physiographic Provinces

Mississippi has been divided into 11 principal physiographic provinces (Fig. 3.9). The largest of these is the Southern Pine Hills (68,900 km^2), followed by the North Central Hills (66,900 km^2), Mississippi Delta (66,300 km^2), Loess Hills (50,400 km^2), Tombigbee and Tennessee River Hills (32,400 km^2), Black Prairie (28,000 km^2), Jackson Prairie (26,000 km^2), Flatwoods (23,100 km^2), Pontotoc Ridge (19,400 km^2), Coastal Meadows (15,200 km^2), and Paleozoic Bottoms (12,400 km^2).

3.6 Geologic Structure

The geologic structure of Mississippi is strongly related to the physiographic provinces. Quaternary sediments include alluvium in the Mississippi Delta, coastal deposits in the Coastal Meadows, loess in the Loess Hills, and clayey sands of the Citronelle Formation in the Coastal Plain (Fig. 3.10). Tertiary sediments include the Pascagoula-Hattiesburg Formations (clay and interbedded sands) and the Catahoula Formation (mudstones and sandstones) in the Southern Pine Hills; the Vicksburg Group-Forest Hill Formation (laminated sands, silts, and clays) and Jackson Group (clay) in the Jackson Prairie; the Claiborne Group (glauconitic clays) and Wilcox Group (mudstones and sandstones) in the North Central Hills; and the Midway Group (bedded clays and sands) in the Flatwoods Province. The oldest sediments are the Cretaceous Selma Group (chalk and glauconitic sandstone) in the Black Prairie; the Cretaceous Eutaw-Tuscaloosa Groups in the Tombigbee and Tennessee River Hills; and the Mississippian Devonian in the Paleozoic Bottoms.

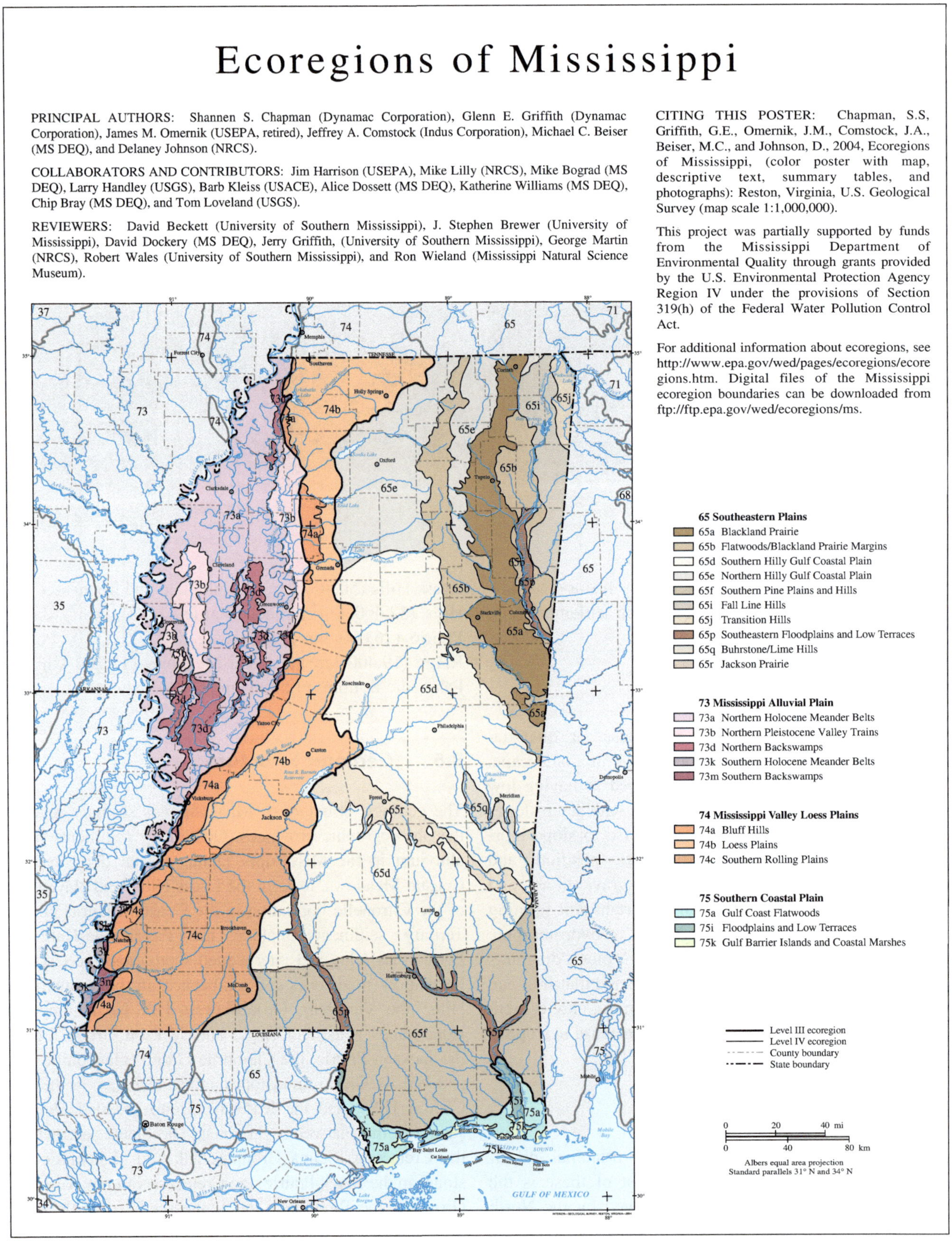

Fig. 3.7 Ecoregions of Mississippi. *Source* US Environmental Protection Agency

Table 3.2 A comparison of ecoregions and general soil regions of Mississippi

Ecoregions of Mississippi[a]	General Soil Map of Mississippi (MLRA)[b]
Southeastern Plains (65)	Alabama and Mississippi Blackland Prairie (135A) and Southern Coastal Plain (133A)
Blackland Prairie (65a)	Blackland Prairie (135A.1) and Southern Pontotoc Ridge (135A.2p)
Flatwoods/Blackland Prairie Margins (65b)	Interior Flatwoods (135A.2f) and Blackland Prairie Margins (135A.2b)
Southern Hilly Gulf Coastal Plain (65d)	Central Hilly Coastal Plain (135A.1c) and Southern Hilly Gulf Coastal Plain (133A.1s)
Northern Hilly Gulf Coastal Plain (65e)	Northern Hilly Gulf Coastal Plain (133A.1n) and Northern Pontotoc Ridge (133A.1p)
Southern Pine Plains and Hills (65f)	Southern Pine Plains and Hills (133A.2)
Fall Line Hills (65i)	Fall Line Hills (133A.1fl)
Transition Hills (65j)	Transition Hills (133A.1t)
Southeastern Floodplains and Low Terraces (65p)	Southeastern Floodplains and Low Terraces (133A.7)
Buhrstone/Lime Hills (65q)	Buhrstone/Lime Hills (133A.1b)
Jackson Prairie (65r)	Jackson Prairie (135A.3)
Mississippi Alluvial Plain (73)	Southern Mississippi River Alluvium (131A)
Northern Holocene Meander Belts (73a)	Northern Holocene Meander Belts (131A.1)
Northern Pleistocene Valley Trains (73b)	Northern Pleistocene Valley Trains (131A.9)
Northern Backswamps (73d)	Northern Backswamps (131A.4)
Southern Holocene Meander Belts (73k)	Southern Holocene Meander Belts (131A.1s)
Southern Backswamps (73m)	Southern Backswamps (131A.4s)
Mississippi Valley Loess Plains (74)	Southern Mississippi Valley Loess (134)
Bluff Hills (74a)	Bluff Hills (134.4)
Loess Plains (74b)	Loess Plains (134.3)
Southern Rolling Plains (74c)	Southern Rolling Plains (134.5)
Southern Coastal Plain (75)	Eastern Gulf Coast Flatwoods (152A) and Gulf Coast Marsh (151)
Gulf Coast Flatwoods (75a)	Gulf Coast Flatwoods (152A.1)
Floodplains and Low Terraces (75i)	Floodplains and Low Terraces (152A.7)
Gulf Barrier Islands and Coastal Marshes (75k)	Gulf Barrier Islands and Coastal Marshes (152A.1c) and Coastal Marshes (151.1)

[a] Chapman et al. (2004)
[b] USDA, Natural Resour. Conserv. Serv. (2015) General soil map of Mississippi

3.7 Surficial Geology

The surficial geology of Mississippi is depicted in Fig. 3.11. The yellow pattern is alluvium; the cream-colored pattern is loess; the violet-colored pattern is marine sediments from sedimentary rocks; the light purple pattern is residuum from fine-grained rocks such as shale that contain smectitic (i.e., "swelling") materials; the gray zone is residuum from limestone; the purple pattern is marine sediments from alluvium adjacent to coastal zone deposits; and the blue is the coastal zone. Figure 3.12 shows the distribution (% of land area) of parent materials in soil series of Mississippi, with alluvium accounting for 50% of the parent materials, followed by marine (25%), and loess (21%).

Salt Domes of Mississippi

Salt Domes are unique geologic features that occur in south central Mississippi as part of the Mississippi Interior Salt Basin (Figs. 3.13 and 3.14). Salt domes are of interest because they contain salt and petroleum, may be a source limestone, sulfur, or gypsum, may be used for nuclear weapons test denotations, and may be used for storage of petroleum or hazardous and nuclear wastes. The Richton Salt Dome, near Hattiesburg, has received attention for potential expansion of the US Strategic Petroleum Reserve. The salt domes are sometimes visible from the surface as broad domes containing sulfur springs or natural gas vents, but they generally require seismic

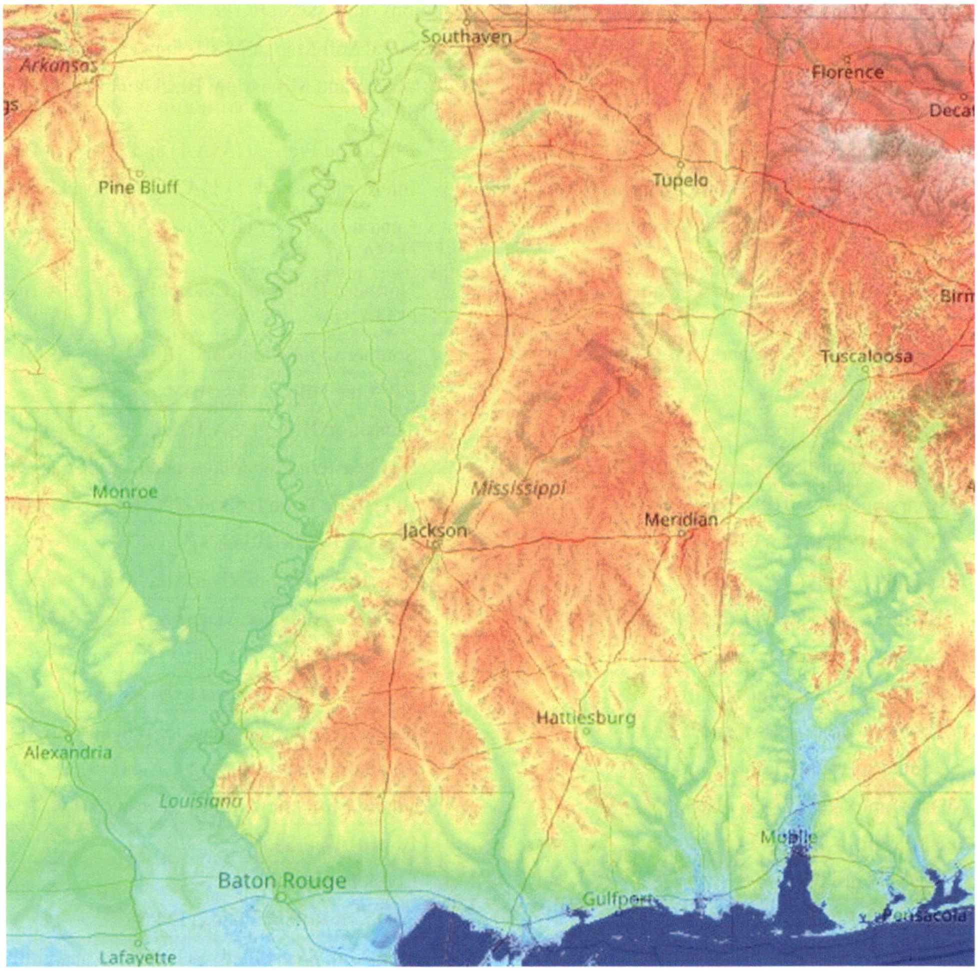

Fig. 3.8 A colored relief map of Mississippi showing the highest elevations in white and red, intermediate elevations in yellow and green, and lowest elevations in light blue. Mean sea level is depicted in dark blue. *Source* http://en-us.topographic-map.com/maps/siu9/Mississippi-River/

techniques or drilling to verify. The salt-enriched zone is forced into a salt dome by the weight of thousands of meters of overlying sediment.

3.8 Time

The soils of Mississippi range from late Holocene (Fluvaquents, Udifluvents, and Udorthents) to the Pliocene or earlier. The oldest soils are Ultisols (Paleaquults, Kandiudults, and Paleudults).

3.9 Humans

Humans have influenced soil development in Mississippi through urbanization, cultivation, irrigation, logging, and other practices that accelerated soil erosion, flooding, and mass wasting. Wood and Pettry (1989) showed the impacts of surface mining on a prime farmland soil in the Southern Mississippi Valley Alluvium. Drastic disturbance created an "open" soil structural matrix that led to subsidence. This topic will be considered more fully in Chap. 15.

3.10 Summary

The expression of a soil results from five factors operating collectively: climate, organisms, relief, parent material, and time. According to the Köppen climate system, Mississippi has a Humid Subtropical (Cfa) climate type characterized by temperate winters and long, hot summers. The mean annual air temperature ranges from 16 °C in northern Mississippi to 20.5 °C along the Gulf Coast. The mean annual precipitation varies from less than 1320 mm in the northern Mississippi River valley to more than 1725 mm along the Gulf Coast and is fairly evenly distributed through the year. Prevailing northerly winds from the Gulf Coast create a strong gradient in moisture from the coast to about Jackson and enables locally violent and destructive thunderstorms, hurricanes, and tornadoes particularly from May through September.

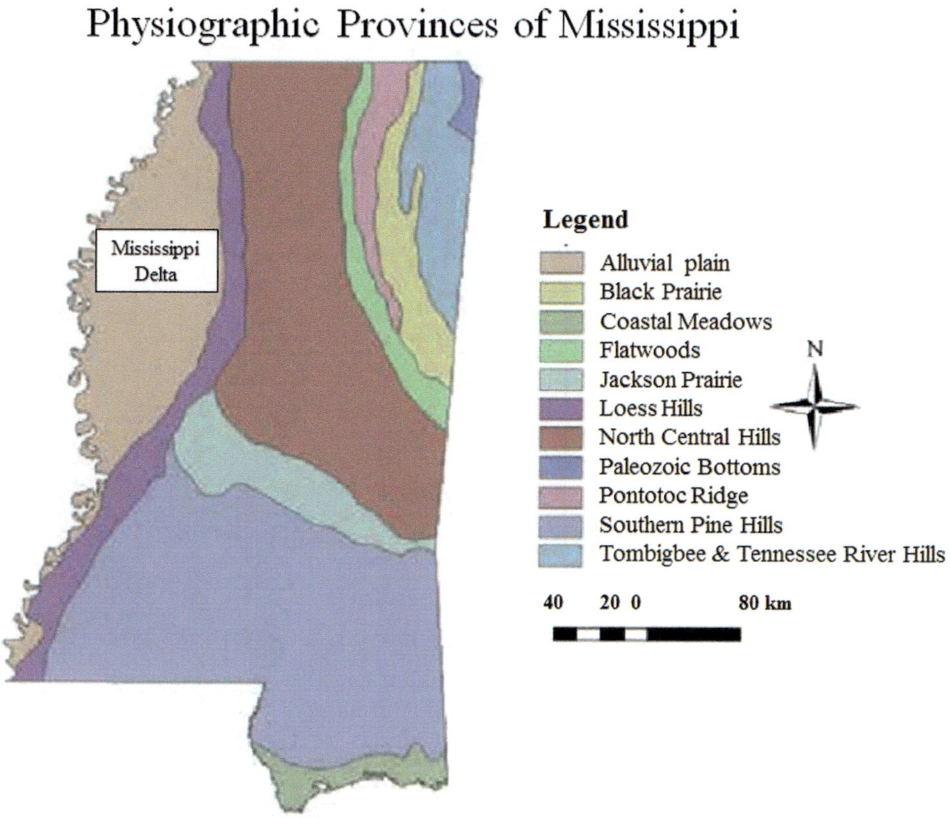

Fig. 3.9 Physiographic provinces of Mississippi. *Source* Hossain (2014)

Heavy snowfall is uncommon in Mississippi, particularly in the south. Droughts and flooding are common in the state. Historically, nearly 90% of Mississippi was covered by forest; forest currently covers about 62% of the state. There are six dominant forest types in Mississippi, including from greatest to least in area, loblolly pine-shortleaf pine, oak-hickory, oak-gum-cypress, oak-pine, longleaf pine-slash pine, and elm-ash-cottonwood. There are 21 level IV ecoregions in the state. Mississippi has been divided into 11 principal physiographic provinces, including from largest to smallest, the Southern Pine Hills, North Central Hills, Mississippi Delta, Loess Hills, Tombigbee and Tennessee River Hills, Black Prairie, Jackson Prairie, Flatwoods, Pontotoc Ridge, Coastal Meadows, and Paleozoic Bottoms. The geological materials of Mississippi range from unconsolidated sediments of Quaternary age to sedimentary rocks of Paleozoic age. The surficial geology of Mississippi is dominated by alluvium, followed by loess, marine deposits, and residuum. Humans have influenced soil development in Mississippi through urbanization, cultivation, irrigation, logging, and other practices that accelerated soil erosion, flooding, and mass wasting.

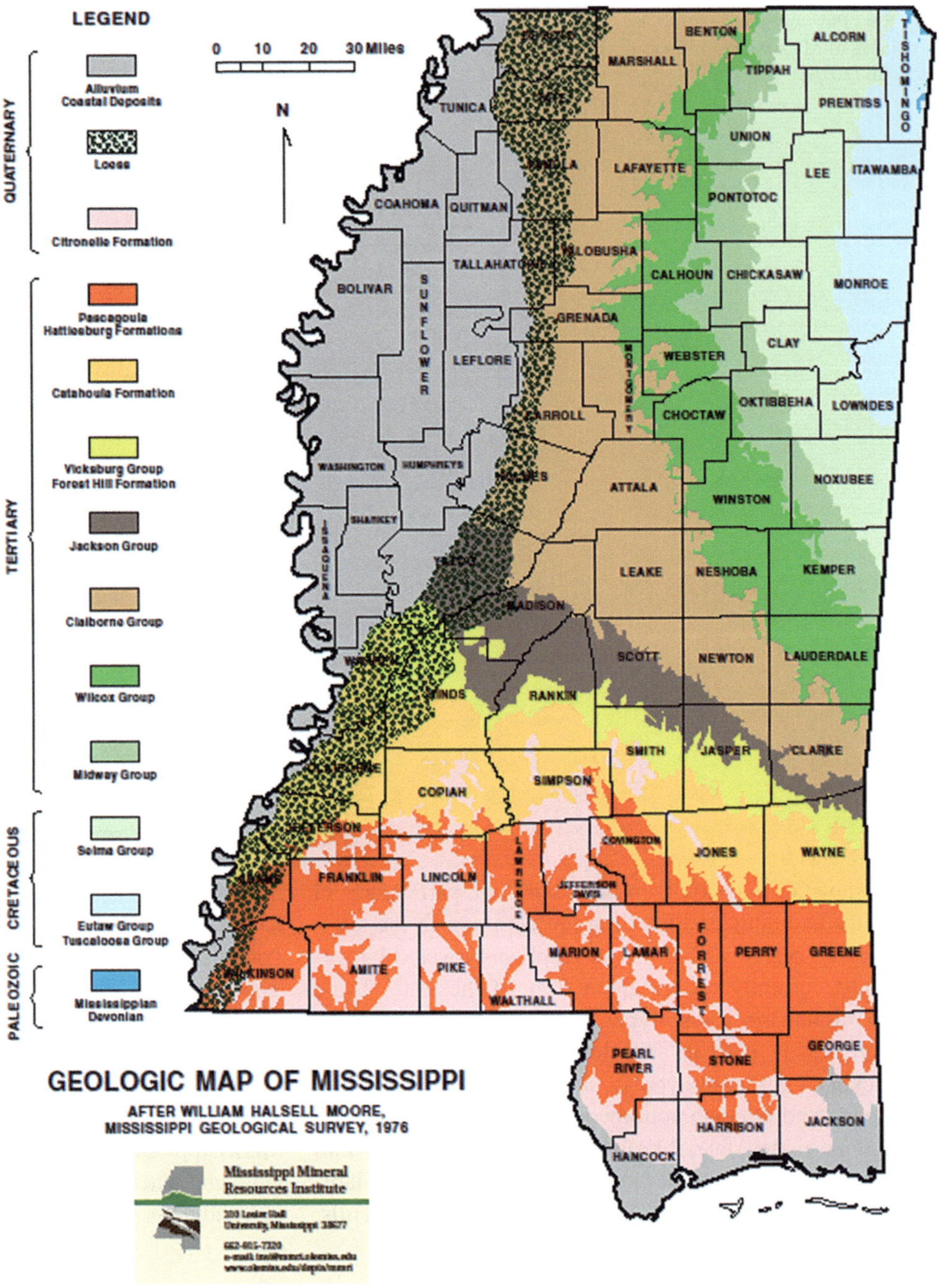

Fig. 3.10 Geologic map of Mississippi. *Source* Mississippi Geological Survey

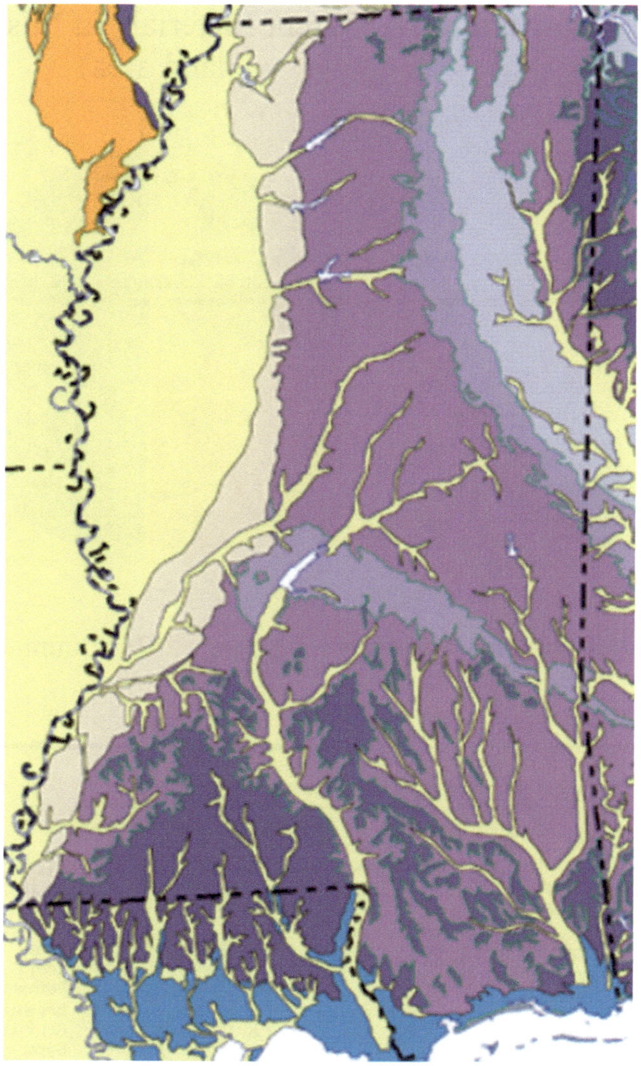

Fig. 3.11 Surficial geology of Mississippi. The yellow pattern is alluvium; the cream-colored pattern is loess; the violet-colored pattern is marine sediments from sedimentary rocks; the light purple pattern is residuum from fine-grained rocks such as shale that contain smectitic (i.e., "swelling") materials; the gray zone is residuum from limestone; the purple pattern is marine sediments from alluvium adjacent to coastal zone deposits; and the blue is the coastal zone. *Source* Soller and Reheis, US Geological Survey

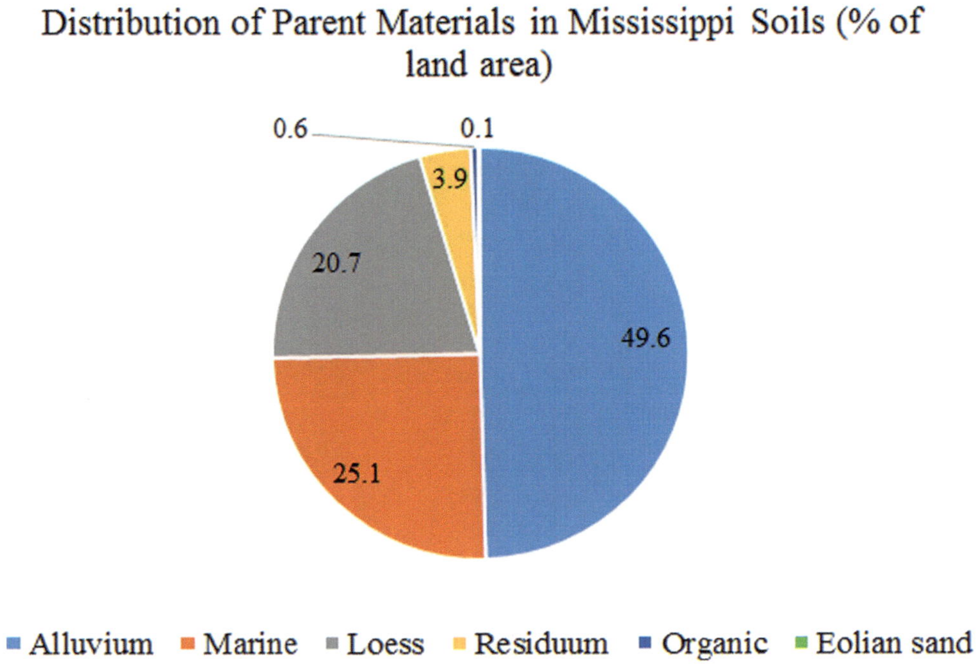

Fig. 3.12 Distribution of parent materials in soil series of Mississippi (% of land area) *Source* by J. Bockheim

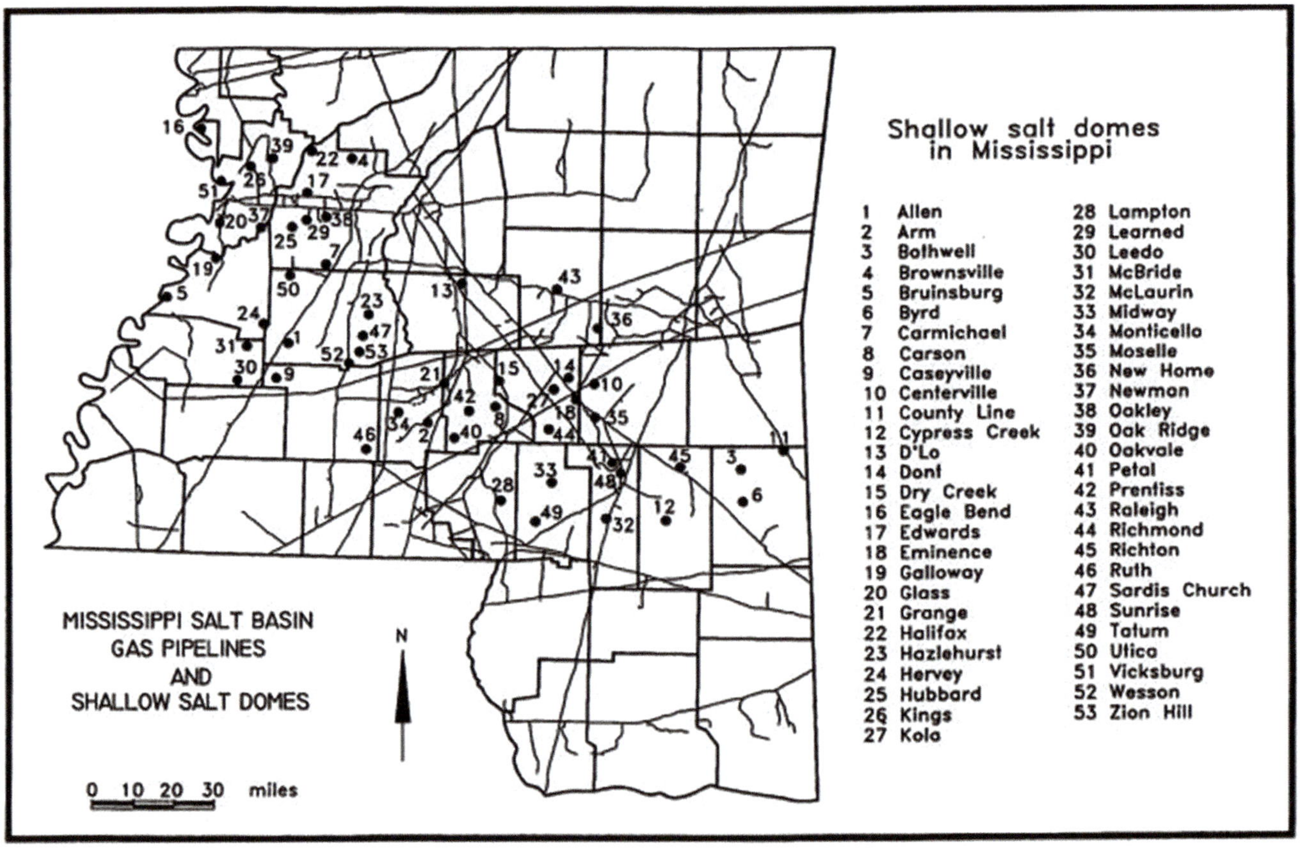

Fig. 3.13 Location of known salt domes in Mississippi (Thieling and Moody 1997) *Source* by J. Bockheim

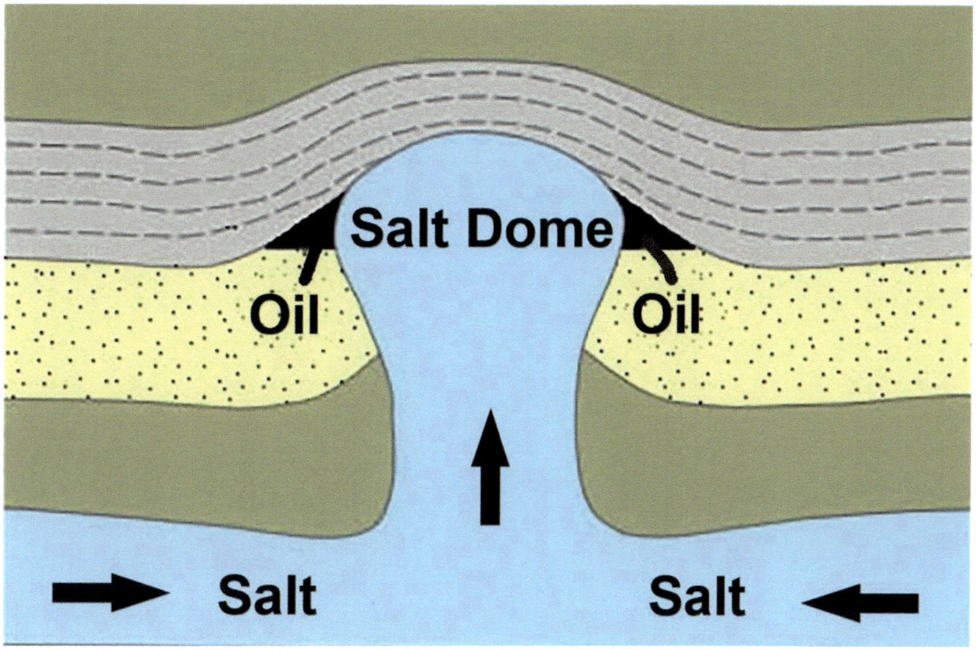

Fig. 3.14 Cross section of a salt dome bearing oil reserves. *Source* Texas Coop. Power

References

Chapman et al. (2004) Ecoregions of Mississippi. US Geol Surv, Reston, VA

Environmental Protection Agency (2016) What climate change means for Mississippi. EPA 430-F-16-026

Fickle JE (2001) Mississippi forests and forestry. Univ Press of Mississippi, Jackson, MS, p 347

Green WP, Pettry DE, Switzer RE (1998) Impact of imported fire ants on the texture and fertility of Mississippi soils. Comm Soil Sci Plant Anal 29:447–457

González JL, Törnqvist TE (2009) A new late Holocene sea-level record from the Mississippi Delta: evidence for a climate/sea level connection? Quat Sci Rev 28:1737–1749

Hossain A (2014) Groundwater depletion in the Mississippi delta as observed by the Gravity Recovery and Climate Experiment (GRACE) satellite system. Mississippi Water Resources Conf., Jackson, MS (https://www.researchgae.net/publication/262067496)

Jenny H (1941) Factors of soil formation. McGraw-Hill Book Co

Russell C, Waters CN, Himson S, Holmes R, Burns A, Zalasiewicz J, Williams M (2021) Geological evolution of the Mississippi River into the Anthropocene. Anthrop Rev 8:115–140

Soil Survey Staff (1999) Soil Taxonomy: a basic system of soil classification for making and interpreting soil surveys. 2nd edit. USDA, NRCS, Agric Handb No. 436. 871 pp

Switzer RE, Pettry DE (1992) Phosphorus distribution in Loessial Fragiudults as affected by relief. Soil Sci Soc Am J 56:849–856

Thieling SC, Moody JS (1997) Atlas of shallow Mississippi salt domes. Bull 131, Office of Geology, Miss. Dep Environ Quality, 328 pp

Welch SM, Waldron JL, Eversol AG, Simoes JC (2008) Seasonal variation and ecological effects of Camp Shelby burrowing crayfish (*Fallicambarus gordoni*) burrows. Am Midl Nat 159:378–384

Wood CW, Pettry DE (1989) Initial pedogenic progression in a drastically disturbed prime farmland soil. Soil Sci 147:196–207

4 General Soil Regions of Mississippi

4.1 Introduction

The Natural Resources Conservation Service has identified 270 Major Land Resource Areas (MLRAs) in the United States, the Caribbean, and the Pacific Basin (Natural Resources Conservation Service, 2020). MLRAs are geographically associated land areas that serve as a framework for organizing soil surveys and information about land as a resource for farming, ranching, forestry, recreation, and other uses. There are six MLRAs in Mississippi, including from greatest to least in area, the Southern Coastal Plain (MLRA 133A), the Southern Mississippi Valley Loess (MLRA 134), the Southern Mississippi Alluvium (MLRA 131A), the Alabama and Mississippi Blackland Prairie (MLRA 135A), the Eastern Gulf Coast Flatwoods (MLRA 152A), and the Gulf Coast Marsh (MLRA 151) (Fig. 4.1, Table 4.1).

A variation of the MLRAs—Soil Resource Areas—divides the state into ten units (Fig. 4.2). The main differences between the two systems are (i) the Southern Mississippi Valley Loess (MLRA 134) is divided into four units, including the Lower Thick Loess, the Lower Thin Loess, the Upper Thick Loess, and the Upper Thin Loess; (ii) the Southern Coastal Plain (MLRA 133A) is divided into Upper Coastal Plain, Lower Coastal Plain, (iii) an Interior Flatwoods region is recognized in the Upper Coastal Plain; and (iv) the Gulf Coast Marsh (MLRA 151) and Eastern Gulf Coast Flatwoods (MLRA 152A) are combined into the Coastal Flatwoods region.

The 2015 general soil map of Mississippi subdivides the six MLRAs into 26 ecoregions (Fig. 2.7 and Table 3.2). The following discussion pertains to the physiography, geology, climate, vegetation, and soils of each of these soil ecoregions.

4.2 Gulf Coastal Plain (MLRA 133C)

The Gulf Coastal Plain (MLRA 133C) is the largest MLRA in Mississippi, comprising 59,100 km^2, which is 47% of the state area (Table 4.1). About 40% of this MLRA is in Mississippi, with the remainder in Alabama (43%), Tennessee (6%), Florida (6%), Louisiana (3%), and Georgia (2%). The Gulf Coastal Plain is divided into nine soil ecoregions, including the Buhrstone/Lime Hills (133C.1b), the Central Hilly Coastal Plain (133C.1c), the Fall Line Hills (133C.1fl), the Northern Hilly Gulf Coast Plain (133C.1n) and Northern Pontotoc Ridge (133C.1p), the Southeastern Floodplains and Low Terraces (133C.7), the Southern Hilly Gulf Coastal Plain (133C.1s), the Southern Pine Plains and Hills (133C.2), and the Transition Hills (133C.1t) (Table 4.2).

Elevations range from 75 to 200 m (Table 4.2) maximum slopes average 15 ± 16%. Most of the Gulf Coastal Plain is underlain by unconsolidated coastal plain sediments. Figure 4.3 shows the Gulf Coastal Plain, as seen from Woodall Hill, and straddles the fall line from the Appalachian Mountains to the Gulf Coastal Plain. The mean annual air temperature ranges between 13.7 and 20 °C, and the mean annual precipitation ranges between 1400 and 1725 mm. The area supports oak-pine vegetation composed primarily of southern and northern red oak, white oak, loblolly pine, longleaf pine, slash pine, shortleaf pine, sweetgum, and yellow poplar.

Figure 4.4 shows the Providence-Savannah-Chenneby soil association in northwest Prentiss County. The Providence soil series is a Fragiudalfs derived from a loess cap that is 2 ft (0.6 m) thick over loamy upland deposits; the Savannah is a Fragiudults formed in loamy marine or fluvial terrace deposits; and the Chenneby is a Dystrudepts formed in loamy and silty alluvium on floodplains.

Common soil orders in MLRA 133C are Ultisols, Entisols, and Inceptisols (Table 4.3). The soils in the area typically have a thermic soil temperature regime, an udic or aquic soil moisture regime, and a siliceous or mixed mineralogy. The soils generally are very deep, somewhat excessively drained to poorly drained, and loamy. Hapludults have formed in marine sediments (Luverne and Sweatman series) and mixed marine sediments and alluvium (Smithdale series) on hills and ridges. Fragiudults (Ora and Savannah series)

D. Johnson et al., *The Soils of Mississippi*, World Soils Book Series,
https://doi.org/10.1007/978-3-031-36235-4_4

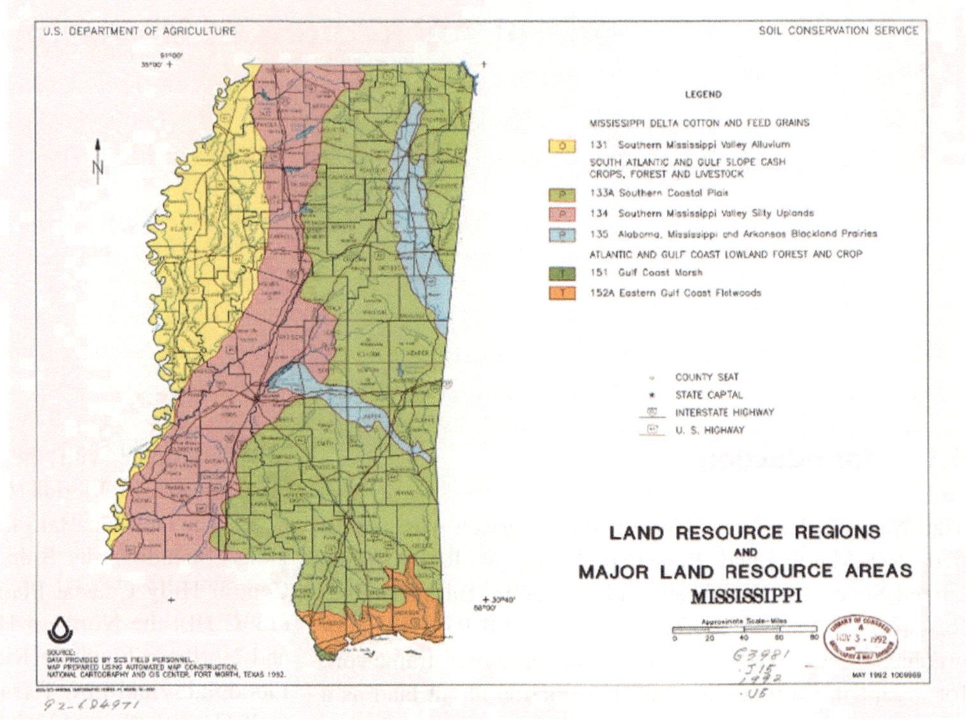

Fig. 4.1 Major land resource areas of Mississippi. MLRA 131 is now identified as 131A, 133A is 133C, and 135 is 135A. *Source* USDA, Soil Conservation Service

Table 4.1 Areas of major land resource areas in Mississippi

MLRA	Description	Total area (km²)	MS (%)	MS area (km²)	MS (%)
131A	Southern Mississippi Alluvium	80,245	25	20,061	16
133C	Gulf Coastal Plain	147,690	40	59,076	47
134	Southern Mississippi Valley Loess	69,705	41	28,272	23
135A	Alabama and Mississippi Blackland Prairie	26,330	57	15,008	12
151	Gulf Coast Marsh	19,635	1	196	0
152A	Eastern Gulf Coast Flatwoods	25,580	11	2814	2
	Total			125,428	100

and Paleudults (Heidel and Wadley series) have formed in mixed marine and fluvial sediments on uplands and stream terraces. Fluvaquents (Bibb series) and Endoaquepts (Mantachie series) have formed in alluvium on flood plains.

4.3 Southern Mississippi Valley Loess (MLRA 134)

The Southern Mississippi Valley Loess (MLRA 134) is the second largest MLRA in Mississippi, comprising nearly 28,300 km^2, which is 22% of the state area (Table 4.1). About 41% of this MLRA is in Mississippi, with the remainder in Tennessee (23%), Louisiana (15%), Arkansas (10%), Kentucky (9%), Missouri (3%), and Illinois (1%). The Southern Mississippi Valley Loess corresponds with the Loess Hills physiographic province. Rutledge and others (1996) provided an historical summary of loess stratigraphy and properties in the Lower Mississippi Valley. Muhs and others (2001) studied the impact of climate and parent material on chemical weathering in loess-derived soils of the Mississippi River Valley from Wisconsin to Louisiana. Based on elemental ratios, there was greater chemical weathering in the southern part of the valley. However, clay contents were less in the south than in the north, because soils in the north received fine-grained loess from source to the west of the Mississippi River valley, whereas soils in the south were unaffected by such additions.

The Southern Mississippi Valley Loess region is subdivided into the Bluff Hills (134.4), the Loess Plains (134.3), and the Southern Rolling Hills (134.5) (Table 4.2). Elevations range from 18 to 190 m. Maximum slopes average 11 ± 17%. The loess within the region ranges from 5 m near the Mississippi River to less than 1 m away from the river that is underlain by alluvium. The loess originated from the glaciated landscapes of the North during the Pleistocene Ice Age, was washed down the Mississippi River and deposited on flood plains, and subsequently picked up by the wind and deposited a second time on the high eastern bluffs of the river (Fisk 1944; Wascher et al. 1947). Figure 4.5 shows the Loess Hills near Yazoo City.

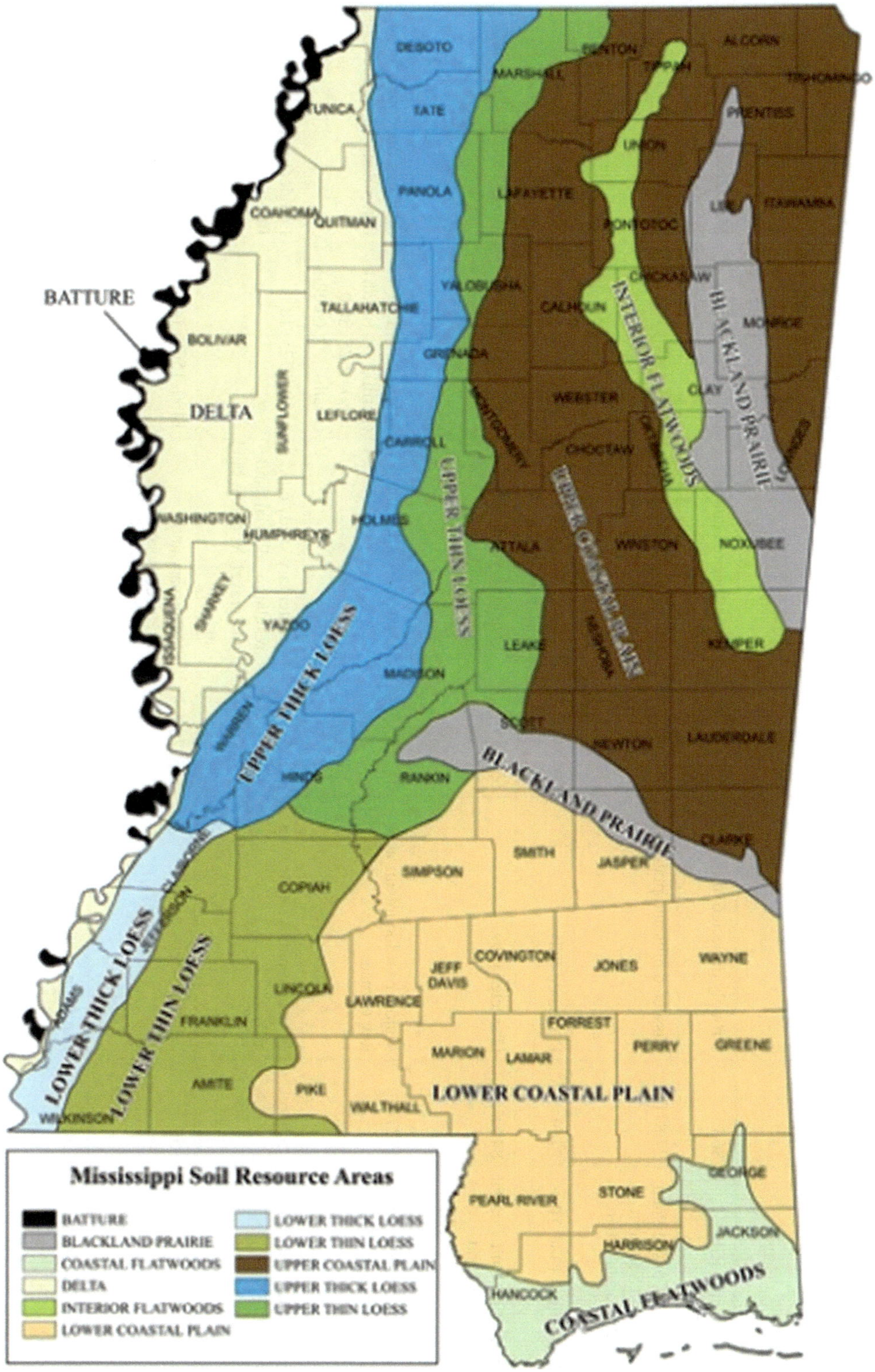

Fig. 4.2 Soil resource areas of Mississippi. *Source* Kushla and Londo 2014; courtesy of Mississippi Department of Wildlife, Fishers, and Parks

Table 4.2 Area and site factors for soil ecoregions in Mississippi

Major land resource area (MLRA)	Soil ecoregions (IV)	Area (km^2)	Physiography	Elevation (m)	Geology	Mean annual precipitation (mm)	Mean annual air temperature (°C)	Potential natural vegetation
Southern Coastal Plain (133C)	Buhrstone/Lime Hills (133.C.4)	868	Strongly dissected hills	75–200	Limestone residuum	1450–1575	17	Mixed oak-pine forest
	Central Hilly Coastal Plain (133C.1c) and Southern Hilly Gulf Coastal Plain (133C.1s)	27,871	Dissected hills and plains	45–200	Quaternary sand residuum	1400–1500	17	Mixed oak-pine forest
	Fall Line Hills (133C.1fl)	2769	Dissected hills	60–245	Quaternary sand and gravel residuum	1400–1500	15.5	Mixed oak-pine forest
	Northern Hilly Gulf Coastal Plain (133C.1n) and Northern Pontotoc Ridge (133C.1p)	8400	Dissected hills	65–225	Quaternary and older sandy residuum	1400–1500	15.5	Mixed oak and oak-pine forests
	Southeastern Floodplains and Low Terraces (133C.7)	1777	Major river floodplains and low terraces	3–315	Quaternary alluvium	1450 (north), 1500–1625 (south)	13.7 (north), 20.0 (south)	Southern floodplain forest, bottomland hardwood forest
	Southern Pine Plains and Hills (133C.2)	15,866	Dissected plains and rolling hills	6–155	Quaternary sandy clay residuum; alluvial sand and gravel; Miocene sand and clay	1500–1725	18	Longleaf pine; pine-oak forests
	Transition Hills (133A.1t)	650	Dissected hills	125–200	Quaternary and Cretaceous sand and gravel; Devonian shale, sandstone, limestone	1400–1500	14	Mixed oak-pine forest
Southern Mississippi River Alluvium (131A)	Northern Backswamps (131A.4)	2567	Flat plains with depressions	20–60	Quaternary alluvium	1320–1420	16.5	Baldcypress-water tupelo; mixed oaks, hockory-elm-ash-sweetgum
	Northern Holocene Meander Belts (131A.1)	14,058	Flat plains and river meander belts with levees, point bars, and oxbows	9–70	Holocene alluvium	1320–1425	16.5	Bottomland hardwoods; riverfront forests; baldcypress-water tupelo
	Northern Pleistocene Valley Trains (131A.9)	2852	Flat plains with rellct branching channels	30–60	Quaternary alluvium	1370–1420	17	Bottomland hardwoods; baldcypress-water tupelo
	Southern Backswamps (131A.4s)	263	Flat plains with depressions	10–30	Q uaternary alluvium	1475–1550	18	Baldcypress-water tupelo; oak-hickory-elm-ash
	Southern Holocene Meander Belts (131A.1s)	367	Flat plains and river meander belts with levees, point bars, and oxbows; large rivers; smaller low-gradient streams	7–30	Holocene alluvium	1475–1550	18	Bottomland hardwoods; riverfront forests; baldcypress-water tupelo

(continued)

Table 4.2 (continued)

Major land resource area (MLRA)	Soil ecoregions (IV)	Area (km^2)	Physiography	Elevation (m)	Geology	Mean annual precipitation (mm)	Mean annual air temperature (°C)	Potential natural vegetation
Southern Mississippi Valley Loess (134)	Bluff Hills (134.4)	4123	Dissected hills and ridges; irregular plains	18–110	Quaternary loess; Tertiary sand, silt, clay	1300–1400 (north), 1400–1575 (south)	16 (north), 20 (south)	Oak-hickory; mixed hardwoods
	Loess Plains (134.3)	13,383	Dissected, irregular plains; low-gradient, bottomed streams	20–190	Quaternary loess; some alluvium, residuum	1320–1420	16	Oak-hickory, oak-hickory-pine; some floodplain forests and bottomland hardwoods
	Southern Rolling Hills (134.5)	10,062	Dissected, irregular plains; low-gradient, bottomed streams	18–155	Quaternary loess and loessal colluvium; some alluvium	1420–1625	18	Oak-hickory-pine; some loblolly pine
Alabama dnd Mississippi Blackland Prairie (135A)	Blackland Prairie (135A.1)	5460	Undulating, irregular plains; low-gradient streams	40–160	Quaternary and Tertiary clayey residuum; Cretaceous chalk, marl, and calcareous clay	1400–1500	16	Oak-cedar; bluestem prairie
	Blackland Prairie Margins (135A.2b), Interior Flatwoods (135A.2f), and Southern Pontotoc Ridge (135A.2p)	6840	Smooth lowland plains; undulating, irregular plains; low-gradient, bottomed streams	40–180	Quaternary and Tertiary clayey residuum; Cretaceous chalk, marl, and calcareous clay	1400–1500	16	Mixed oak; oak-hickory-pine; flatwoods oak-pine; mixed hardwoods; some bottomland hardwoods
	Jackson Prairie (135A.3)	2473	Undulating, irregular plains; low-gradient streams	64–200	Quaternary and Tertiary smectiticlayey residuum; Tertiary chalk, marl, and calcareous clay	1400–1500	17	Oak-hickory-pine; scattered little bluestem prairies
Gulf Coast Marsh (151) and Eastern Gulf Coast Flatwoods (152A)	Coastal Marshes (151.1) and Gulf Coast Islands and Coastal Marshes (152A.1c)	570	Tidal marshes, bays, river deltas, lagoons, barrier islands, dunes, beaches	0–8	Quaternary quartz sand, shell fragments, silt, clay, muck, peat	1575–1650	19	Salt marshes, tidal fresh-water marshes, maritime shrbus, slash pine savannas, maritime evergreen forests
	Floodplains and Low Terraces (152A.7)	453	Major river floodplains and associated low terraces; low-gradient streams with oxbow lakes	0.5–15	Quaternary quartz sand, shell fragments, silt, clay, muck, peat	1575–1650	19	Southern floodplain forests; bottomland hardwoods
	Gulf Coast Flatwoods (152A.1)	1725	Flat to gently undulating plains; low-gradient streams	0.5–23	Quaternary quartz sand, shell fragments, silt, clay, muck, peat	1575–1650	19	Slash pine flatwoods/savannas, longleaf pine savannas, wet (slash) pine-pond pine-cypress savanna

Fig. 4.3 Tombigbee and Tennessee Hills, as viewed from Woodall Hill, the highest point in Mississippi (246 m), are part of the Southern Coastal Plain (MLRA 133A). Photo by J. Bockheim

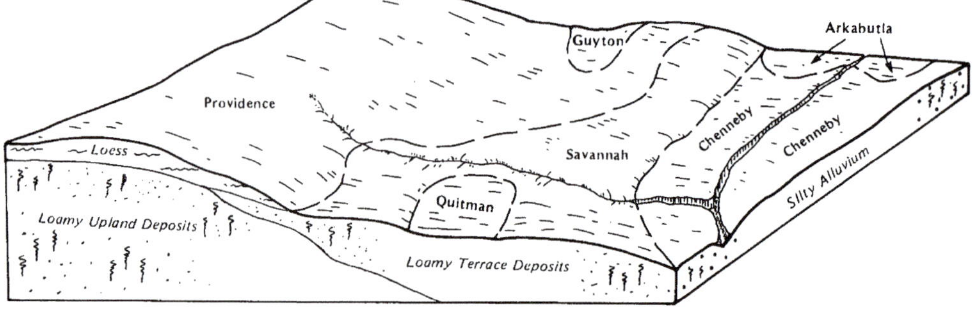

Fig. 4.4 Prentice-Savannah-Chenneby soil association in MLRA 133C, the Southern Coastal Plain. *Source* Soil Survey of Prentiss County, Mississippi

Table 4.3 Dominant great groups and soil series by Major Land Resource Area and Soil ecoregion in Mississippi

Major land resource area (MLRA)	Soil ecoregions (IV)[a]	Eco-region[b]	Dominant great groups (soil series)
Southern Coastal Plain (133C)	Buhrstone/Lime Hills (133.C.4)	65q	Fragiudults (Arundel, Lauderdale, Sweatman, Smithdale); Fluvaquents (Bibb); Dystrudepts (Jena, Kirkville); Endoaquepts (Mantachie)
	Central Hilly Coastal Plain (133C.1c) and Southern Hilly Gulf Coastal Plain (133C.1s)	65d	Hapludults (Smithdale, Luverne); Fragiudults (Ora, Savannah, Prentiss); Paleudults (Ruston); Hapludalfs (Lexington); Fluvaquents (Bibb); Udifluvents (Iuka); Epiaquepts (Urbo)
	Fall Line Hills (133C.1fl)	65c	Hapludults (Smithdale, Sweatman, Luverne); Fragiudults (Savannah); Hapludalfs (Okeelala, Brantley); Fluvaquents (Bibb); Udifluvents)(Iuka); Endoaquepts (Mantachie)
	Northern Hilly Gulf Coastal Plain (133C.1n) and Northern Pontotoc Ridge (133C.1p)	65e	Hapludults (Smithdale, Luverne); Paleudults (Ruston); Fragiudults (Ora, Savannah, Prentiss); Fragiudalfs (Providence); Hapludalfs (Lexington); Fluvaquents (Bibb); Udifluvents (Iuka); Epiaquepts (Urbo)
	Southeastern Floodplains and Low Terraces (133C.7)	65p	Hapludults (Cahaba, Latonia, Columbus); Albaquults (Leaf); Dystrudepts (Jena, Kirkville); Endoaquepts (Mantachie, Rosebloom); Fluvaquents (Bibb); Udifluvents (Nugent); Quartzipsamments (Bigbee); Glossaqualfs (Guyton)
	Southern Pine Plains and Hills (133C.2)	65F	Paleudults (McLaurin, Heidel, Benndale, Malbis, Ruston); Hapludults (Smithdale); Fragiudults (Prentiss); Paleaquults (Smithton, Trebloc); Paleudalfs (Susquehanna, Freest); Fluvaquents (Bibb)
	Transition Hills (133C.1t)	65j	Hapludults (Saffell, Smithdale, Luverne); Paleudults (Ruston); Fluvaquents (Bibb); Endoaquepts (Mantachie); Dystrudepts (Kirkville, Jena)
Southern Mississippi River Alluvium (131A)	Northern Backswamps (131A.4)	73d	Dystraquerts (Alligator); Epiaquerts (Sharkey); Endoaquepts (Dowling); Epiaquepts (Tunica); Hapludolls (Bowdre)
	Northern Holocene Meander Belts (131A.1)	73a	Endoaquepts (Commerce, Convent); Dystrudepts (Beulah); Udifluvents (Robinsonville); Udipsamments (Crevasse); Hapludalfs (Dubbs); Endoaqualfs (Dundee); Epiaquerts (Sharkey)
	Northern Pleistocene Valley Trains (131A.9)	73b	Endoaqualfs (Dundee, Forestdale); Hapludalfs (Askew, Dubbs); Epiaquerts (Sharkey); Epiaquepts (Tunica); Endoaquepts (Waverly); Eutrudepts (Adler); Udifluvents (Bruno)
	Southern Backswamps (131A.4s)	73m	Epiaquerts (Sharkey); Epiaquepts (Tunica); Endoaquepts (Dowling); Eutrudepts (Adler, Bruin); Hapludolls (Bowdre)
	Southern Holocene Meander Belts (131A.1s)	73k	Endoaquepts (Commerce, Convent); Dystrudepts (Beulah); Udifluvents (Robinsonville); Udipsamments (Crevasse); Hapludalfs (Dubbs); Endoaqualfs (Dundee); Epiaquerts (Sharkey); Epiaquepts (Newellton)
Southern Mississippi Valley Loess (134)	Bluff Hills (134.4)	74a	Hapludalfs (Memphis); Fragiudalfs (Loring); Eutrudepts (Natchez, Adler); Udifluvents (Collins)
	Loess Plains (134.3)	74b	Fraglossudalfs (Grenada, Calloway); Fragiudalfs (Loring, Providence), Hapludalfs (Memphis); Udifluvents (Collins); Fluvaquents (Falaya); Dystrudepts (Oaklimeter, Ariel)

(continued)

Table 4.3 (continued)

Major land resource area (MLRA)	Soil ecoregions (IV)[a]	Eco-region[b]	Dominant great groups (soil series)
	Southern Rolling Hills (134.5)	74c	Fragiudalfs (Providence, Loring); Hapludalfs (Memphis, Lexington, Lorman); Hapludults (Smithdale); Udifluvents (Collins); Fluvaquents (Gillsburg, Falaya); Dystrudepts (Ariel, Oaklimeter)
Alabama dnd Mississippi Blackland Prairie (135A)	Blackland Prairie (135A.1)	65a	Dystruderts (Vaiden, Oktibbeha); Hapluderts (Okolona, Brooksville, Griffith); Eutrudepts (Sumter, Marietta); Epiaquepts (Leeper); Paleudalfs (Kipling); Udorthents (Demopolis); Fluvaquents (Belden); Hapludolls (Catalpa)
	Blackland Prairie Margins (135A.2b), Interior Flatwoods (135A.2f), and Southern Pontotoc Ridge (135A.2p)	65b	Paleudalfs (Falkner, Tippah); Endoaqualfs (Adaton); Hapludalfs (Longview); Dystrudepts (Wilcox); Dystraquerts (Mayhew); Epiaquepts (Leeper, Urbo, Una); Fragiudults (Prentiss); Paleudults (Quitman); Fluvaquents (Mathiston)
	Jackson Prairie (135A.3)	65r	Dystruderts (Vaiden, Llouin, Ichusa, Griffith); Hapluderts (Okolona, Maytag); Paleudalfs (Boswell, Kipling, Freest); Hapludalfs (Okeelala, Brantley); Hapludolls (Catalpa); Eutrudepts (Marietta)
Gulf Coast Marsh (151) and Eastern Gulf Coast Flatwoods (152A)	Coastal Marshers (151.1) and Gulf Coast Islands and Coastal Marshes (152A.1c)	75k	Hapludults (Latonia), Paleudults (Ocilla); Paleaquults (Plummer); Sulfaquents (Bohicket, Axis); Quartzipsamments (Newhan); Psammaquents (Duckston), Sulfihemists (Handsboro); Alaquods (Leon)
	Floodplains and Low Terraces (152A.7)	75i	Paleaquults (Plummer); Haplosaprists (Maurapas); Endoaquepts (Kinston, Chastain, Arkabutla, Rosebloom); Dystrudepts (Jena), Hydraquents (Arat)
	Gulf Coast Flatwoods (152A.1)	75a	Paleaquults (Atmore, Plummer, Daleville, Smithton); Paleudults (Harleston, Poarch, Escambia); Umbraquults (Hyde); Glossaqualfs (Guyton)

[a] General Soil Map of Mississippi (USDA-NRCS 2022)
[b] Ecoregions of Mississippi (USGS 2003)

Fig. 4.5 Loess Hills near Yazoo City. Photo by J. Bockheim

Brown Loam Region of Mississippi

The Brown Loam Region or Bluff Hills is a highly fertile soil belt in Mississippi that extends roughly north–south from Memphis, TN to southern Louisiana that corresponds to the Southern Mississippi Valley Loess region (Fig. 4.6). The topsoil of many of the loess-derived soils in the region have a brown (10YR 4/3) or dark grayish brown (10YR 4/2) moist color, including the Memphis, Loring, Grenada, Calloway, and Bude soil series. The Brown Loam Branch of the Mississippi State University Extension Service was established in Raymond, MS to focus on beef cattle, forages, cotton, soybeans, corn, and conservation management.

The mean annual air temperature ranges between 16 and 20 °C, and the mean annual precipitation ranges between 1300 and 1625 mm (Table 4.2). The area supports oak-pine vegetation, including cherrybark oak, Shumard oak, white oak, post oak, southern red oak, loblolly pine, shortleaf pine, and southern magnolia.

Figure 4.7 shows the Natchez-Memphis soil association on deep loess in Tate County. The Natchez soil series (Eutrudepts) is formed in thick loess on dissected uplands; and the Memphis (Hapludalfs) occurs on stable alluvial terraces composed of loess that exceeds 1.5 m in depth.

The dominant soil orders in this MLRA are Alfisols, Entisols, Inceptisols, and Ultisols (Table 4.3). The soils in the area are very deep or deep, are medium textured, and have a thermic soil temperature regime, an udic soil moisture regime, and mixed mineralogy. Well drained, nearly level to very steep Hapludalfs (Memphis series) are on uplands. Nearly level to steep, well drained Hapludalfs (Memphis series), moderately well drained and somewhat poorly drained Fraglossudalfs (Grenada and Calloway series), moderately well drained Fragiudalfs (Loring series), and well drained Eutrudepts (Natchez and Adler series) have formed in thick deposits of loess. In the eastern part of the area, where the loess mantle thins, moderately well drained Fragiudalfs (Providence series) are on gently sloping to steep and are on ridgetops and side slopes. Well drained Dystrudepts (Ariel series), moderately well drained Udifluvents (Collins series), moderately well drained Dystrudepts (Oaklimeter series), and somewhat poorly drained Fluvaquents (Falaya and Gillsburg series) are on flood plains.

4.4 Southern Mississippi Valley Alluvium (MLRA 131A)

The Southern Mississippi Valley Alluvium (MLRA 131A), also called the Delta, is the third largest MLRA in Mississippi, comprising 20,000 km^2, which is 16% of the state area (Table 4.1). About 25% of this MLRA is in Mississippi, with the remainder in Louisiana (32%), Arkansas (26%), Missouri (12%), Tennessee (3%), and Kentucky (1%). The Southern Mississippi Valley Alluvium corresponds with the Mississippi Delta Alluvial Plain physiographic province.

Fig. 4.6 Brown Loam near Oxford, MS, was described by Mabry in 1898. *Source* Mabry, 1898

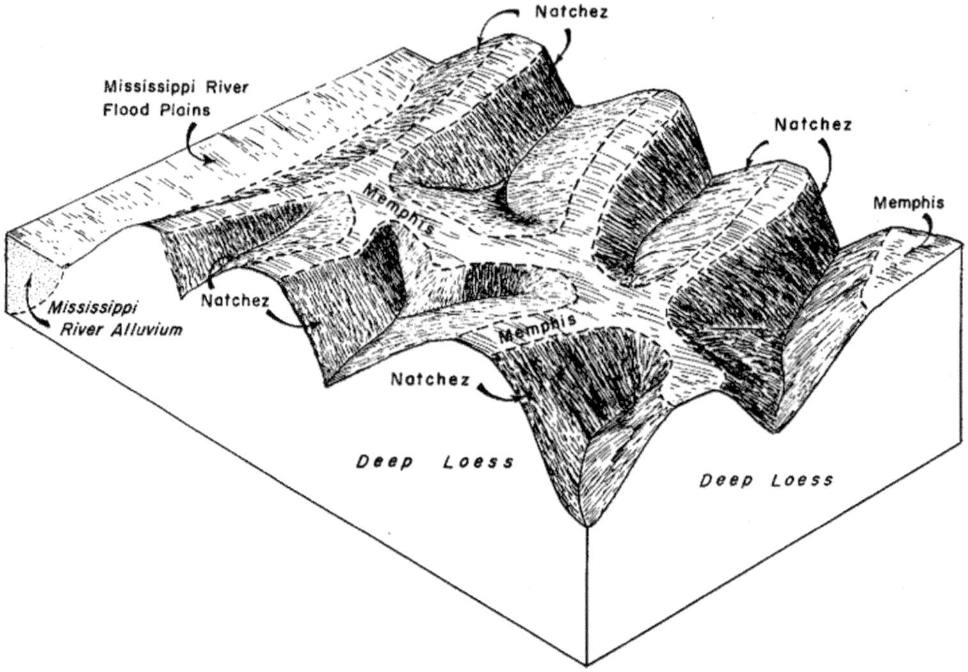

Fig. 4.7 Natchez-Memphis soil association in Tate County is derived from deep loess in the Southern Mississippi Valley Loess (MLRA 134). *Source* USDA Natural Resources Conservation Service

This region is divided into five subregions, including the Northern Backswamps (131A.4), the Northern Holocene Meander Belts (131A.1), the Northern Pleistocene Valley Trains (131A.9), the Southern Backswamps (131A.4s), and the Southern Holocene Meander Belts (131A.1s) (Table 4.2).

The Delta features thick deposits of alluvium from flooding and lateral migration of the Mississippi River. Elevations range from 7 to 70 m (Table 4.2). Maximum slopes average $4.3 \pm 2.6\%$. Figure 4.8 shows the Southern Mississippi Valley Alluvium (MLRA 134) near Sarah, Mississippi.

Fig. 4.8 Southern Mississippi Valley Alluvium (MLRA 131A) near Sarah, Mississippi. Photo by J. Bockheim

Buckshot Soils of Mississippi

Buckshot soils are clayey soils in the Delta that upon drying contain small, round aggregates (very-fine and fine granular structure), resembling shotgun buckshot (Fig. 4.9). Farmers around the turn of the last century differentiated between "blue buckshot" and "black buckshot" land. Apparently, the terms were used to characterize the general way in which the Yazoo clay responded to cultivation under specific moisture conditions. The term may have been first used in the 1902 Soil Survey of the Smedes Area–parts of Yazoo, Madison, Issaquena, and Sharkey Counties–and was applied primarily to the Yazoo clay, which is no longer recognized. No mention of buckshot is made in any of the Official Soil Descriptions of clay-rich soil series in the Delta.

The mean annual air temperature ranges between 16.5 and 18 °C, and the mean annual precipitation ranges from 1320 to 1550 mm (Table 4.2). The area originally supported bottomland hardwood and hardwood-cypress vegetation, including water oak, Nuttall oak, cherrybark oak, native pecan, red maple, sweetgum, eastern cottonwood, and hickory. In the swamps, the major species were and currently are baldcypress, water tupelo, water oak, green ash, red maple, and black willow.

Figure 4.10 shows the Alligator-Dowling soil association in Tate County. The Dundee soil series (Endoaqualfs) is formed in silty materials on alluvial terraces; and the Dowling (Endoaquepts) and Alligator (Dystraquerts) soil series are formed in clayey alluvium in oxbows and backswamps. Figure 4.11 shows the Dundee-Dubbs-Tensas soil association in Holmes County. Whereas the Dundee, Dubbs, and Tensas soil series are Udalfs or Aqualfs formed in silty and loamy materials on alluvial terraces, the Alligator soil series, a Dystraquerts, is derived from clayey alluvium in floodplains and backswamps. The Fausse soil series is no longer mapped in Mississippi.

Crawdad Land in Mississippi

Crawdad land is a term that was used for land in the Mississippi Yazoo River Basin that contained crawfish. In Mississippi, crawfish aquaculture is integrated with agricultural crop rotations (Miss. State Univ. Ext.) (Fig. 4.12). Although crayfish are normally associated with Louisiana, Mississippi has one of the most diverse crayfish faunas in the world (US Forest Service; http://srs.fs.usda.gov/crayfish/info.pp). The crayfish live in a variety of habitats including seasonally wet and upland habitats. Most crayfishes burrow down during times of drought, creating mud chimneys to seal their burrow entrances (Fig. 4.13).

The dominant soil orders in MLRA 131A are Alfisols, Vertisols, Inceptisols, and Entisols (Table 4.3). The soil temperature regime is thermic in most of the MLRA. The

Fig. 4.9 Buckshot in a clayey soils near Shaw, MS. Photo by Dr. Neal Smith, Serene Fox Farm Photo by J. Bockheim

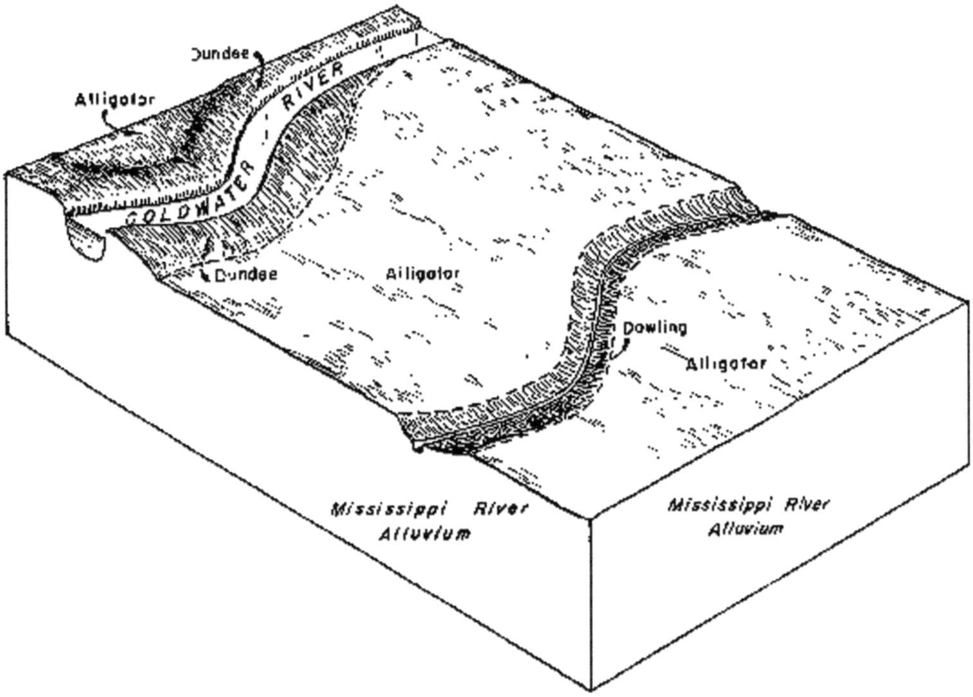

Fig. 4.10 Alligator-Dowling soil association in Tate County is formed in Mississippi River alluvium in the Southern Mississippi Valley Alluvium (MLRA 131A). *Source* USDA Natural Resources Conservation Service

soils usually have an aquic soil moisture regime, smectitic clay mineralogy, and mixed sand and silt fraction mineralogy. The soils are very deep, dominantly poorly drained and somewhat poorly drained, and dominantly loamy or clayey. Nearly level Epiaquerts (Sharkey series), Dystraquerts (Alligator series), Vertic Epiaquepts (Tunica and Newellton series), and Vertic Endoaquepts (Dowling series) dominate the alluvial flats and backswamps of Holocene to late Pleistocene age. Nearly level to gently sloping Endoaquepts (Commerce and Waverly series), Udifluvents (Robinsonville series), Dystrudepts (Beulah series), and Fluvaquents (Convent series) dominate the natural levees of Holocene age. Nearly level to gently undulating, sandy Udifluvents (Bruno series) and Udipsamments

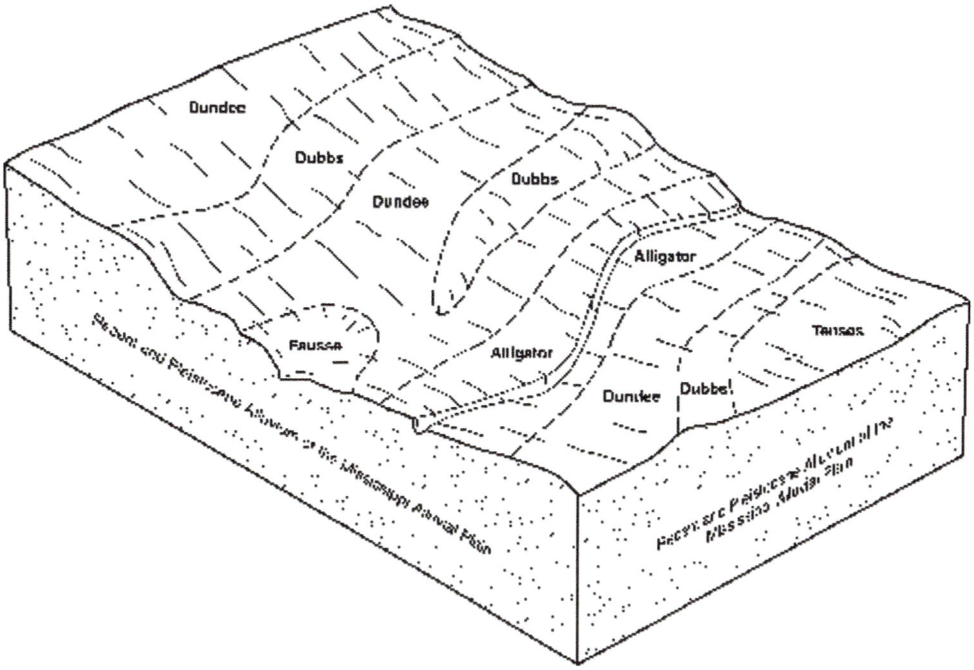

Fig. 4.11 Dundee-Dubbs-Tensas soil association in Holmes County is derived from Mississippi River alluvium in the Southern Mississippi Valley Alluvium (MLRA 131A). *Source* USDA Natural Resources Conservation Service

Fig. 4.12 Loose red swamp crawfish (*Procambarus clarkia*) in Mississippi. *Source* Miss. State Univ. Extension

(Crevasse series) dominate the levee splays and point bars of Holocene age. Nearly level to gently undulating Endoaqualfs (Dundee series), Hapludalfs (Dubbs and Askew series), and Epiaqualfs (Forestdale series) dominate the terraces of Pleistocene age. Flood plains along streams and alluvial fans contain Eutrudepts (Adler and Bruin series) and Hapludolls (Bowdre series). Aslan and Autins (1998) studied Holocene flood plain soil formation on meander belts and in backswamps of the southern Mississippi River Valley. Inceptisols and Alfisols were most common on point bars, ridges, and natural levee crests, and Vertisols were most common in backswamps.

4.5 Alabama and Mississippi Blackland Prairie (MLRA 135A)

The Alabama and Mississippi Blackland Prairie (MLRA 135A) is the fourth largest MLRA in Mississippi, comprising nearly 7760 km^2, which is 12% of the state area (Table 4.1). About 57% of this MLRA is in Mississippi, with the remainder in Alabama (43%). The Alabama and Mississippi Blackland Prairie corresponds with the Black Prairie and Jackson Prairie physiographic provinces. This region is underlain by Cretaceous-age clay, marl, soft limestone, or chalk of the Selma and Jackson Groups. Elevations range

Fig. 4.13 Crayfish chimney. *Source* Dr. Carrie Stevenson, Institute of Food and Agricultural Sciences, University of Florida, Escambia, FL

from 40 to 200 m. Maximum slopes average 17 ± 18%. Figure 4.14 shows Blackland Prairie near Booneville, Mississippi.

The mean annual air temperature ranges between 16 and 17 °C, and the mean annual precipitation ranges from 1400 to 1500 mm (Table 4.2). The area supports hardwoods and pines, including northern and southern red oak, white oak, sweetgum, blackgum, loblolly pine, and shortleaf pine, oak-cedar, and scattered little bluestem prairies.

Figure 4.15 shows the Wilcox-Dulac-Falaya soil association in Tippah County. The Falkner, Dulac, and Bude soil series (Udalfs) are derived from loess over marine clay or alluvium; the Wilcox soil series, a Dystruderts, is formed in residuum derived from clayey shale; and the Falaya (Fluvaquents) and Chastain (Endoaquepts) soil series are derived from alluvium. The Tickfaw soil series is no long recognized.

Figure 4.16 shows the Kipling-Sumter association in Prentiss County. At the highest elevations, the Dulac soil series, an Oxyaquic Fragiudalfs, is derived from a thin layer of loess of acidic marine-clay residuum. The Kipling soil series, a Vertic Paleudalfs, is derived entirely from the marine clay. Whereas the Sumter soil series (Rendollic Eutrudepts) is derived from the Demopolis Chalk Formation, the Catalpa soil series (Fluventic Hapludolls) is formed in clayey alluvium.

The dominant soil orders in this MLRA are Inceptisols and Vertisols (Table 4.3). The soils have a thermic soil temperature regime, a udic or aquic soil moisture regime, and smectitic or carbonatic mineralogy. They are shallow to very deep, generally well drained to somewhat poorly drained, and loamy or clayey. Epiaquepts (Leeper and Urbo series), Epiaquerts (Sucarnoochee and Houlka series), and Hapludolls (Catalpa series) formed in clayey alluvium on flood plains. Eutrudepts formed in loamy alluvium on flood plains (Marietta series) and in clayey sediments and residuum on uplands (Sumter series). Dystruderts (Oktibbeha, Watsonia, and Vaiden series), Hapluderts (Brooksville,

Fig. 4.14 Alabama and Mississippi Blackland Prairie (MLRA 135A) near Booneville, Mississippi. Photo by J. Bockheim

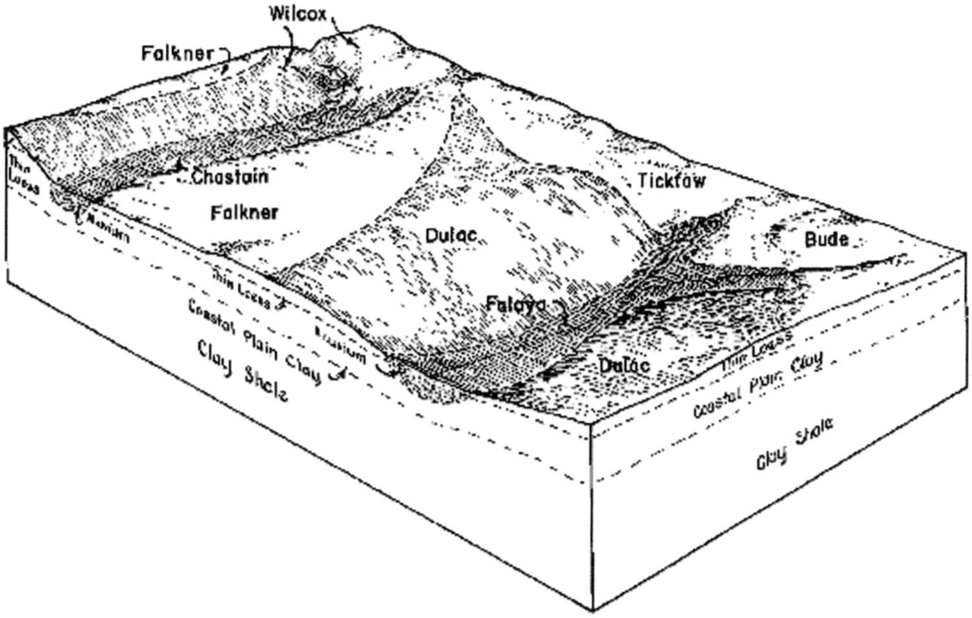

Fig. 4.15 Wilcox-Dulac-Falkner soil association in Tippah County, representative of MLRA 135A, the Alabama and Mississippi Blackland Prairie. *Source* USDA Natural Resources Conservation Service

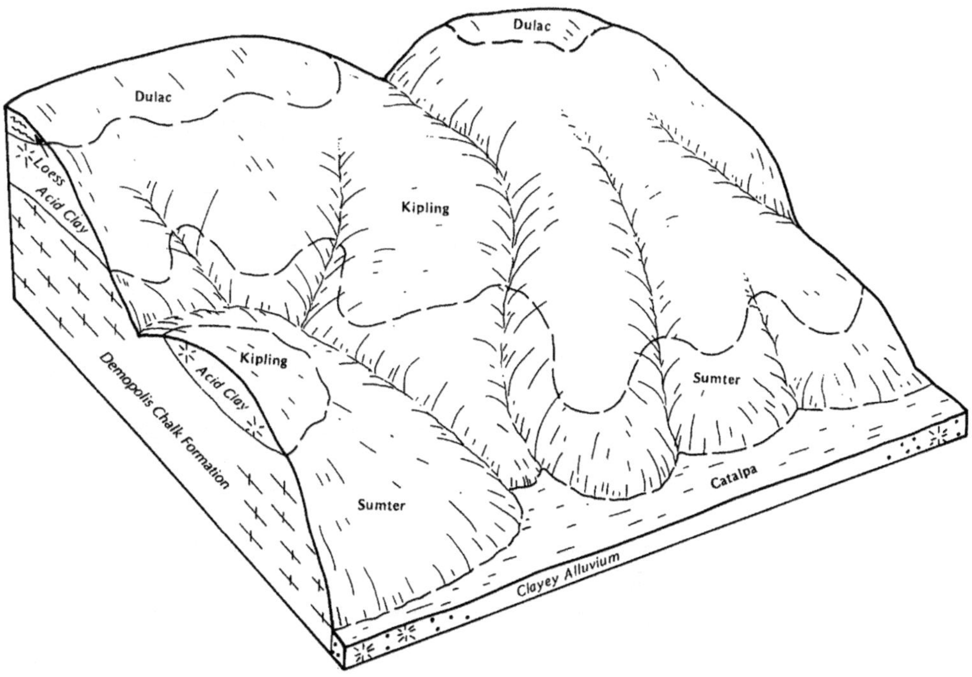

Fig. 4.16 Kipling-Sumter association in Prentiss County is derived from chalk residuum and clayey alluvium in the Alabama and Mississippi Blackland Prairie (MLRA 135A). *Source* USDA Natural Resources Conservation Service

Okolona, and Houston series), and Paleudalfs (Kipling and Searcy series) formed in clayey sediments on uplands. Udorthents (Demopolis series) formed in residuum on ridges and hills.

Campbell and Seymour (2011) related soil taxa and series to parent materials in MLRA 135A. Uplands and high terraces featured Ultisols (mainly Fragiudults and Paleudults); clayey uplands contained Alfisols (Hapludalfs and Paleudalfs) and Vertisols (Hapluderts and Dystruderts); chalky uplands had mainly Vertisols (Hapluderts and Epiaquerts); and floodplains featured most commonly featured Hapluderts and Fluvaquents (Fig. 4.17).

4.6 Eastern Gulf Coast Flatwoods (MLRA 152A)

The Eastern Gulf Coast Flatwoods (MLRA 152A) is the fifth largest MLRA in Mississippi, comprising nearly 3,067 km^2, which is 2.2% of the state area (Table 4.1). About 11% of this MLRA is in Mississippi, with the remainder in Florida (71%), Alabama (9%), and Louisiana (8%). The Eastern Gulf Coast Flatwoods is part of the Coastal Meadows physiographic province. This region features Pleistocene-age terraces consisting of ancient Mississippi River deposits of unconsolidated fine sand grading to coarser sand and gravel

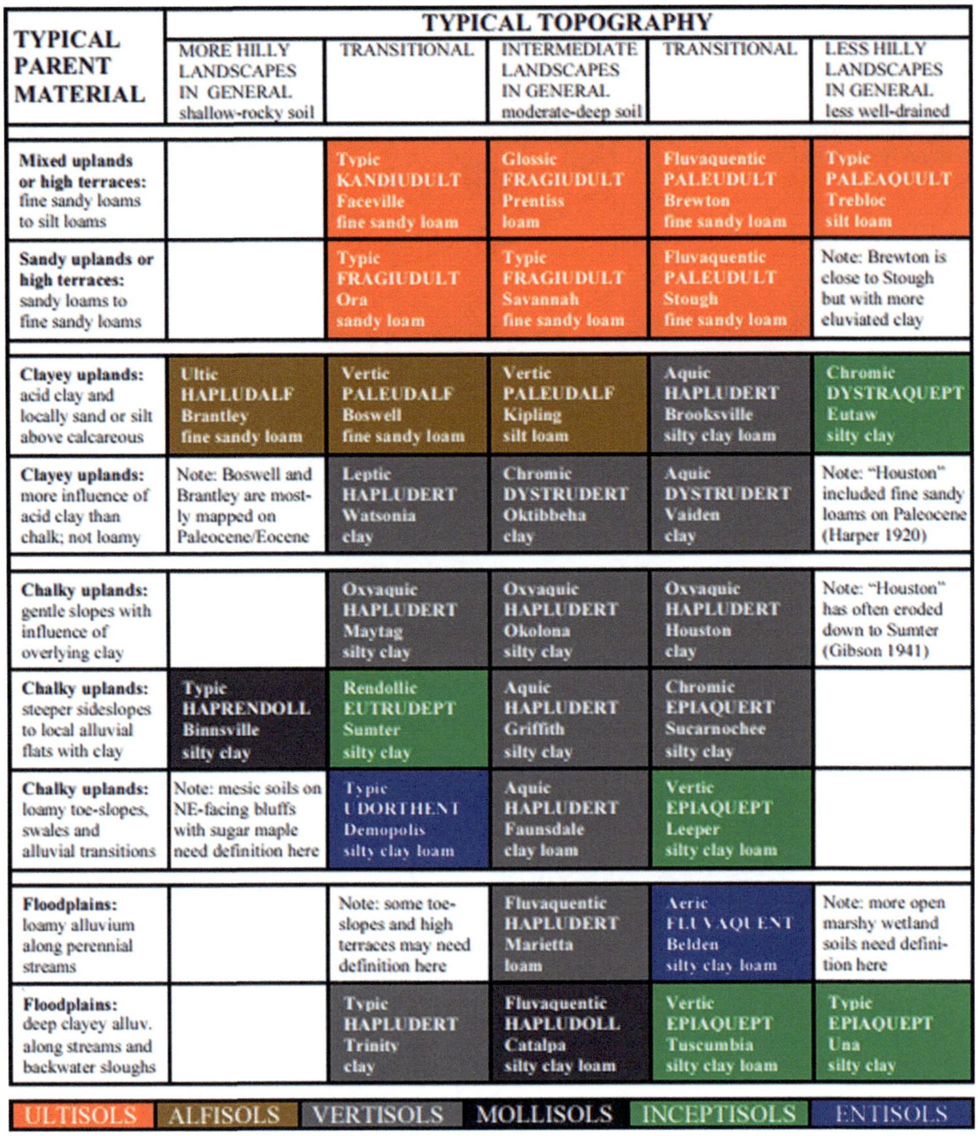

TYPICAL PARENT MATERIAL	TYPICAL TOPOGRAPHY: MORE HILLY LANDSCAPES IN GENERAL shallow-rocky soil	TRANSITIONAL	INTERMEDIATE LANDSCAPES IN GENERAL moderate-deep soil	TRANSITIONAL	LESS HILLY LANDSCAPES IN GENERAL less well-drained
Mixed uplands or high terraces: fine sandy loams to silt loams		Typic KANDIUDULT Faceville fine sandy loam	Glossic FRAGIUDULT Prentiss loam	Fluvaquentic PALEUDULT Brewton fine sandy loam	Typic PALEAQUULT Trebloc silt loam
Sandy uplands or high terraces: sandy loams to fine sandy loams		Typic FRAGIUDULT Ora sandy loam	Typic FRAGIUDULT Savannah fine sandy loam	Fluvaquentic PALEUDULT Stough fine sandy loam	Note: Brewton is close to Stough but with more eluviated clay
Clayey uplands: acid clay and locally sand or silt above calcareous	Ultic HAPLUDALF Brantley fine sandy loam	Vertic PALEUDALF Boswell fine sandy loam	Vertic PALEUDALF Kipling silt loam	Aquic HAPLUDERT Brooksville silty clay loam	Chromic DYSTRAQUEPT Eutaw silty clay
Clayey uplands: more influence of acid clay than chalk; not loamy	Note: Boswell and Brantley are mostly mapped on Paleocene/Eocene	Leptic HAPLUDERT Watsonia clay	Chromic DYSTRUDERT Oktibbeha clay	Aquic DYSTRUDERT Vaiden clay	Note: "Houston" included fine sandy loams on Paleocene (Harper 1920)
Chalky uplands: gentle slopes with influence of overlying clay		Oxyaquic HAPLUDERT Maytag silty clay	Oxyaquic HAPLUDERT Okolona silty clay	Oxyaquic HAPLUDERT Houston clay	Note: "Houston" has often eroded down to Sumter (Gibson 1941)
Chalky uplands: steeper sideslopes to local alluvial flats with clay	Typic HAPRENDOLL Binnsville silty clay	Rendollic EUTRUDEPT Sumter silty clay	Aquic HAPLUDERT Griffith silty clay	Chromic EPIAQUERT Sucarnochee silty clay	
Chalky uplands: loamy toe-slopes, swales and alluvial transitions	Note: mesic soils on NE-facing bluffs with sugar maple need definition here	Typic UDORTHENT Demopolis silty clay loam	Aquic HAPLUDERT Faunsdale clay loam	Vertic EPIAQUEPT Leeper silty clay loam	
Floodplains: loamy alluvium along perennial streams		Note: some toe-slopes and high terraces may need definition here	Fluvaquentic HAPLUDERT Marietta loam	Aeric FLUVAQUENT Belden silty clay loam	Note: more open marshy wetland soils need definition here
Floodplains: deep clayey alluv. along streams and backwater sloughs		Typic HAPLUDERT Trinity clay	Fluvaquentic HAPLUDOLL Catalpa silty clay loam	Vertic EPIAQUEPT Tuscumbia silty clay loam	Typic EPIAQUEPT Una silty clay

ULTISOLS | ALFISOLS | VERTISOLS | MOLLISOLS | INCEPTISOLS | ENTISOLS

Fig. 4.17 Relation between soil taxa and series and parent materials in the Alabama and Mississippi Blackland Prairie (MLRA 135A) (Campbell and Seymour 2011) Photo by J. Bockheim

at depth. Gal and others (2021) described the influence of antecedent geology on the Holocene formation of stable barriers along the Mississippi-Alabama chain. Elevations range from 0 to 23 m. Maximum slopes average $8 \pm 12\%$. Figure 4.18 shows Longleaf-slash pine in the Eastern Gulf Coast Flatwoods near Gautier, Mississippi.

The mean annual air temperature averages 19 °C, and the mean annual precipitation ranges between 1575 and 1650 mm (Table 4.2). The area supports longleaf-slash pine, saltwater marsh vegetation.

Figure 4.19 shows the Eustis-Latonia-Lakeland soil association in Harrison County. The soils are mainly Udults on dissected sandy and loamy alluvial and marine terraces. The Lakeland soil series, a Quartzipsamments, is formed in wind-reworked sandy marine deposits.

The dominant soil orders in this MLRA are Alfisols, Ultisols, and Entisols (Table 4.3). The soils in the area dominantly have a thermic soil temperature regime, an aquic or udic soil moisture regime, and siliceous mineralogy. They generally are deep or very deep; are somewhat poorly drained to very poorly drained; and are loamy, mucky, or sandy. Haplosaprists formed in organic deposits in swamps and depressions (Dorovan and Pamlico series) and in marshes and swamps (Maurepas series). Sulfihemists (Handsboro series) have formed in saltwater and brackish water marshes. Quartzipsamments (Newhan series) and Psammaquents (Duckston series) formed on dunes and in interdunal swales on barrier islands. Glossaqualfs (Guyton series) and Hydraquents (Arat series) formed in alluvium on flood plains. Endoaquults (Myatt series) and Paleudults

Fig. 4.18 Eastern Gulf Coast Flatwoods Region (MLRA 152) near Gautier, Mississippi. Photo by J. Bockheim

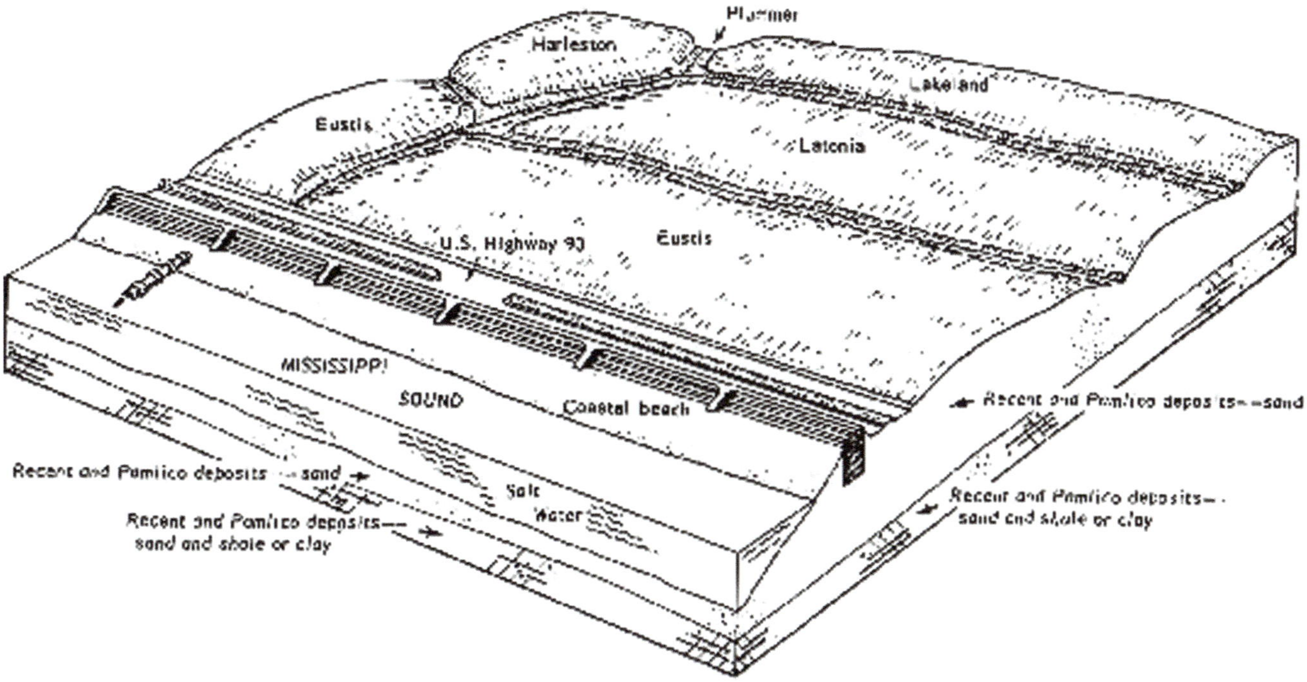

Fig. 4.19 Eustis-Latonia-Lakeland soil association in Harrison County is formed in recent and older marine terrace deposits in the East Gulf Coast Flatwoods (MLRA 152). *Source* USDA Natural Resources Conservation Service

Fig. 4.20 Salt marsh vegetation along the Mississippi Gulf Coast near Pascagoula (MLRA 151). Photo by J. Bockheim

(Stough series) formed in mixed fluvial and marine sediments on flats and stream terraces. Paleaquults (Plummer and Bayou series) and Paleudults (Escambia and Ocilla series) formed in loamy and sandy sediments on marine terraces.

4.7 Gulf Coast Marsh (MLRA 151)

The Gulf Coast Marsh (MLRA 151) is the smallest MLRA in Mississippi, comprising only 196 km^2, which is 0.1% of the state area (Table 4.1). About 1% of this MLRA is in Mississippi, with the remainder in Louisiana (95%) and Texas (4%). The Gulf Coast Marsh is part of the Coastal Meadows physiographic province (Table 4.3). This region features primarily Mississippi River alluvium composed of clay, silt, and fine sand of Pleistocene age. Elevations range from 0 to 3 m. Maximum slopes average $3 \pm 2\%$. Figure 4.20 shows Gulf Coast saltmarsh near Pascagoula. The mean annual air temperature averages 19 ± 1.3 °C, and the mean annual precipitation averages 1410 ± 120 mm. The area supports saltwater marsh vegetation.

Figure 4.21 shows the Handboro soil series, a Typic Sulfihemists, that is derived from freshwater and tidewater peat.

The dominant soil orders in this MLRA are Alfisols and Histosols. The soils in the area dominantly have a hyperthermic soil temperature regime, an aquic soil moisture regime, and smectitic mineralogy. They generally are very deep, very poorly drained, and clayey. The two predominant soil series in the Mississippi portion of the MLRA 151 are the Guyton, a Glossaqualfs formed in loamy flood plain deposits, and the Handsboro, a Sulfihemist formed in decomposed herbaceous plant remains in flooded salt marshes.

4.8 Summary

Mississippi contains six Major Land Resource Areas (MLRAs), including from largest to smallest, the Gulf Coastal Plain (MLRA 133C), the Southern Mississippi Valley Loess (MLRA 134), the Southern Mississippi Alluvium (MLRA 131A), the Alabama and Mississippi Blackland Prairie (MLRA 135A), the Eastern Gulf Coast Flatwoods (MLRA 152), and the Gulf Coast Marsh (MLRA 151). The MLRAs vary distinctly in mean annual air temperature, mean annual precipitation, elevation range, average maximum slope, vegetation, parent materials, landforms, and soils.

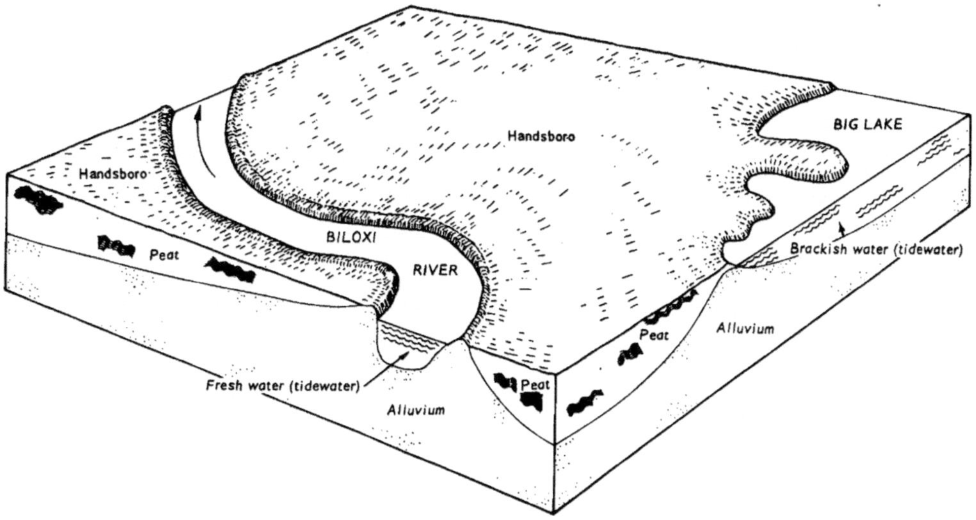

Fig. 4.21 Handsboro soil series is derived from peat along the Gulf Coast Marsh (MLRA 151) near Biloxi. *Source* USDA Natural Resources Conservation Service

References

Aslan A, Autin WJ (1998) Holocene flood-plain soil formation in the southern lower Mississippi Valley: implications for interpreting alluvial paleosols. Geol Soc Am Bull 110:443–449

Campbell JJN, Seymour WR Jr (2011) A review of native vegetation types in the Black Belt of Mississippi and Alabama, with suggested relationships to the catenas of soil series. J Miss Acad Sci 56:166–184

Fisk HN (1944) Geological investigation of the alluvial valley of the lower Mississippi River. US Army Corps of Engineers, Mississippi River commission, Vicksburg, MS 78 pp

Gal NS, Wallace DJ, Miner MD, Hollis RJ, Dike C, Flocks JG (2021) Influence of antecedent geology on the Holocene formation and evolution of Horn Island, Mississippi, USA. Marine Geol 431:106375

Muhs DR, Bettis EA III, Been J, McGeechin JP (2001) Impact of climate and parent material on chemical weathering in loess-derived soils of the Mississippi River Valley. Soil Sci Soc Am J 65:1761–1777

Rutledge EM, Guccione MJ, Markewich HW, Wysocki DA, Ward LB (1996) Loess stratigraphy of the lower Mississippi Valley. Engin Geol 45:167–183

United States Department of Agriculture, Natural Resources Conservation Service (2022) Land resource regions and major land resource areas of the United States, the Caribbean, and the Pacific Basin. USDA Agric Handbook 296

United States Geological Survey (2003) Ecoregions of Mississippi

Wascher HL, Humbert RP, Cady JG (1947) Loess in the Southern Mississippi Valley. Identification and distribution of loess sheets. Soil Sci Soc Am Proc 12:389–399

Diagnostic Horizons and Taxonomic Structure of Mississippi Soils

5

5.1 Introduction

Four of the eight epipedons (diagnostic surface horizons) in *Soil Taxonomy* (1999, 2014) and eight of the 20 diagnostic subsurface horizons occur in soils of Mississippi. Mississippi contains soils representative of eight of the 12 soil orders, 17 of the 67 suborders, 42 of the 270 great groups, 98 subgroups, 187 families, and 232 soil series. Data on diagnostic horizon thicknesses are provided for all soil series in Appendix C. Soil classification data for all established soil series in Mississippi are provided in Appendix D.

5.2 Diagnostic Horizons

Diagnostic surface horizons, or epipedons, and subsurface horizons are important in classifying the soils of Mississippi. Based on occurrence in soil series, diagnostic surface horizons can be ranked: ochric (94%), mollic (3.2%), histic (1.9%), and umbric (0.9%) (Table 5.1). Mississippi has a higher proportion of ochric epipedons and a lower proportion of mollic epipedons than for the US as a whole (Bockheim 2014). Thicknesses of epipedons for Mississippi soil series are ranked: histic 117 ± 51 cm, mollic 51 ± 30 cm, umbric 51 ± 22, and ochric 24 ± 21 cm (Table 5.1). Whereas the mollic epipedon is most common in Vertisols and Mollisols, the umbric epipedon is most common in Ultisols and Inceptisols.

Diagnostic subsoil horizons can be ranked by occurrence in soil series: argillic (61%), cambic (22%), glossic (10%), albic (9.7%), fragipan (9.3%), and natric (2.8%) (Table 5.1). Kandic and spodic horizons are limited to two or less soil series each. About 17% of the soil series in Mississippi lack a diagnostic subsurface horizon. Thickness of diagnostic subsurface horizons of Mississippi soil series (more than two occurrences) can be ranked: argillic (119 ± 48 cm), cambic (104 ± 51 cm), natric (101 ± 21 cm), fragipan (81 ± 38 cm), glossic (67 ± 49 cm), and albic (30 ± 25 cm).

Argillic horizon occurrence can be ranked: Ultisols > Alfisols > Vertisols. Argillic horizons are most common in Hapludults, Paleudults, Hapludalfs, and Paleudalfs, but they occur in many other soil great groups. Cambic horizon occurrence can be ranked: Inceptisols >> Vertisols. Cambic horizons are most common in Endoaquepts, Dystrudepts, Eutrudepts, Epiaquepts, and Hapluderts, but they occur in other soil great groups. Glossic horizon occurrence can be ranked: Alfisols >> Ultisols, and they occur most commonly in the Glossaqualfs, Natraqualfs, Glossudalfs, and Fraglossudalfs great groups. Albic horizons can be ranked: Alfisols >> Ultisols, and they occur commonly in Natraqualfs, Fraglossudalfs, Glossaqualfs, and Paleaquults. Fragipans can be ranked: Ultisols > Alfisols, and they occur commonly in Fragiudults, Fragiudalfs, and Fraglossudalfs. Great groups most often lacking a diagnostic subsurface horizon include the Udifluvents, Quartzipsamments, Fluvaquents, and Haplosaprists.

5.3 Orders

Although soil series representing eight orders occur in Mississippi, five account for 99% of the soil series, including Ultisols (31%), Alfisols (30%), Inceptisols (18%), Vertisols (11%), and Entisols (9%). The remainder is made up by Histosols (0.6%), Mollisols (0.5%), and a Spodosol (0%) (Fig. 5.1).

5.4 Suborders

Mississippi contains soils representative of 17 of the 67 suborders recognized in Soil Taxonomy. Seven suborders account for two-thirds (89%) of the soil series of Mississippi, including the Udults (28%), Udalfs (24%), Aquepts (11%), Aquerts (7.9%), Udepts (7.6%), Aqualfs (5.2%), and Aquents (5.0%) (Fig. 5.2).

D. Johnson et al., *The Soils of Mississippi*, World Soils Book Series,
https://doi.org/10.1007/978-3-031-36235-4_5

Table 5.1 Diagnostic horizon thickness for Mississippi soil series

Horizon	Mean (cm)	Std Dev (cm)	% of soil series
Surface			
Ochric	24	21	94.0
Mollic	51	30	3.2
Histic	117	51	1.9
Umbric	51	22	0.9
Subsurface			
Argillic	119	48	61
Cambic	104	51	22
None			17
Glossic	67	49	10.0
Albic	30	25	9.7
Fragipan	81	38	9.3
Natric	101	27	2.8
Sulfuric	114		0.9
Spodic	107		0.5
Kandic	85	22	0.2

Note many Mississippi soil series contain multiple subsurface horizons

5.5 Great Groups

Mississippi contains soils representative of 42 of the 270 great groups. Twelve great groups account for 77% of the soil series in Mississippi, including the Paleudults (9.8%), Fragiudalfs (9.8%), Hapludults (9.7%), Fragiudults (8.3%), Hapludalfs (8.3%), Endoaquepts (8.3%), Paleudalfs (4.9%), Fluvaquents (4.8%), Dystrudepts (4.7%), Epiaquerts (4.0%), and Endoaqualfs (4.0%) (Fig. 5.3).

5.6 Subgroups

There are 98 soil subgroups in Mississippi. Excluding those in the Typic category, the most common subgroups in Mississippi are aquic (15%), Fluvaquentic (7.4%), Vertic (4.8%), Aeric (4.3%), Chromic (3.9%), Oxyaquic (3.5%), Fluventic (3.0%), and Plinthic (3.0%) (Fig. 5.4a). The frequent use of Typic (35%) is common, in that it delineates the central concept of a great group. Aquic implies restricted drainage; Fluvaquentic refers to a transition to soils in the Fluvaquentic great group; Vertic implies

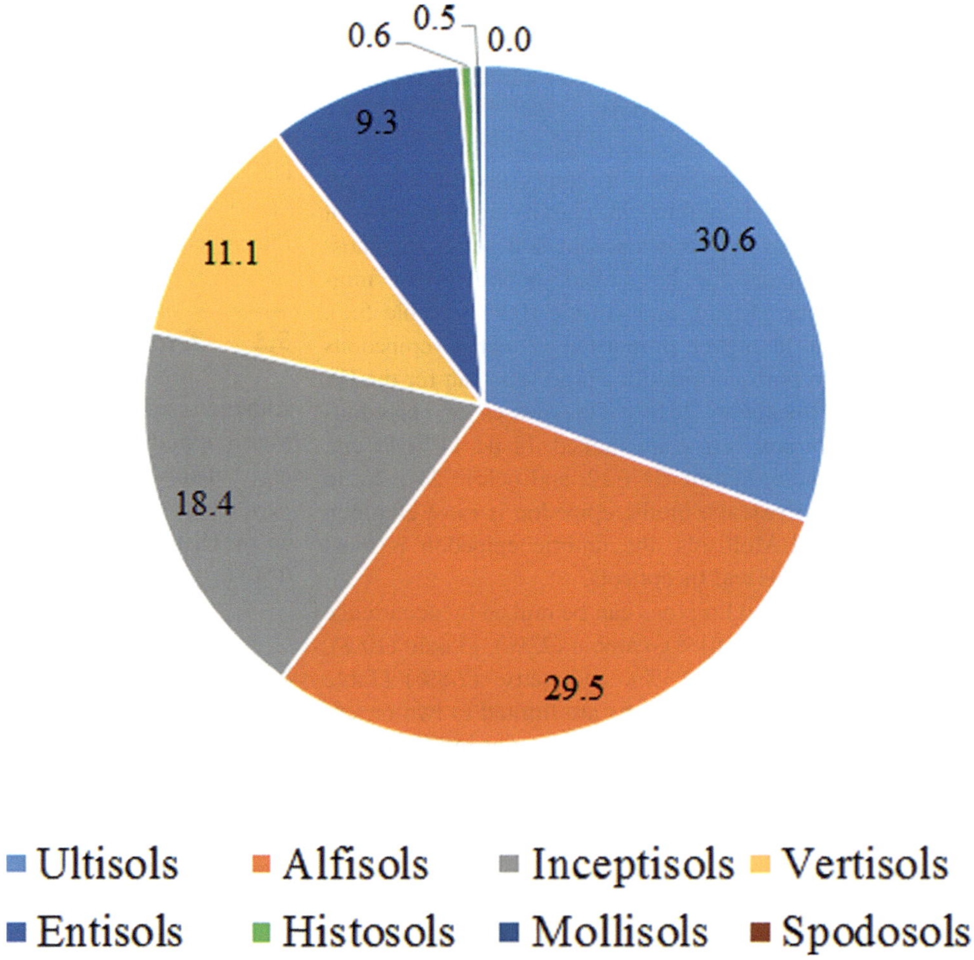

Fig. 5.1 Distribution (%) of soil orders in Mississippi by area *Source* by J. Bockheim

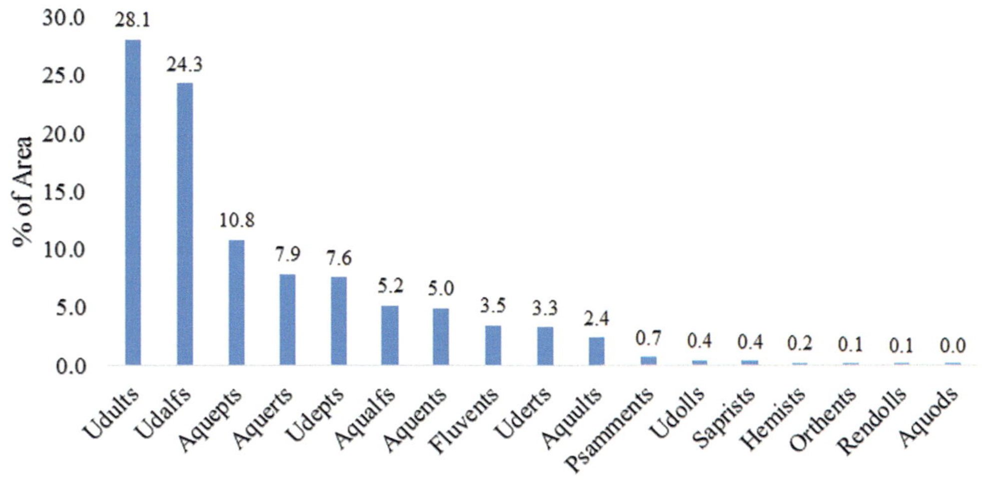

Fig. 5.2 Distribution (%) of suborders in Mississippi by area *Source* by J. Bockheim

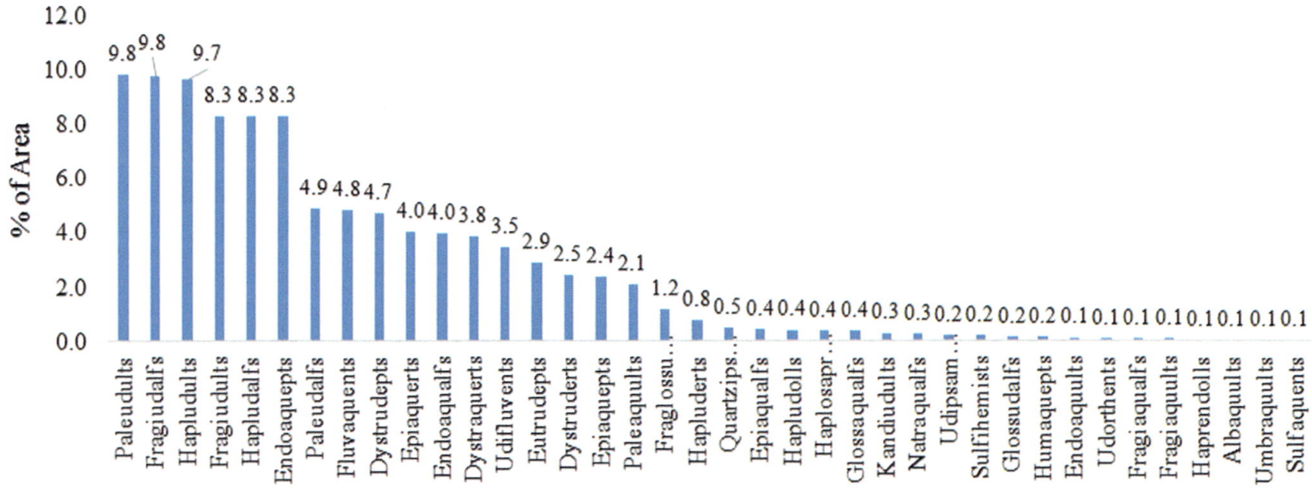

Fig. 5.3 Distribution (%) of great groups in Mississippi by area *Source* by J. Bockheim

seasonal cracking and a tendency toward Vertisols; Aeric refers to a coarse-textured layer at the surface; Chromic refers to a high-chroma color; Oxyaquic refers to seasonally restricted drainage; Fluventic refers to a transition with the Fluvent suborder; and Plinthic implies that the soil contains plinthite.

Subgroups follow one of four options: the central concept (e.g., Typic and Haplic); intergrades that have specific properties that differ from the Typic or Haplic and have one or more characteristics similar to another order, suborder, or great group; extragrades, which identify soils with properties that do not clearly intergrade toward specifically defined categories (e.g., Lithic or Calcic), and intragrades reflect variations within a great group. More than one-third (36%) of the soil series follow the central concept, 18% are intergrades, 40% are intragrades, and 6% are extragrades (Fig. 5.4b).

About 3.9% of the soil series in Mississippi have a lithic or paralithic contact within the upper 100 cm.

5.7 Families

There are 187 soil families in Mississippi. More than three-quarters (80%) of the soils in Mississippi are in five particle-size classes, including fine-silty (21%), fine (20%), fine-loamy (17%), coarse-loamy (13%), and coarse-silty (9.5%) (Fig. 5.5a). The predominant mineral classes are siliceous (40%), mixed (38%), and smectitic (15%) (Fig. 5.5b). Two-thirds (67%) of the soil series are in the active, semiactive, and superactive cation-exchange activity classes (Fig. 5.5c). Only 17% of the soil series in Mississippi are in a reaction class (Fig. 5.5d). All but one of the soil series in Mississippi has a thermic soil

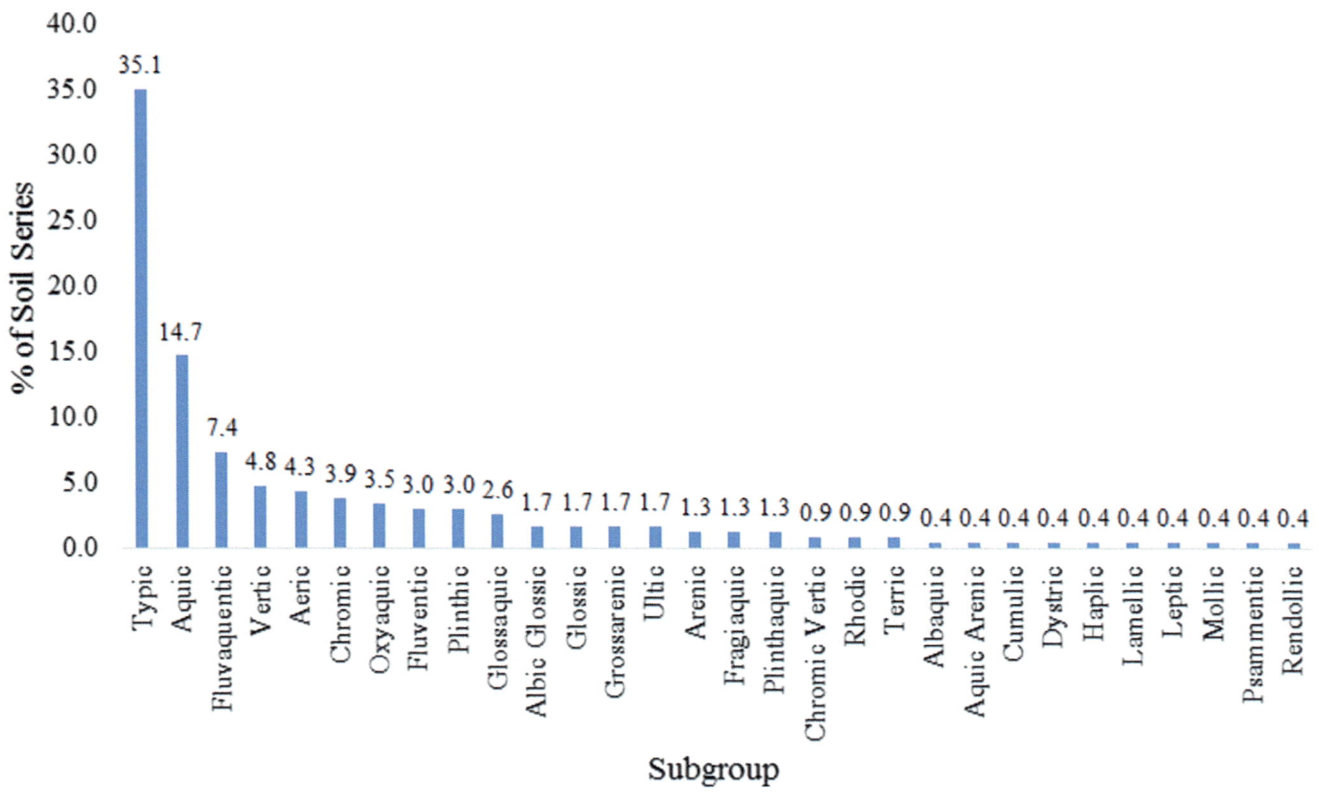

Fig. 5.4 **a** Distribution of subgroups in Mississippi by area. **b** Distribution (%) of soil series by subgroup category *Source* by J. Bockheim

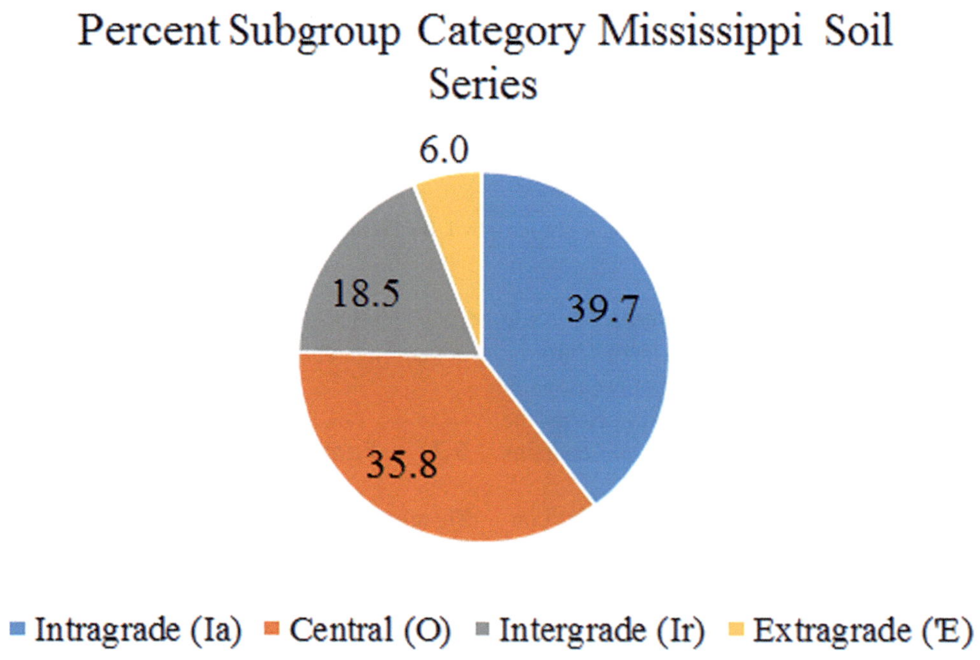

(continued)

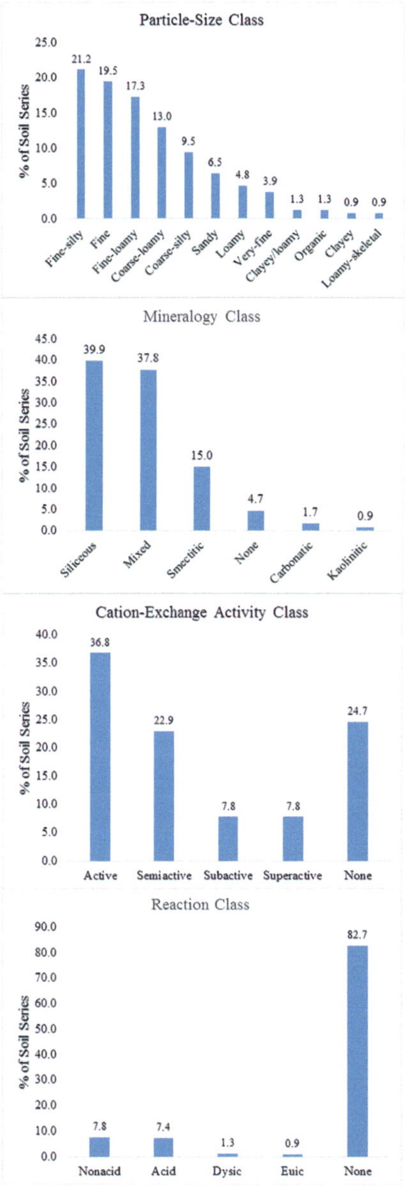

Fig. 5.5 Distribution of family classes of Mississippi soil series *Source* by J. Bockheim

temperature regime (Fig. 5.6a). More than two-thirds (68%) of Mississippi's soils have an udic soil moisture regime; the remaining 32% of the soils have an aquic SMR (Fig. 5.6b).

5.8 Soil Series

Mississippi contains 232 established soil series (Table 5.2), of which 10% occur only in Mississippi (Fig. 5.7). The majority (81%) of soil series recognized in Mississippi

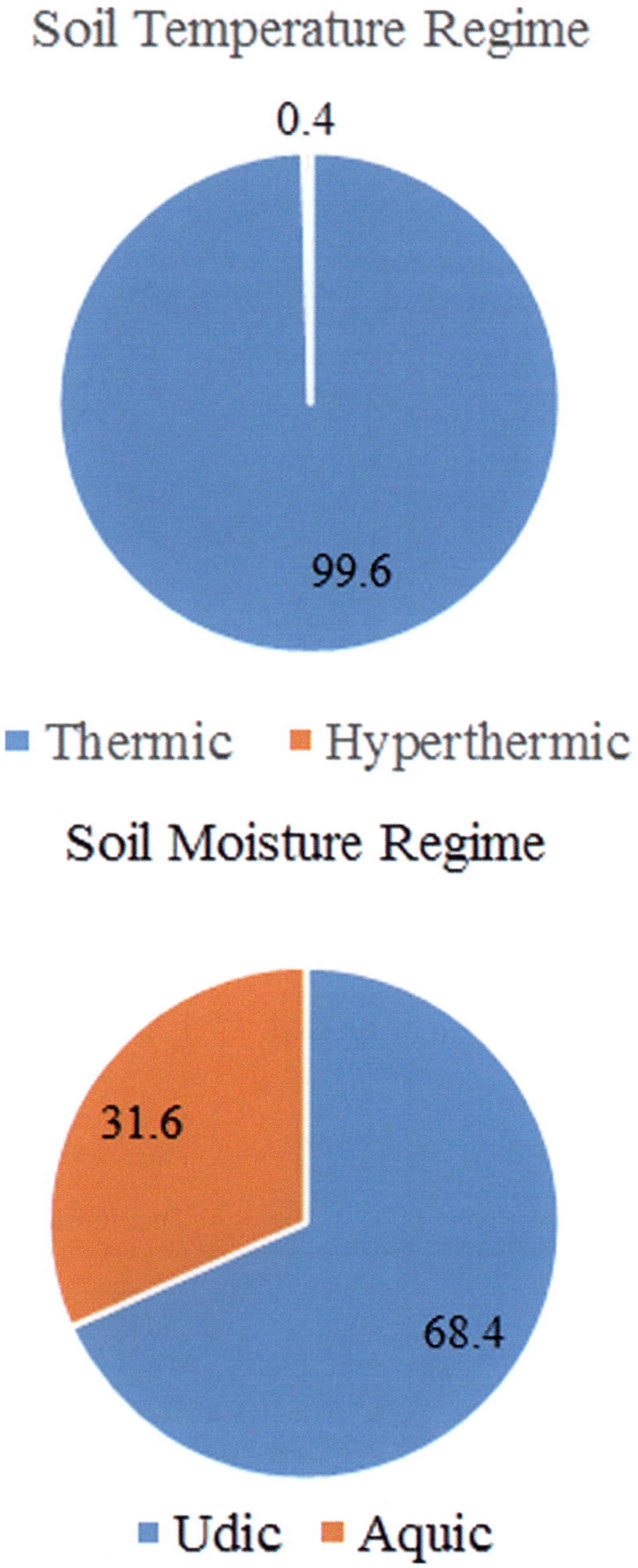

Fig. 5.6 Distribution of soil temperature regimes and soil moisture regimes of soil series in Mississippi *Source* by J. Bockheim

Table 5.2 Soil orders, suborders, and great groups of Mississippi

Order	Suborder	Great group	No. soil series	Area (km^2)
Alfisols	Aqualfs	Endoaqualfs	4	3424
Alfisols	Aqualfs	Epiaqualfs	1	370
Alfisols	Aqualfs	Fragiaqualfs	2	83
Alfisols	Aqualfs	Glossaqualfs	5	346
Alfisols	Aqualfs	Natraqualfs	6	246
Alfisols	Udalfs	Fragiudalfs	5	8432
Alfisols	Udalfs	Fraglossudalfs	5	1007
Alfisols	Udalfs	Glossudalfs	4	153
Alfisols	Udalfs	Hapludalfs	18	7144
Alfisols	Udalfs	Natrudalfs	1	17
Alfisols	Udalfs	Paleudalfs	11	4219
		Total	62	25,441
Entisols	Aquents	Fluvaquents	5	4146
Entisols	Aquents	Hydraquents	1	29
Entisols	Aquents	Psammaquents	2	42
Entisols	Aquents	Sulfaquents	2	47
Entisols	Fluvents	Udifluvents	9	3002
Entisols	Orthents	Udorthents	1	89
Entisols	Psamments	Quartzipsamments	6	413
Entisols	Psamments	Udipsamments	1	204
		Total	27	7972
Histosols	Hemists	Sulfihemists	1	177
Histosols	Saprists	Haplosaprists	4	350
		Total	5	527
Inceptisols	Aquepts	Endoaquepts	11	7143
Inceptisols	Aquepts	Epiaquepts	7	2039
Inceptisols	Aquepts	Humaquepts	1	132
Inceptisols	Udepts	Dystrudepts	10	4046
Inceptisols	Udepts	Eutrudepts	7	2474
		Total	36	15,834
Mollisols	Rendolls	Haprendolls	2	55
Mollisols	Udolls	Hapludolls	2	361
		Total	4	416
Spodosols	Aquods	Alaquods	1	1
Ultisols	Aquults	Albaquults	1	51
Ultisols	Aquults	Endoaquults	1	117
Ultisols	Aquults	Fragiaquults	1	72
Ultisols	Aquults	Paleaquults	8	1812
Ultisols	Aquults	Umbraquults	1	48
Ultisols	Udults	Fragiudults	8	7158
Ultisols	Udults	Hapludults	26	8312
Ultisols	Udults	Kandiudults	4	262
Ultisols	Udults	Paleudults	27	8475
		Total	77	26,307
Vertisols	Aquerts	Dystraquerts	4	3305
Vertisols	Aquerts	Epiaquerts	3	3469
Vertisols	Uderts	Dystruderts	6	2122
Vertisols	Uderts	Hapluderts	7	698
		Total	20	9594
		Sum	232	86,092

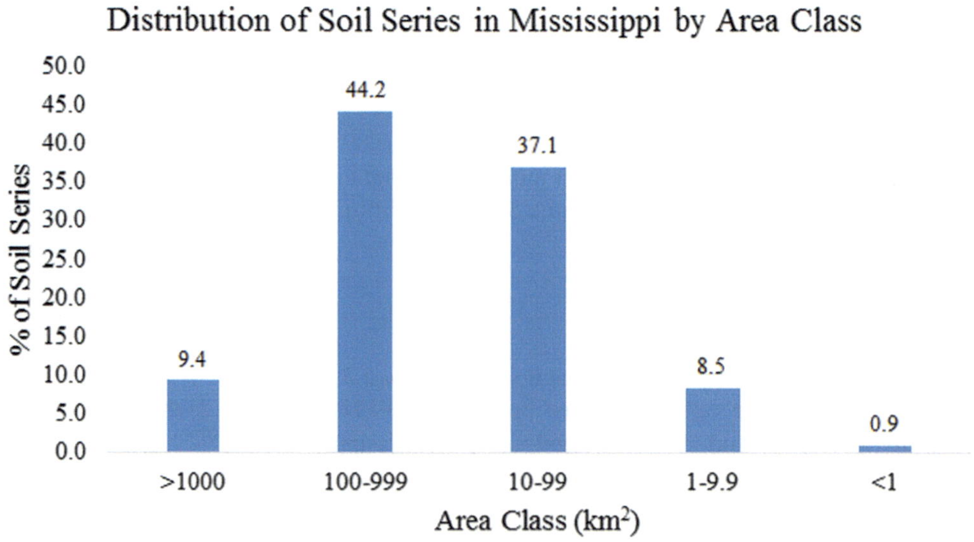

Fig. 5.7 Distribution of soil series in Mississippi by area class *Source* by J. Bockheim

are in the 100–999 km^2 (44%) or 10–99 km^2 (37%) area classes, followed by the > 1000 km^2 (9%) and 1–9.9 km^2 (8.5%) classes. This contrasts with the area distribution of soil series in the US as a whole, where 56% of the soil area contains "mega" (> 1000 km^2) soil series.

5.9 Summary

Four of the eight epipedons (diagnostic surface horizons) in *Soil Taxonomy* (ochric, umbric, histic, and mollic) and eight (argillic, cambic, glossic, albic, fragipan, natric, kandic, and spodic) of the 20 diagnostic subsurface horizons occur in soils of Mississippi. Mississippi contains soils representative of eight of the 12 soil orders, 17 of the 67 suborders, 42 of the 270 great groups, 98 subgroups, 187 families, and 232 soil series. The dominant orders are Ultisols, Alfisols, Inceptisols, Vertisols, and Entisols, with Histosols, Mollisols and a Spodosol accounting for a small proportion of the soils.

Reference

Bockheim JG (2014) Soil geography of the USA: a diagnostic horizon approach. Springer International, New York

6 Taxonomic Soil Regions of Mississippi

6.1 Introduction

Mississippi is divided here into 19 soil regions based on the relative abundance of great groups. Site factors are summarized in Table 6.1, and thicknesses of diagnostic horizons are provided in Table 6.2. The methods for preparing soil great group maps in this chapter and soil order maps in Chaps. 7 through 15 are described below.

Soil mapping in Mississippi has one level of status: SSURGO (Soil Survey Geographic Database) certified. Soil survey areas that are SSURGO certified cover 100% of Mississippi. The mapping generally is at a scale between 1:15,000 and 1:24,000, and all soil information is contained in Web Soil Survey located at https://websoilsurvey.sc.egov.usda.gov. All soil series identified in these areas have an established status.

Soil order and soil great group maps were developed using all soil mapping from October 2021 SSURGO certified areas. Certified soil map units and initial soil survey map units, as previously mentioned, can have up to three major components although many map units have only one major component. Each component has an area percentage and each component has a soil classification. Minor components, i.e., less than 10% of a map unit, were filtered out. Soil order, and great group maps were prepared by James Curtis, Mississippi State Scientist.

6.2 Fragiudalfs (Soil Region 1)

Fragiudalfs are well-developed soils that contain an ochric epipedon over argillic and fragipan horizons. They have formed in loess over alluvium or marine deposits on uplands and alluvial terraces (Table 6.1). Fragiudalfs on maximum slopes averaging $13 \pm 5.5\%$. These soils are very deep and generally are moderately well drained. The native vegetation on Fragiudalfs commonly is mixed hardwoods, often with pines. The mean annual air temperature is 17 ± 1.5 ℃, and the mean annual precipitation is 1360 ± 87 mm.

The Fragiudalfs great group in Mississippi has five soil series, which cover 8400 km^2. Fragiudalfs are most common in the Southern Mississippi Loess (MLRA 134) (Fig. 6.1). Major Fragiudalfs include the Providence and Loring soil series, each of which occupies more than 4300 km^2.

Fragiudalfs are in the fine-silty particle-size class, the mixed mineral class; and 80% are in the active cation-exchange activity class. Fragiudalfs have a udic soil moisture regime and a thermic soil temperature regime.

Fragiudalfs have an ochric epipedon that averages 20 ± 7.6 cm over an argillic horizon that averages 149 ± 41 cm and a fragipan that averages 75 ± 38 cm in thickness (Table 6.2). A few Fragiudalfs have a glossic horizon or plinthite. The Providence soil series, a fine-silty, mixed, active, thermic Oxyaquic Fragiudalfs, is formed in a mantle of loess about 60 cm thick over sandy and loamy sediments (Fig. 6.2). This soil as a dark gray A horizon from 0 to 7.5 cm, a grayish brown albic horizon from ssss7.5 to 18 cm, and an argillic horizon from 18 cm to over 200 (80 in.). The upper argillic is formed in loess and has a strong brown color; the lower argillic is yellowish red in color and has a fragipan from 58 to 135 cm (23–53 in.).

6.3 Paleudults (Soil Region 2)

Paleudults are strongly developed soils with an ochric epipedon over a deep argillic horizon. They have developed on loamy or sandy alluvium on alluvial terraces and marine deposits on marine terraces in uplands. Maximum slopes average $12 \pm 6.3\%$. The soils are deep to very deep and well drained to somewhat poorly drained. The vegetation is pines and hardwoods and southern pines (flatwoods). The mean annual air temperature is 19 ± 0.7 ℃, and the mean annual precipitation is 1465 ± 139 mm.

D. Johnson et al., *The Soils of Mississippi*, World Soils Book Series,
https://doi.org/10.1007/978-3-031-36235-4_6

Table 6.1 Soil-forming factors of major great groups in Mississippi

Soil region	Soil great group	Area (km^2)	No. soil series	MAP (mm/yr)	MAAT (°C)	Soil-moisture regime	Soil temperature regime	Vegetation type	Upper slope (%)	Parent materials	Composition	Landforms
1	Fragiudalfs	8432	5	17 ± 1.5	1360 ± 87	Udic	Thermic	Hardwoods; hardwoods and pines	13 ± 5.5	Loess; loess/alluvium or marine	Silty; silty/clayey or loamy	Uplands; alluvial terraces
2	Paleudults	8475	27	19 ± 0.9	1465 ± 134	Udic	Thermic	Pines; oak-pine	12 ± 7.7	Alluvium, marine	Loamy; clayey; sandy	Uplands; marine terraces; alluvial terraces
3	Hapludults	8312	26	18 ± 1.2	1365 ± 123	Udic	Thermic	Pines; oak-pine	18 ± 18	Alluvium, marine, residuum	Clayey, loamy, sandy	Alluvial terraces; marine terraces
4	Hapludalfs	7144	18	18 ± 1.5	1295 ± 68	Udic	Thermic	Hardwoods; oak-pine	31 ± 22	Alluvium, residuum, loess	Loamy, clayey, silty	Uplands; alluvial terraces
5	Fragiudults	7158	8	18 ± 2.0	1485 ± 128	Udic	Thermic	Oak-pine; hardwoods	11 ± 5.7	Loess; marine; alluvium	Loamy, silty	Uplands; alluvial terraces
6	Endoaquepts	7143	11	18 ± 1.7	1340 ± 175	Aquic	Thermic	Bottomland hardwoods; hardwoods	2.6 ± 1.3	Alluvium	Clayey/sandy; loamy	Flood plains
7	Endoaqualfs	3434	4	18 ± 1.8	1310 ± 25	Aquic	Thermic	Hardwoods	3.5 ± 3.0	Alluvium	Silty	Alluvial terraces
8	Paleudalfs	4219	11	18 ± 1.3	1320 ± 119	Udic	Thermic	Oak-pine	14 ± 12	Marine; loess/marine; alluvium	Loamy/clayey; clayey	Marine terraces; uplands; alluvial terraces
9	Dystrudepts	4046	10	18 ± 0.7	1330 ± 89	Udic	Thermic	Bottomland hardwoods; hardwoods	2.9 ± 2.3	Alluvium	Loamy; silty	Flood plains; alluvial terraces
10	Fluvaquents	4148	5	18 ± 1.2	1365 ± 64	Aquic	Thermic	Bottomland hardwoods	2.0 ± 0.0	Alluvium	Silty	Flood plains
11	Epiaquerts	3469	3	18 ± 0.7	1410 ± 18	Aquic	Thermic	Hardwoods; bottomland hardwoods	3.0 ± 1.7	Alluvium	Clayey	Flood plains
12	Dystraquerts	3305	4	17 ± 1.9	1290 ± 63	Aquic	Thermic	Hardwoods; oak-pine	5.0 ± 5.0	Marine, alluvium, residuum	Clayey	Uplands
13	Udifluvents	3002	9	18 ± 2.1	1355 ± 93	Udic	Thermic	Hardwoods; oak-pine	2.7 ± 1.2	Alluvium	Sandy; silty; loamy	Flood plains
14	Eutrudepts	2474	7	18 ± 0.8	1345 ± 63	Udic	Thermic	Hardwoods; oak-pine	18 ± 25	Alluvium	Silty; loamy	Flood plains; uplands
15	Dystruderts	2122	6	17 ± 0.3	1365 ± 76	Udic	Thermic	Oak-pine	18 ± 24	Alluvium; residuum	Clayey; clayey/chalk	Uplands
16	Epiaquepts	2039	7	17 ± 1.5	1335 ± 47	Aquic	Thermic	Bottomland hardwoods; hardwoods	3.0 ± 1.2	Alluvium	Clayey; clayey/loamy	Flood plains; natural levees
17	Paleaquults	1812	8	18 ± 1.1	1375 ± 137	Aquic	Thermic	Oak-pine; pines	3.5 ± 2.5	Marine, alluvium	Loamy; sandy	Uplands; alluvial terraces; marine terraces
18	Fraglossudalfs	1007	5	17.5 ± 1.7	1325 ± 58	Udic	Thermic	Hardwoods	7.2 ± 4.6	Loess; loess/alluvium	Silty; silty/loamy	Alluvial terraces; uplands
19	Hapluderts	698	7	18 ± 0.74	1265 ± 55	Udic	Thermic	Redcedar, hackberry, grasses	8.6 ± 8.0	Residuum; alluvium	Clayey; clayey/chalk	Uplands; flood plains

Table 6.2 Diagnostic horizon thicknesses of dominant great groups in Mississippi

					Thickness (cm)						(%)
No	Great group	Mollic	Ochric	Umbric	Cambic	Argillic	Fragipan	Glossic	Albic	Plinthite	None (%)
1	Fragiudalfs		20±7.6			149±41	75±38			0	0
2	Paleudults		36±37			144±38	137±9.9 (7.7%)			79±30 (27%)	0
3	Hapludults		24±15			88±37				0	0
4	Hapludalfs		19±12			89±38				0	0
5	Fragiudults		20±7.6			157±44	84±42	31 (25%)		145 (25%)	0
6	Endoaquepts		15 ± 5.9		128±55					0	0
7	Endoaqualfs		20±12			119±40				0	0
8	Paleudalfs		16±5.4			170±26		37 (18%)		0	0
9	Dystrudepts		23±12		102±29					0	0
10	Fluvaquents		20±7.9		47±38 (40%)					0	60
11	Epiaquerts		22±3.5		166±30					0	0
12	Dystraquerts		16±2.9		147 (25%)	120±37 (50%)				0	25
13	Udifluvents		17±4.0							0	100
14	Eutrudepts		21±6.9		72±61					0	14
15	Dystruderts		7.9±3.3		151±17 (50%)	58±32 (83%)				0	0
16	Epiaquepts		16±5.6		104±52					0	14
17	Paleaquults		46±39			133±30			45±52 (43%)	127 (12%)	0
18	Fraglossudalfs		21±7.9		41 (20%)	119±29	71±16	32±24 (80%)	19±19 (60%)	0	0
19	Hapluderts	70±32 (57%)	19±19 (43%)		87±42 (81%)					0	19

There are 27 soil series in the Paleudults great group that cover an area of 8475 km^2. Paleudults occur most commonly in the Eastern Gulf Coast Flatwoods (MLRA 152A) and the Gulf Coastal Plain (MLRA 133C) (Fig. 6.3). The most extensive Paleudults in Mississippi are the Ruston (2018 km^2), McLaurin (1767 km^2), and Benndale (950 km^2) soil series.

Paleudults are most commonly in the coarse-loamy, fine-loamy, and loamy particle-size classes and the siliceous mineral class; they have a udic soil moisture regime and a thermic soil temperature regime.

Paleudults have an ochric epipedon that averages 36 ± 37 cm over an argillic horizon that averages 144 ± 38 cm in thickness. Seven of the 27 Paleudults have plinthite and these soil series occur in the southern part of the state. The McLaurin soil series, a coarse-loamy, siliceous, subactive, thermic Typic Paleudults, has formed in loamy marine or stream sediments (Fig. 6.4). This soil as a very dark grayish brown A horizon from 0 to 12 cm, a dark grayish brown albic horizon from 12 to 20 cm, a yellowish brown transitional BE horizon from 20 to 36 cm, and a yellowish red to red argillic horizon from 36 cm to beyond 150 cm.

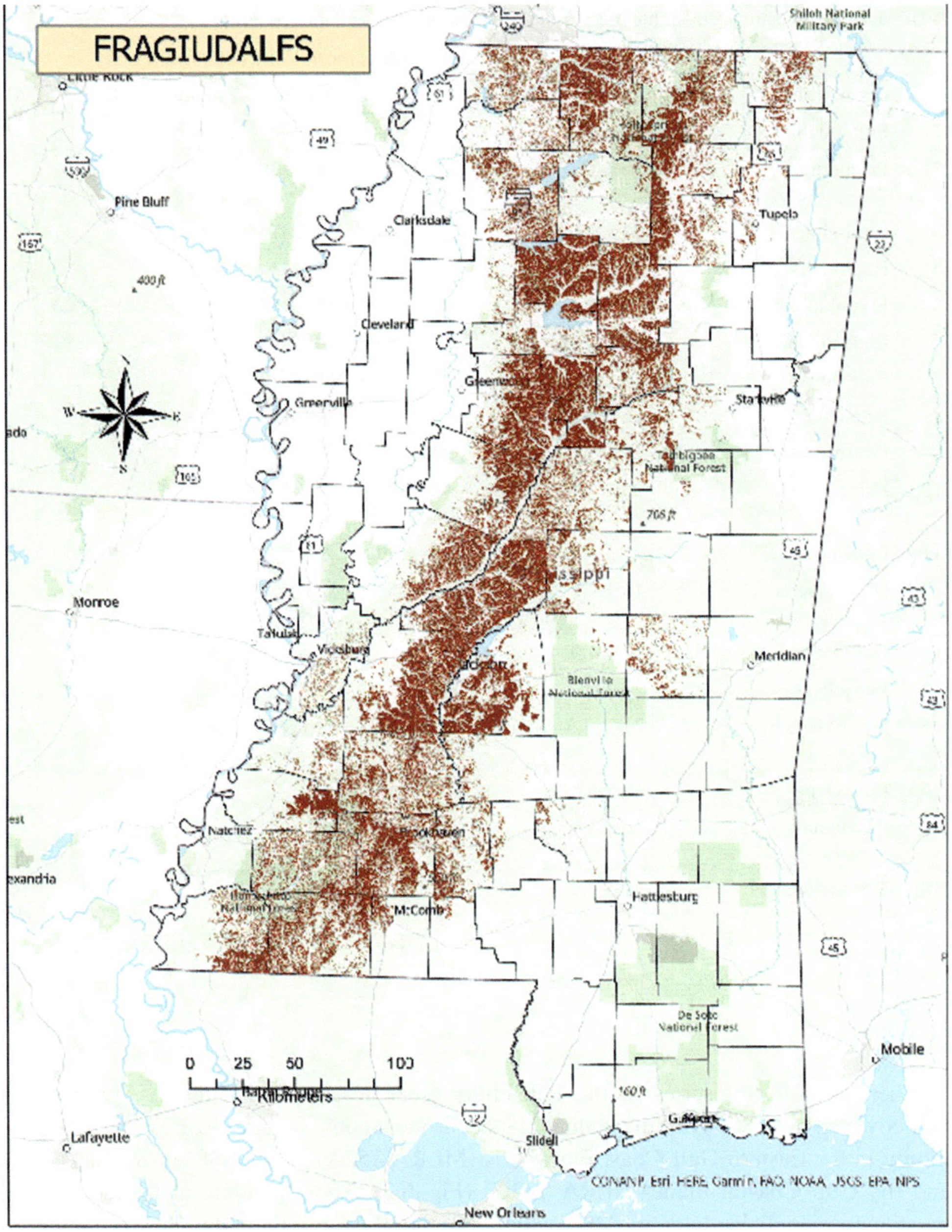

Fig. 6.1 Distribution of Fragiudalfs in Mississippi. *Source* USDA Natural Resources Conservation Service, Mississippi

6.4 Hapludults (Soil Region 3)

Hapludults are well-developed soils that contain an ochric epipedon and a deep argillic horizon. They have formed in clayey or loamy alluvium on alluvial terraces and marine deposits on marine terraces in uplands. Maximum slopes average $20 \pm 22\%$. These soils tend to be very deep and well drained to moderately well drained. The native vegetation on Hapludults is hardwoods and pines. The mean annual air temperature is 18 ± 1.4 ℃, and the mean annual precipitation is 1375 ± 95 mm.

There are 26 soil series in the Hapludults great group in Mississippi that cover an area of 8300 km^2. They occur primarily in the Gulf Coast Plain (MLRA 133C) (Fig. 6.5). The most extensive Hapludults are the Smithdale (3190 km^2), Sweatman (2410 km^2) soil series.

Hapludults in Mississippi are most commonly in the fine and fine-loamy particle-size classes, the siliceous or mixed mineral class, and in the semiactive cation-exchange activity class. Hapludults have a udic soil moisture regime and a thermic soil temperature regime.

Fig. 6.2 Providence soil series, a fine-silty, mixed, active, thermic Oxyaquic Fragiudalfs, is formed in a mantle of loess about 60 cm thick over sandy and loamy sediments. The scale is in inches. *Source* USDA Natural Resources Conservation Service

Hapludults have an ochric epipedon that averages 24 ± 15 cm over an argillic horizon that averages 88 ± 37 cm in thickness. The Cuthbert soil series, a fine, mixed, semiactive, thermic Typic Hapludults, is formed in residuum from weakly consolidated sandstone and shale (Fig. 6.6). This soil has a very dark gray A horizon from 0 to 10 cm, a gray albic horizon from 10 to 20 cm, and a dark red argillic horizon from 20 to 85 cm. The underlying C horizon has common shale plates.

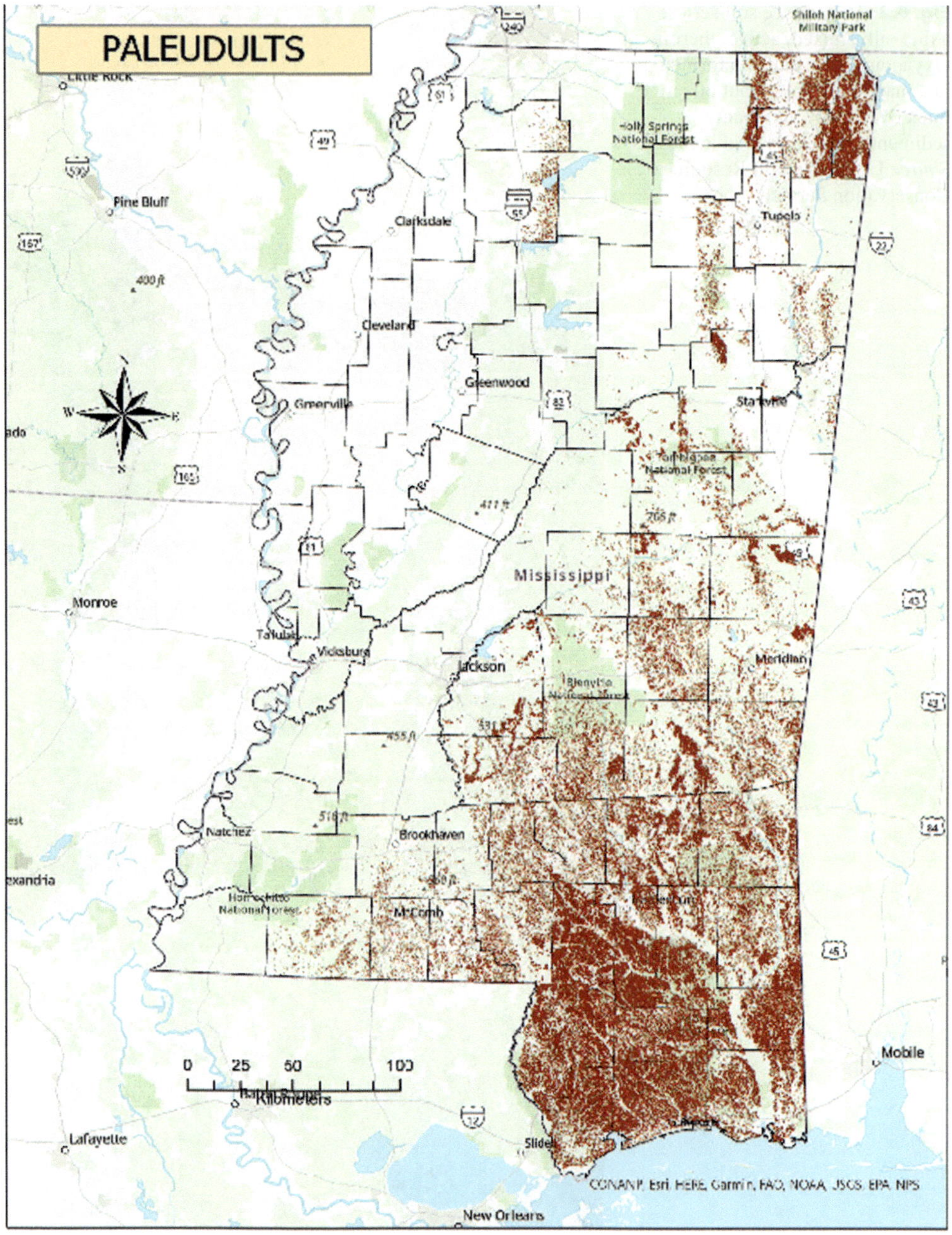

Fig. 6.3 Distribution of Paleudults in Mississippi. *Source* USDA Natural Resources Conservation Service, Mississippi

6.5 Hapludalfs (Soil Region 4)

Hapludalfs are well-developed soils that contain an ochric epipedon over an argillic horizon. They have formed in loamy, clayey, and silty alluvium, marine deposits, and residuum on low alluvial terraces and in uplands. Maximum slopes average 26 ± 21%. These soils tend to be deep to very deep and well drained to somewhat poorly drained. The native vegetation is hardwoods and pines. The mean annual air temperature is 18 ± 1.5 °C, and the mean annual precipitation is 1300 ± 71 mm.

There are 18 soil series in the Hapludalfs great group that occupy 7100 km^2, primarily in the Southern Mississippi Valley Loess (MLRA 134) but also occur in the Gulf Coastal Plain (MLRA 133C) and the Southern Mississippi Valley Alluvium (MLRA 131A) (Fig. 6.7). The

Fig. 6.4 McLaurin soil series, a coarse-loamy, siliceous, subactive, thermic Typic Paleudults, has formed in loamy marine or stream sediments. This pedon shows a 2.5 m section near Wiggins, MS. Photo by J. Bockheim

most extensive Hapludalfs are the Memphis (4275 km^2) and Lorman (1080 km^2) soil series.

Hapludalfs generally occur in fine-silty and fine particle-size classes, the mixed mineral class, and the active cation-exchange activity class. They have a udic soil moisture regime and a thermic soil temperature regime.

Hapludalfs have an ochric epipedon that averages 19 ± 12 cm over an argillic horizon that averages 89 ± 38 cm in thickness. The Dubbs soil series, a fine-silty, mixed, active, thermic Typic Hapludalfs, has formed in loamy alluvium (Fig. 6.8). This soil has a dark grayish brown plow layer (Ap horizon) from 0 to 12 cm over a yellowish brown argillic horizon from 12 to 100 cm. The Dubbs soil series is one of the classic "brown or yellow loams" common to the Delta region.

6.6 Fragiudults (Soil Region 5)

Fragiudults are strongly developed soils with an ochric epipedon overlying an argillic and fragipan horizons. They are formed in silty loess over gravelly or clayey residuum and loamy marine deposits on alluvial terraces, marine terraces, and uplands. Maximum slopes average $11 \pm 6.2\%$. These soils tend to be very deep and moderately well drained. The native vegetation is hardwoods or pines and hardwoods. The mean annual air temperature is 18 ± 2.2 °C, and the mean annual precipitation is 1500 ± 120 mm.

There are eight soil series in the Fragiudults great group in Mississippi that cover 7160 km^2, primarily in the Gulf Coastal Plain (MLRA 133C) (Fig. 6.9). The most extensive

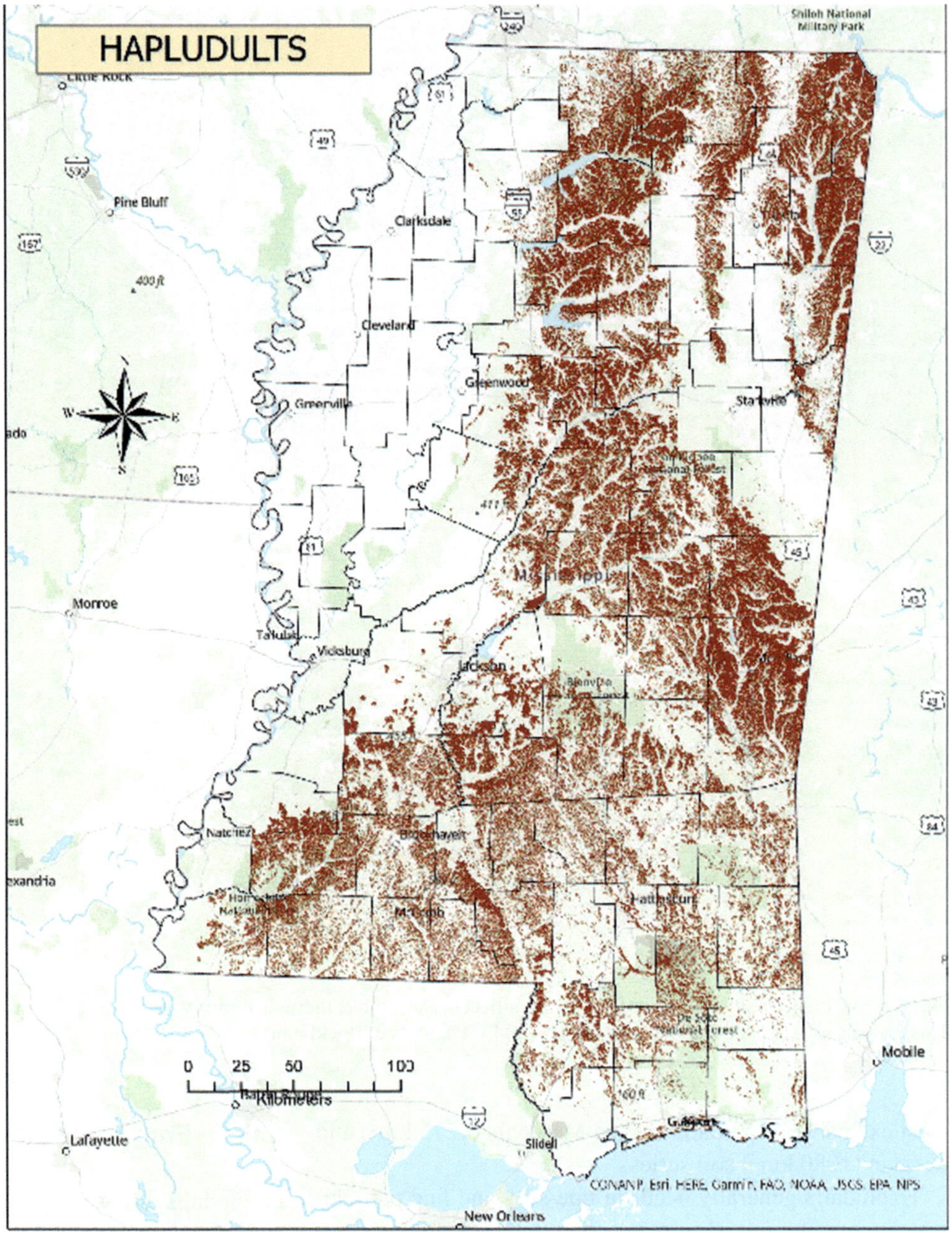

Fig. 6.5 Distribution of Hapludults in Mississippi. *Source* USDA Natural Resources Conservation Service, Mississippi

Fragiudults are the Ora (3065 km^2) and the Savannah (2535 km^2) soil series.

Fragiudults most commonly in the fine-loamy, coarse-silty, or fine-silty particle-size class, the siliceous mineralogy class, and the semiactive cation-exchange activity class. They have a udic soil moisture regime and a thermic soil temperature regime.

Fragiudults have an ochric epipedon that averages 20 ± 7.6 cm over an argillic horizon that averages 157 ± 44 cm and a fragipan that averages 84 ± 42 cm in thickness. Two of the Fragiudults have a glossic horizon and two have plinthite. The Ora soil series, a fine-loamy, siliceous, semiactive, thermic Typic Fragiudults, has formed in loamy marine and alluvial materials (Fig. 6.10). This soil contains a dark grayish brown A horizon from 0 to 7.5 cm (0–3 in.), a grayish brown albic horizon from 7.5 to 18 cm (3–7 in.), and a yellowish argillic horizon from 18 to 140 cm (7–56 in.). A fragipan exists from 65 to 140 cm (26–56 in.).

Fig. 6.6 Cuthbert soil series, a fine, mixed, semiactive, thermic Typic Hapludults, is formed in residuum from weakly consolidated sandstone and shale. The section is about 1.5 m deep. USDA Natural Resources Conservation Service Photo by J. Bockheim

6.7 Endoaquepts (Soil Region 6)

Endoaquepts are moderately well-developed soils that contain an ochric epipedon over a deep cambic horizon. They have formed in loamy or clayey alluvium on flood plains. The average maximum slope is $2.5 \pm 1.4\%$. These soils tend to be very deep and somewhat poorly drained to very poorly drained. The native vegetation is bottomland hardwoods. The mean annual air temperature is 18 ± 1.8 °C, and the mean annual precipitation is 1340 ± 190 mm.

There are 11 soil series in the Endoaquepts great group in Mississippi that cover 7140 km^2, primarily occurring in the Southern Mississippi Valley Alluvium (MLRA 131A) but also in the Gulf Coastal Plain (MLRA 133C)

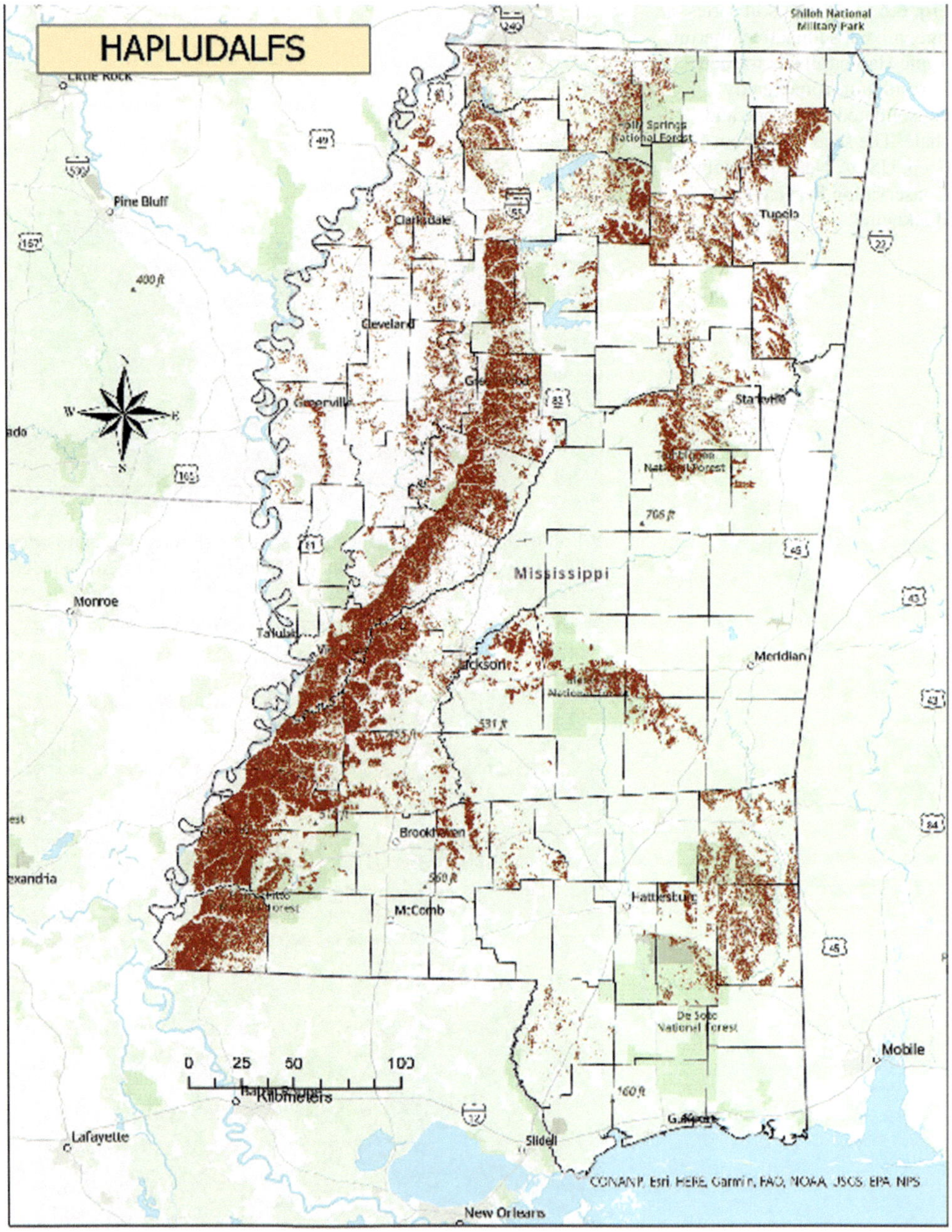

Fig. 6.7 Distribution of Hapludalfs in Mississippi. *Source* USDA Natural Resources Conservation Service, Mississippi

(Fig. 6.11). The most extensive Endoaquepts are the Mantachie (1845 km^2), Dowling (1684 km^2), and Arkabutla (942 km^2) soil series.

Endoaquepts most commonly in the fine-silty, coarse-silty, and fine-loamy particle-size classes, the mixed mineralogy class, the active and superactive cation-exchange activity classes, and the acid and nonacid reaction classes. They have an aquic soil moisture regime and a thermic soil temperature regime.

Endoaquepts have an ochric epipedon that averages 15 ± 5.9 cm over a cambic horizon that averages 128 ± 55 cm in thickness. The Commerce soil series, a fine-silty, mixed, superactive, nonacid, thermic Fluvaquentic Endoaquepts, has formed in loamy alluvial sediments in the Delta (Fig. 6.12). This soil has a dark grayish brown ochric epipedon (Ap horizon) from 0 to 18 cm (0–7 in.) and a dark grayish brown to dark gray, gleyed cambic horizon from 18 to 200 cm (7–80 in.). Slickensides are apparent below a depth of 160 cm (63 in.).

Fig. 6.8 Dubbs soil series, a fine-silty, mixed, active, thermic Typic Hapludalfs, has formed in loamy alluvium. The scale is in meters. *Source* USDA Natural Resources Conservation Service

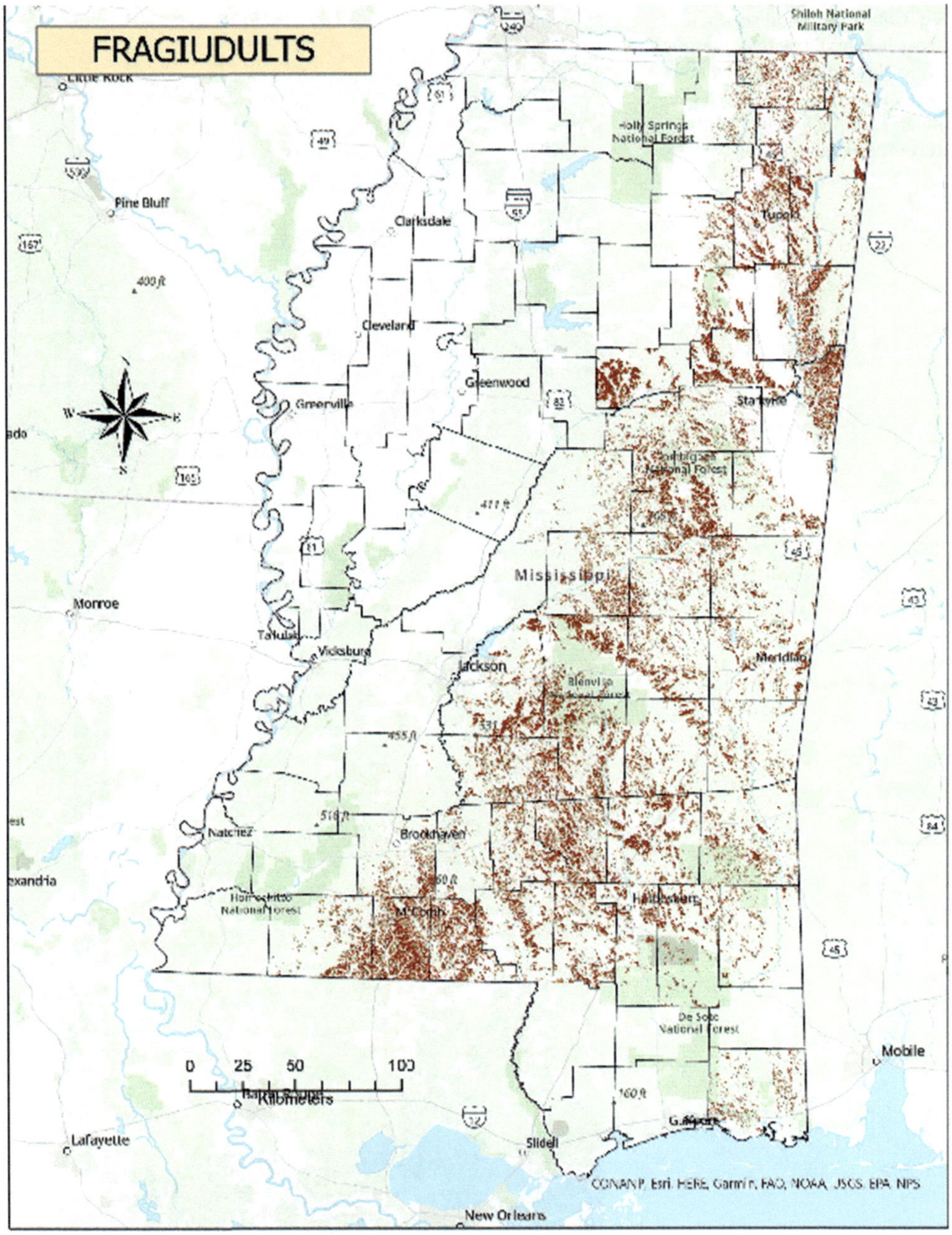

Fig. 6.9 Distribution of Fragiudults in Mississippi. *Source* USDA Natural Resources Conservation Service, Mississippi

6.8 Endoaqualfs (Soil Region 7)

Endoaqualfs are well-developed soils with an ochric epipedon over a deep argillic horizon. They are formed in silty alluvium on alluvial terraces. Maximum slopes average 3.5 ± 3.0%. These soils tend to be very deep and poorly drained. The native vegetation is hardwoods, particularly oaks and sweetgum, and pines. The mean annual air temperature is 18 ± 1.8 ℃, and the mean annual precipitation is 1310 ± 25 mm.

There are four soil series in the Endoaqualfs great group that cover 3430 km^2, primarily in the Southern Mississippi Valley Alluvium (MLRA 131A) (Fig. 6.13). The most extensive Endoaqualfs are the Forestdale (2560 km^2) and the Dundee (1500 km^2) soil series.

Endoaqualfs are mainly in fine-silty and fine particle-size classes, the mixed mineralogy class, and the active and semiactive cation-exchange activity classes. They have an aquic soil moisture regime and a thermic soil temperature regime.

Fig. 6.10 Ora soil series, a fine-loamy, siliceous, semiactive, thermic Typic Fragiudults, has formed in loamy marine and alluvial materials. The scale is in inches. *Source* USDA Natural Resources Conservation Service

Endoaqualfs have an ochric epipedon that averages 20 ± 12 cm over an argillic horizon that averages 119 ± 40 cm in thickness. The Forestdale soil series, a fine, smectitic, thermic Typic Endoaqualfs, has formed in clayey and silty alluvium in Tallahatchie County, the Delta (Fig. 6.14). This soil has a dark grayish brown ochric epipedon (Ap horizon) from 0 to 15 cm over a gray gleyed argillic horizon from 15 to over 150 cm.

6.9 Paleudalfs (Soil Region 8)

Paleudalfs are strongly developed soils with an ochric epipedon over a deep argillic horizon. They have formed in silty loess over loamy or clayey marine sediments or clayey or loamy marine sediments on marine terraces and uplands. Maximum slopes are 13 ± 13%. These soils tend to be very

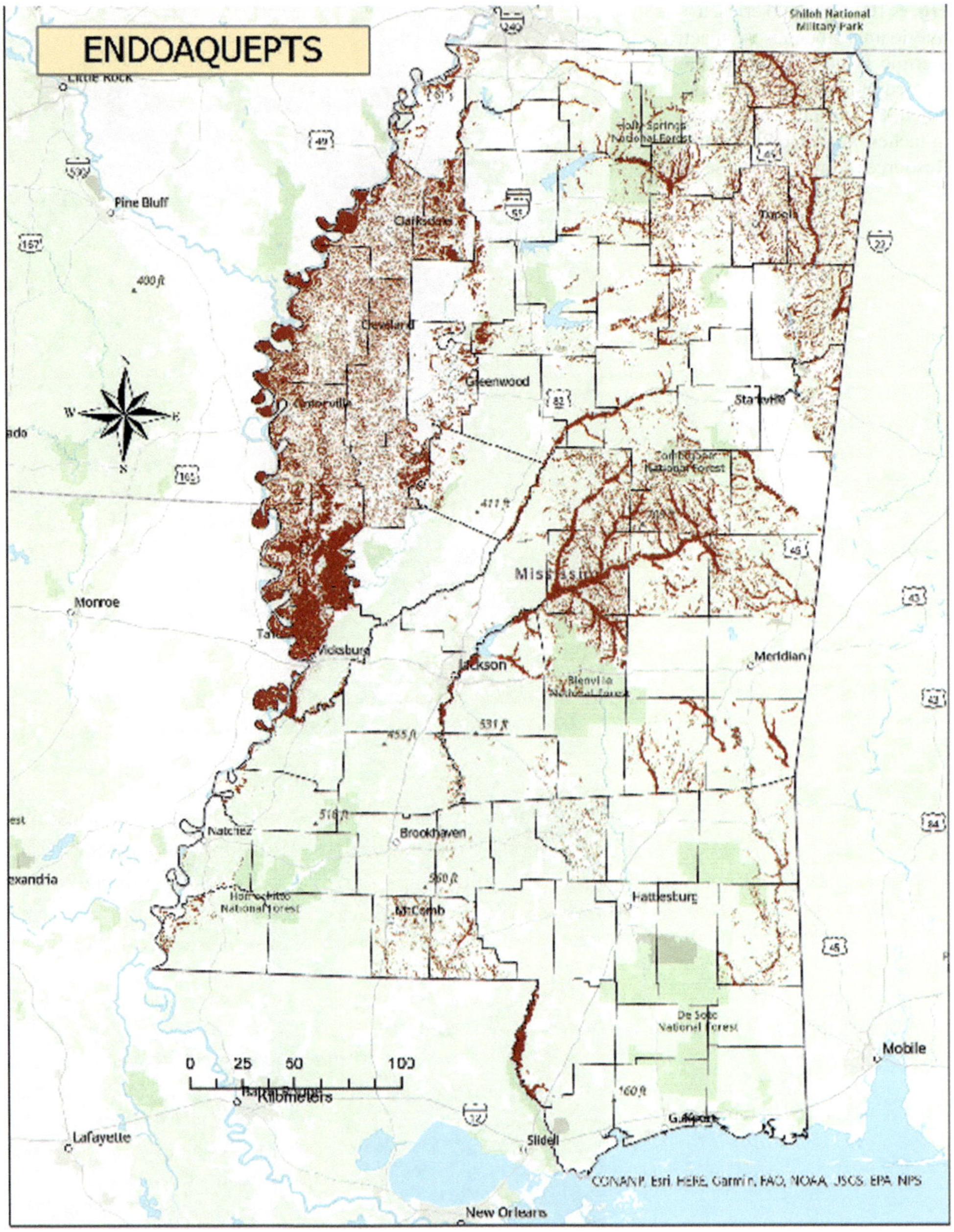

Fig. 6.11 Distribution of Endoaquepts in Mississippi. *Source* USDA Natural Resources Conservation Service, Mississippi

deep and well drained to somewhat poorly drained. The vegetation is hardwoods and pines. The mean annual air temperature is 18 ± 14 ℃, and the mean annual precipitation is 1315 ± 96 mm.

There are 11 Paleudalfs in Mississippi that account for 4200 km^2, primarily in the Gulf Coastal Plain (MLRA 133C) and the Alabama and Mississippi Blackland Prairie (MLRA 135A) (Fig. 6.15). The most extensive Paleudalfs are the Kipling (1020 km^2) and the Falkner (768 km^2) soil series.

Paleudalfs are in fine-loamy, fine-silty, and fine particle-size classes; siliceous, mixed, and smectitic mineralogy classes; and active and semiactive cation-exchange activity classes. They have a udic soil moisture regime and a thermic soil temperature regime.

Paleudalfs have an ochric epipedon that averages 16 ± 5.4 cm over an argillic horizon that averages 170 ± 26 cm in thickness. Two of the Paleudults have a glossic horizon over the argillic horizon.. The Susquehanna soil series, a fine, smectitic, thermic Vertic Paleudalfs, has

Fig. 6.12 Commerce soil series, a fine-silty, mixed, superactive, nonacid, thermic Fluvaquentic Endoaquepts, has formed in loamy alluvial sediments in the Delta. The scale is in inches. *Source* USDA Natural Resources Conservation Service

formed in silty clay or clay marine or alluvial deposits in the Southern Coastal Plain (Fig. 6.16). This soil has a dark gray A horizon from 0 to 3 inches (0–7.5 cm), a yellowish brown albic horizon from 3 to 5 inches (7.5–13 cm), and a thick argillic horizon from 5 to beyond 77 in. (13 to > 195 cm). The color of the argillic horizon varies with depth from yellowish red in the upper part to mottled light brownish gray and red in the lower part, due to restricted drainage.

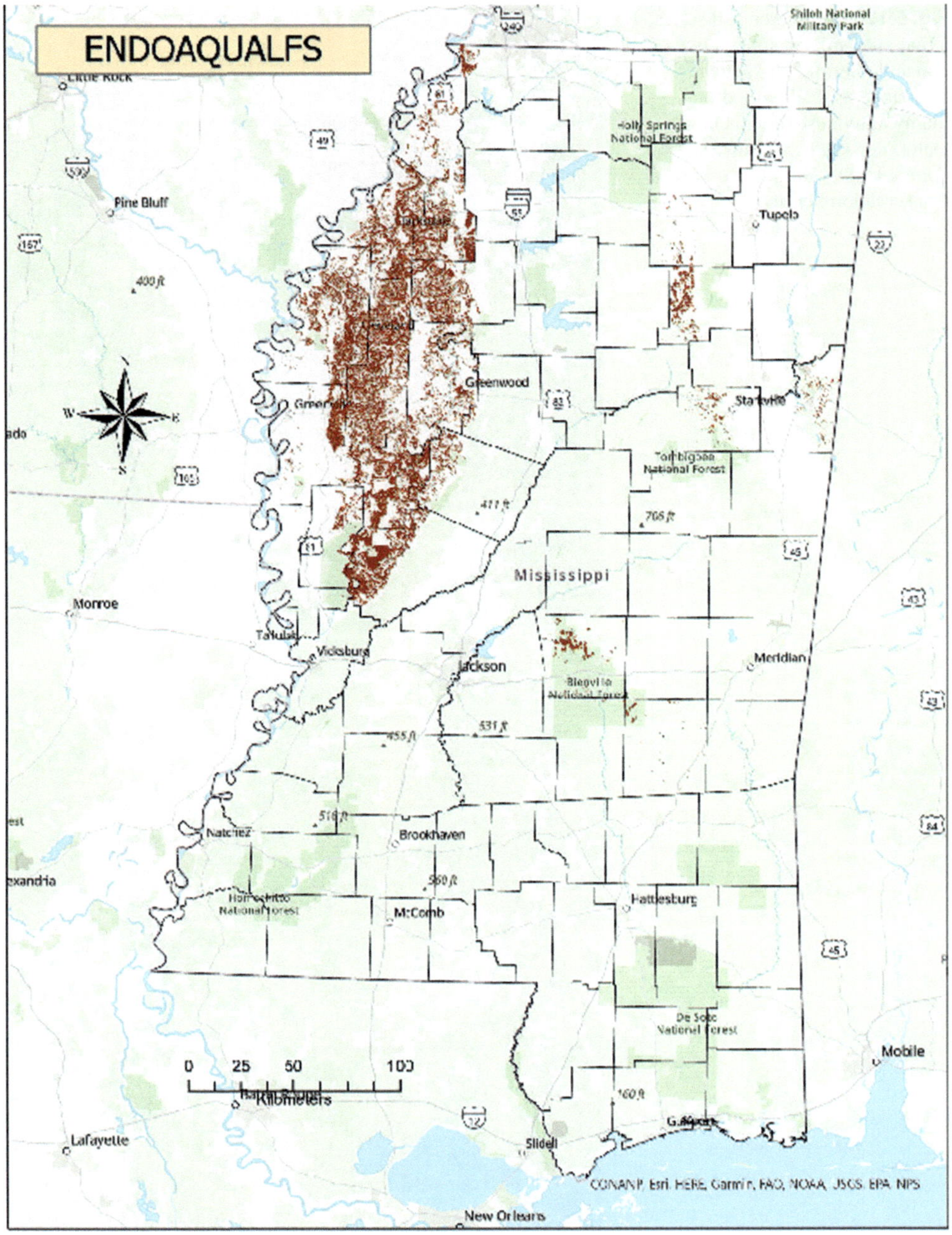

Fig. 6.13 Distribution of Endoaqualfs in Mississippi. *Source* USDA Natural Resources Conservation Service, Mississippi

6.10 Dystrudepts (Soil Region 9)

Dystrudepts are moderately well-developed soils that have an ochric epipedon over a cambic horizon. These soils have formed in silty, loamy, or sandy alluvium on low alluvial terraces, flood plains, and natural levees. Maximum slopes average 3.2 ± 2.8%. Dystrudepts are in deep to very deep materials and are somewhat excessively drained to somewhat poorly drained. The native vegetation is hardwoods, hardwoods and pine, and bottomland hardwoods. The mean annual air temperature is 18 ± 0.7 ℃, and the mean annual precipitation is 1325 ± 108 mm.

There are ten soil series in the Dystrudepts that occupy 4050 km^2, primarily in the Gulf Coastal Plain (MLRS 133C) (Fig. 6.17). The most extensive soil series in the Dystrudepts is the Oaklimeter (1237 km^2).

Dystrudepts are commonly in the coarse-loamy, coarse-silty, and fine-silty particle-size classes, the silicic and mixed mineralogy classes, and the active cation-exchange activity class. They have a udic soil moisture regime and a thermic soil temperature regime.

Dystrudepts have an ochric epipedon that averages 23 ± 12 cm over a cambic horizon that averages 102 ± 29 cm in thickness. The Oaklimeter soil series, a very

Fig. 6.14 Forestdale soil series, a fine, smectitic, thermic Typic Endoaqualfs, has formed in clayey and silty alluvium in Tallahatchie County, the Delta. The scale is in meters. *Source* USDA Natural Resources Conservation Service

deep, moderately well-drained soil formed in silty alluvium on flood plains and lower terraces, is classified as a coarse-silty, mixed, active, thermic Fluvaquentic Dystrudepts.

6.11 Fluvaquents (Soil Region 10)

Fluvaquents are weakly developed soils that have only an ochric epipedon. They form in silty, loamy, and sandy alluvium in flood plains. They have maximum slopes of $2.0 \pm 0\%$. Fluvaquents tend to be deep to very deep and somewhat poorly drained to poorly drained. The vegetation is bottomland hardwoods, hardwoods, or hardwoods and loblolly pine. The mean annual air temperature is 17 ± 1.1 ℃, and the mean annual precipitation is $1,335 \pm 39$ mm.

There are five soil series in the Fluvaquents great group that occupy 4150 km^2, primarily in the Gulf Coastal Plain (MLRA 133C) and the Southern Mississippi Valley Loess (MLRA 134) (Fig. 6.18). The most extensive Fluvaquents

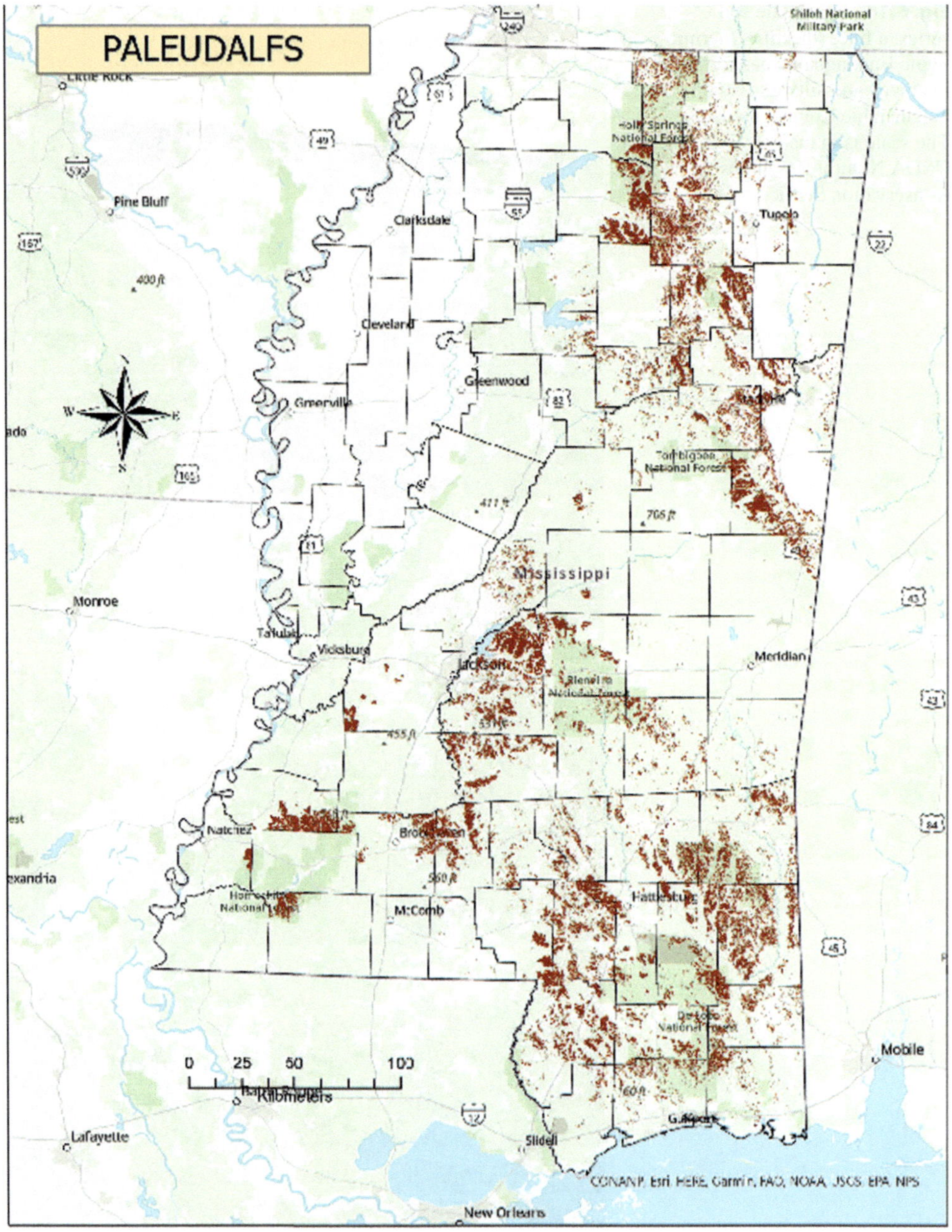

Fig. 6.15 Distribution of Paleudalfs in Mississippi. *Source* USDA Natural Resources Conservation Service, Mississippi

are the Falaya (1574 km^2), Bibb (1020 km^2), and Gillsburg (912 km^2) soil series.

Fluvaquents commonly are in coarse-silty and fine-silty particle-size classes, the mixed or siliceous mineralogy class, the active cation-exchange activity class, and the acid reaction class. They have an aquic soil moisture regime and a thermic soil temperature regime.

Fluvaquents generally have only an ochric epipedon that averages 20 ± 7.9 cm in thickness. Two of the Fluvaquents paradoxically have a cambic horizon that averages 47 ± 38 cm. The Falaya soil series, a coarse-silty, mixed, active, acid, thermic Aeric Fluvaquents, has formed in silty alluvium from loess (Fig. 6.19). This soil has a brown A horizon from 0 to 25 cm and a brown cambic horizon from 25 to 43 cm.

Fig. 6.16 Susquehanna soil series, a fine, smectitic, thermic Vertic Paleudalfs, has formed in silty clay or clay marine or alluvial deposits in the Southern Coastal Plain. The scale is in feet. *Source* USDA Natural Resources Conservation Service

6.12 Epiaquerts (Soil Region 11)

Epiaquerts are moderately well-developed soils with an ochric epipedon over a deep cambic horizon. These soils form from clayey alluvium in flood plains and on low alluvial terraces. Maximum slopes are 3.0+/1 1.7%. Epiaquerts tend to be very deep and somewhat poorly drained to poorly drained. The native vegetation is bottomland hardwoods, hardwood, and hardwoods with loblolly pine. The mean annual air temperature is 18±0.7 ℃, and the mean annual precipitation is 1410±18 mm.

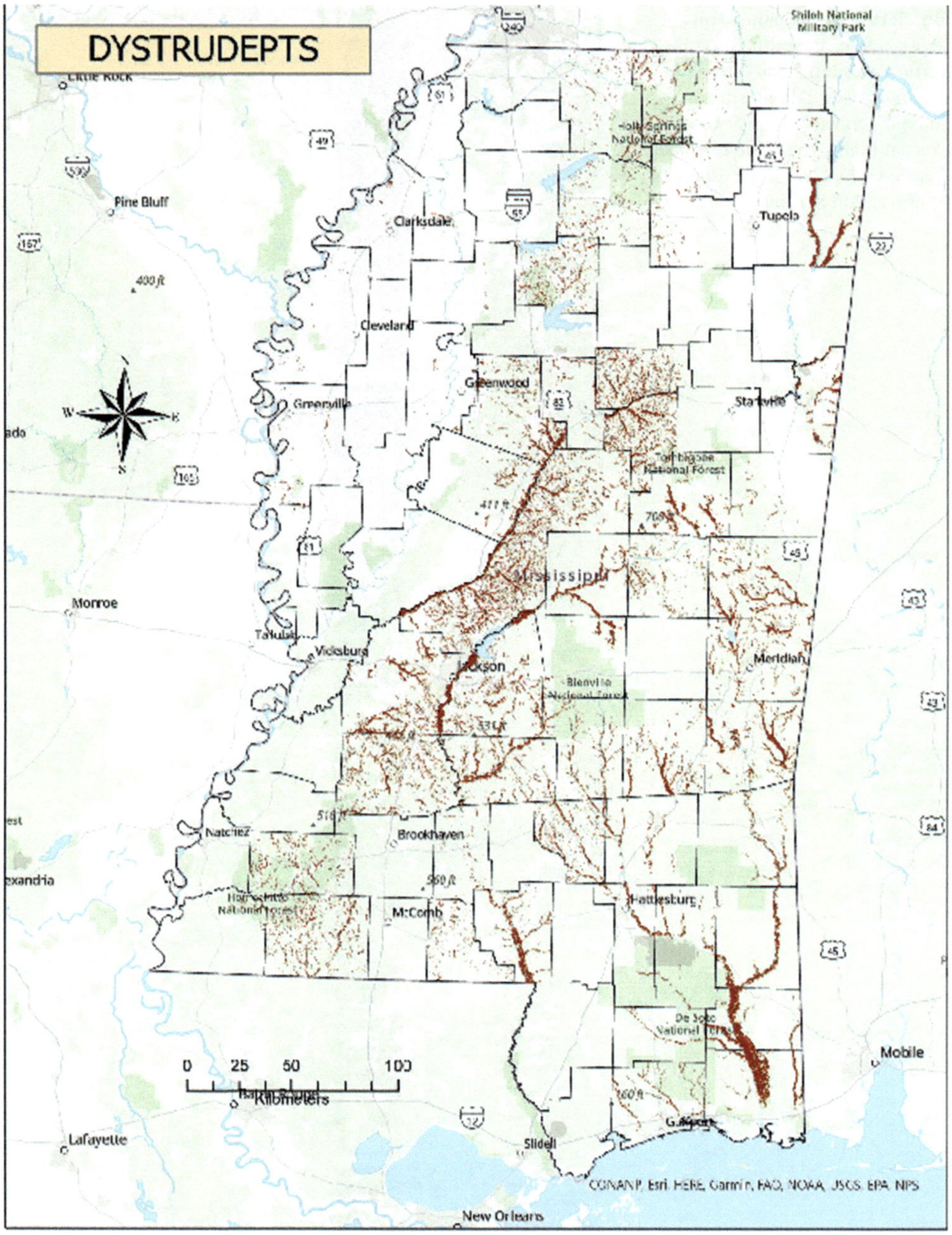

Fig. 6.17 Distribution of Dystrudepts in Mississippi. *Source* USDA Natural Resources Conservation Service, Mississippi

There are three soil series in the Epiaquerts great group that cover 3470 km^2, primarily in the Southern Mississippi Valley Alluvium (MLRA 131A) (Fig. 6.20). The Sharkey soil series (3080 km^2) is the dominant Epiquert in Mississippi.

Epiaquerts are in fine or very-fine particle-size classes, the smectitic mineral class, the aquic soil moisture regime, and a thermic soil temperature regime. Epiaquerts have an ochric epipedon that averages 22 ± 3.5 cm over a cambic horizon that averages 166 ± 30 cm in thickness. The Sharkey soil series, a very-fine, smectitic, thermic Chromic Epiaquerts, has formed in clayey alluvium in flood plains and low terraces of the Mississippi River (Fig. 6.21). This soil has a dark grayish brown plow layer (Ap horizon) to 25 cm over a dark gray to gray cambic horizon to beyond 210 cm. The wedge-shaped aggregates and slickensides (glossy pressure faces) are not readily visible. Figure 6.22 shows surface cracks in the Sharkey soil that represent the prismatic structure.

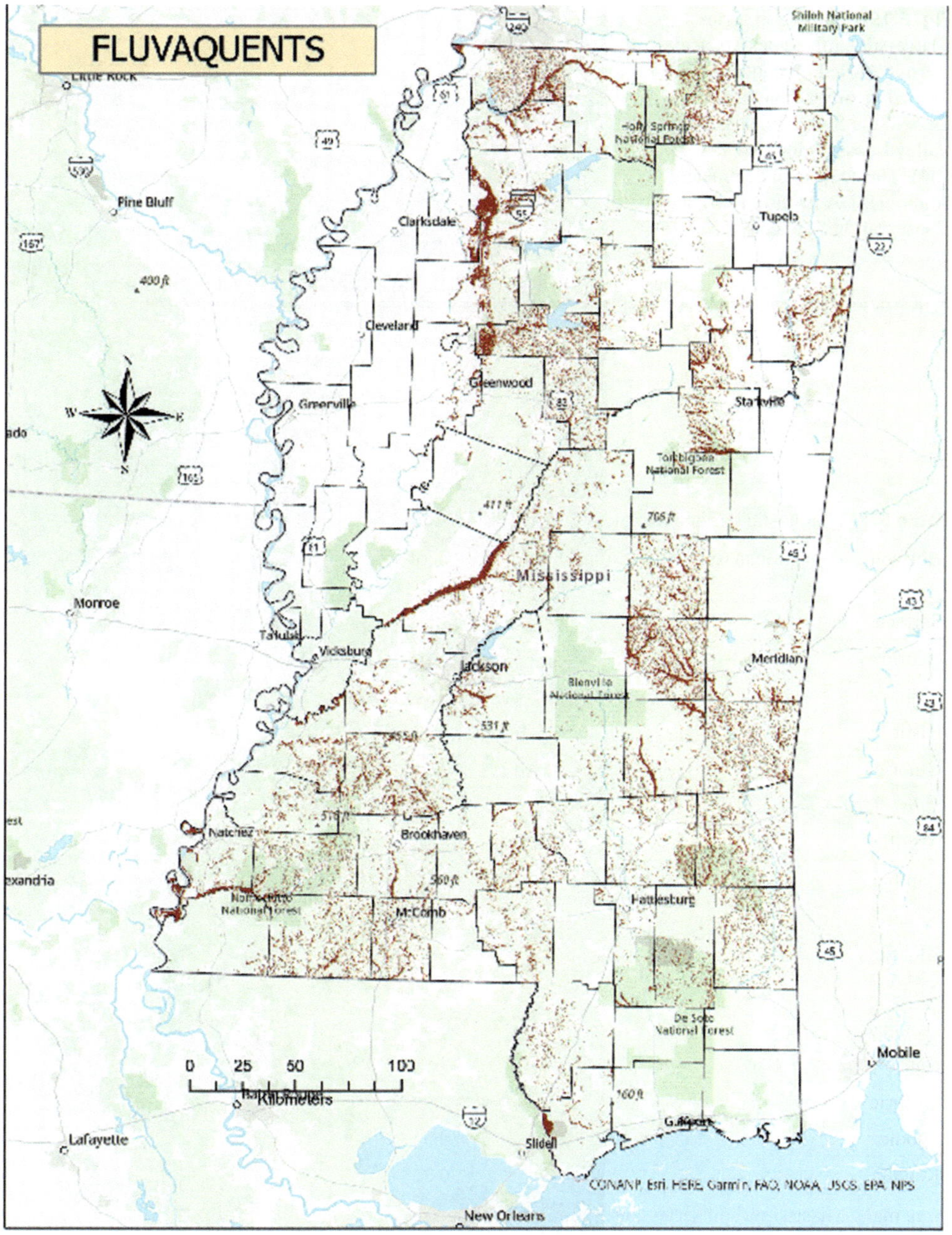

Fig. 6.18 Distribution of Fluvaquents in Mississippi. *Source* USDA Natural Resources Conservation Service, Mississippi

6.13 Dystraquerts (Soil Region 12)

Dystraquerts are moderately well-developed soils with an ochric epipedon over an argillic or cambic horizon. They are formed in clayey alluvium, residuum, or marine deposits in uplands and in floodplains. Maximum slopes are 5.0±5.0%. Dystraquerts are deep to very deep and poorly drained. The native vegetation is mixed hardwoods and pines and bottomland hardwoods. The mean annual air temperature is 17±1.9 ℃, and the mean annual precipitation is 1290±63 mm.

There are four soil series in the Dystraquerts great group that cover 3,300 km^2, primarily in the Southern Mississippi Valley Alluvium (MLRA 131A) (Fig. 6.23). The Alligator (2847 km^2) is the most extensive soil series in the Dystraquerts great group in Mississippi.

Fig. 6.19 Falaya soil series, a coarse-silty, mixed, active, acid, thermic Aeric Fluvaquents, has formed in silty alluvium from loess in the Southern Mississippi Valley Loess region (MLRA 134). The scale is in decimeters. *Source* USDA Natural Resources Conservation Service

Dystraquerts are in fine and very-fine particle-size classes, the smectitic mineral class, and have an aquic soil moisture regime and a thermic soil temperature regime.

Dystraquerts have an ochric epipedon averaging 16 ± 2.9 cm over an argillic horizon that averages 120 ± 37 (50% of soil series), or a cambic horizon with a thickness of 147 cm (25%), or the great group lacks a diagnostic subsurface horizon. The Alligator soil series, a very-fine, smectitic, thermic Chromic Dystraquerts, has formed in clayey alluvium in backswamps and meander scrolls, sloughs, and flood plains of the Mississippi River (Fig. 6.24). This soil has a dark grayish brown plow layer (Ap horizon) in the upper 18 cm that is underlain by a light brownish gray to grayish brown cambic horizon to 180 cm.

6.14 Udifluvents (Soil Region 13)

Udifluvents are weakly developed soils that only contain an ochric epipedon. They have developed in silty, loamy, and sandy alluvium in floodplains. They have maximum slopes of 2.6 ± 1.3%. Udifluvents are deep to very deep and range from excessively drained to moderately well drained. The native vegetation is bottomland hardwoods, hardwoods

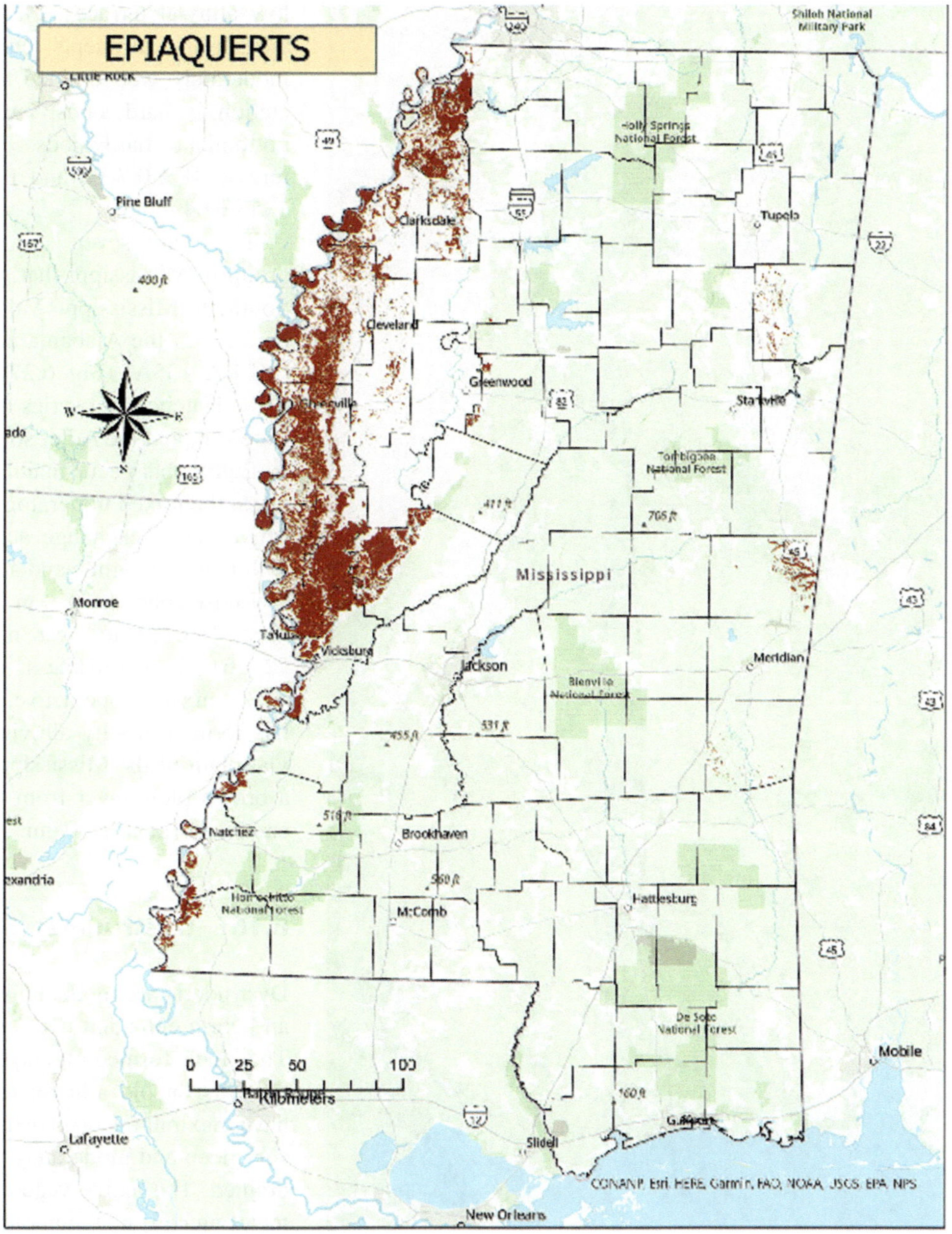

Fig. 6.20 Distribution of Epiaquerts in Mississippi. *Source* USDA Natural Resources Conservation Service, Mississippi

and pines, and hardwoods. The mean annual air temperature is 18 ± 1.7 °C, and the mean annual precipitation is 1345 ± 101 mm.

There are nine soil series in the Udifluvents great group in Mississippi that cover 3000 km^2, primarily in the Southern Mississippi Valley Loess (MLRA 134) (Fig. 6.25). The Collins soil series (1500 km^2) is the most pervasive Udifluvents in Mississippi.

Udifluvents are in coarse-loamy, coarse-silty, and sandy particle-size classes, mixed and siliceous mineral classes, active and superactive cation-exchange activity classes, and acid and nonacid reaction classes. They have a udic soil moisture regime and a thermal soil temperature regime.

Udifluvents have only an ochric epipedon that averages 17 ± 4.0 cm in thickness; they lack a diagnostic subsurface horizon. The Collins soils series, a coarse-silty, mixed, active, acid Aquic Udifluvents, has formed in silty alluvium on flood plains of streams in the Southern Mississippi Valley Alluvium (MLRA 134) (Fig. 6.26). This soil has a brown plow layer (Ap horizon) to 18 cm that is underlain by a yellowish brown to brown C horizon.

Fig. 6.21 Sharkey soil series, a very-fine, smectitic, thermic Chromic Epiaquerts, has formed in clayey alluvium in flood plains and low terraces of the Mississippi River. This soil was photographed in Bolivar County, MS. The scale is in decimeters. *Source* USDA Natural Resources Conservation Service

6.15 Eutrudepts (Soil Region 14)

Eutrudepts are moderately well-developed soils and contain an ochric epipedon over a cambic horizon. They have formed in silty alluvium and loess in flood plains and low alluvial terraces. They occur on maximum slopes of 14 ± 26%. Eutrudepts tend to be deep to very deep and moderately well drained to well drained. The native vegetation is hardwoods, hardwoods and loblolly pine, and bottomland hardwoods. The mean annual air temperature is 18 ± 0.7 ℃, and the mean annual precipitation is 1355 ± 64 mm.

There are seven soil series in the Eutrudepts great group in Mississippi that cover 2475 km^2, primarily in the Southern Mississippi Valley Loess region (MLRA 134) but also in the Alabama and Mississippi Blackland Prairie (MLRA 135A) (Fig. 6.27). The most extensive Eutrudept is the Natchez soil series (733 km^2), which is Mississippi's state soil, and the Adler soil series (612 km^2).

Eutrudepts occur mainly in the coarse-silty particle-size class, the mixed mineralogy class, and the active and superactive cation-exchange activity classes. They have a udic soil moisture regime and a thermic soil temperature regime.

Eutrudepts have an ochric epipedon that averages 21 ± 6.9 cm over a cambic horizon that averages 72 ± 61 cm in thickness. The Bruin soil series, a coarse-silty, mixed, superactive, thermic Oxyaquic Eutrudepts, has formed in silty alluvium on natural levees in the alluvial plain of the Mississippi River (Fig. 6.28). This soil has a brown plow layer from to 28 cm and a yellowish brown cambic horizon to 50 cm.

6.16 Dystruderts (Soil Region 15)

Dystruderts are moderately well-developed soils containing an ochric epipedon over an argillic and/or cambic horizon. They have formed on clayey alluvium, residuum, or marine deposits on old alluvial terraces and uplands. Dystruderts have maximum slopes of 14 ± 12%. They tend to be deep to very deep and moderately well drained to somewhat poorly drained. The native vegetation is pine and hardwoods. The mean annual air temperature is 17 ± 0.0 ℃, and the mean annual precipitation is 1335 ± 15 mm.

There are six soil series in the Dystruderts great group in Mississippi that account for an area of 2100 km^2, mainly in the Alabama and Mississippi Blackland Prairie (MLRA 135A) (Fig. 6.29). The Vaiden (884 km^2) and Wilcox (500 km^2) soil series are the most extensive Dystruderts in the state.

Dystruderts are in very-fine or fine particle-size classes and the smectitic mineralogy class, and they have a udic soil moisture regime and a thermic soil temperature regime.

Dystruderts have an ochric epipedon that averages 7.9 ± 3.3 cm over and argillic horizon that averages 58 ± 33 cm and/or a cambic horizon that averages

Fig. 6.22 Surface cracks in the Sharkey soil near Tensas Parish, Louisiana. *Source* USDA Natural Resources Conservation Service

151 ± 17 cm in thickness. About one-third of the Dystruderts have a paralithic contact with 1.6 m of the surface. The Suggsville soil series, a very-fine, smectitic, thermic Chromic Dystruderts, has formed in clayey sediments overlying limestone and chalk in the Alabama and Mississippi Blackland Prairies (Fig. 6.30). This soil has dark brown A horizon from 0 to 2.5 cm that is underlain by a yellowish red argillic horizon with slickensides that extends to 42 in. (107 cm).

6.17 Epiaquepts (Soil Region 16)

Epiaquepts are moderately well-developed soils with an ochric epipedon over a cambic horizon. They have formed in clayey and clayey over loamy alluvium in flood plains, particularly on natural levees. They occur on maximum slopes of 3.3 ± 1.3%. Epiaquepts tend to be deep to very deep and are from somewhat poorly drained to poorly drained. The native vegetation is hardwoods or bottomland hardwoods. The mean annual air temperature is 17 ± 1.7 °C, and the mean annual precipitation is 1345 ± 48 mm.

There are seven soil series in the Epiaquepts great group of Mississippi that cover 2040 km^2, primarily in the Southern Mississippi Valley Alluvium (MLRA 131A) and the Blackland Prairie (MLRA 135A) (Fig. 6.31). The Leeper (686 km^2) and Urbo (566 km^2) are the major soil series in this great group.

Epiaquepts are in the fine and clayey over loamy particle-size classes, smectitic and mixed mineralogy classes, active and superactive cation-exchange activity classes, and nonacid and acid reaction classes. They have an aquic soil moisture regime and a thermic soil temperature regime.

Epiaquepts have an ochric epipedon that averages 16 ± 5.6 cm over a cambic horizon that averages 104 ± 52 cm in thickness. The Newellton soil series, a clayey over loamy, smectitic over mixed, superactive, nonacid, thermic Fluvaquentic Epiaquepts, has formed in clayey over loamy alluvium on natural levees in the alluvial plain of the Mississippi River and its tributaries (Fig. 6.32). This soil has a dark grayish brown Ap horizon from 0 to 7 in. (18 cm) over a dark grayish brown cambic horizon with subangular blocky structure that extends to 26 in. (65 cm).

6.18 Paleaquults (Soil Region 17)

Paleaquults are strongly developed soils with an ochric epipedon over a deep argillic horizon. They have formed in clayey, loamy, and sandy alluvium and marine deposits

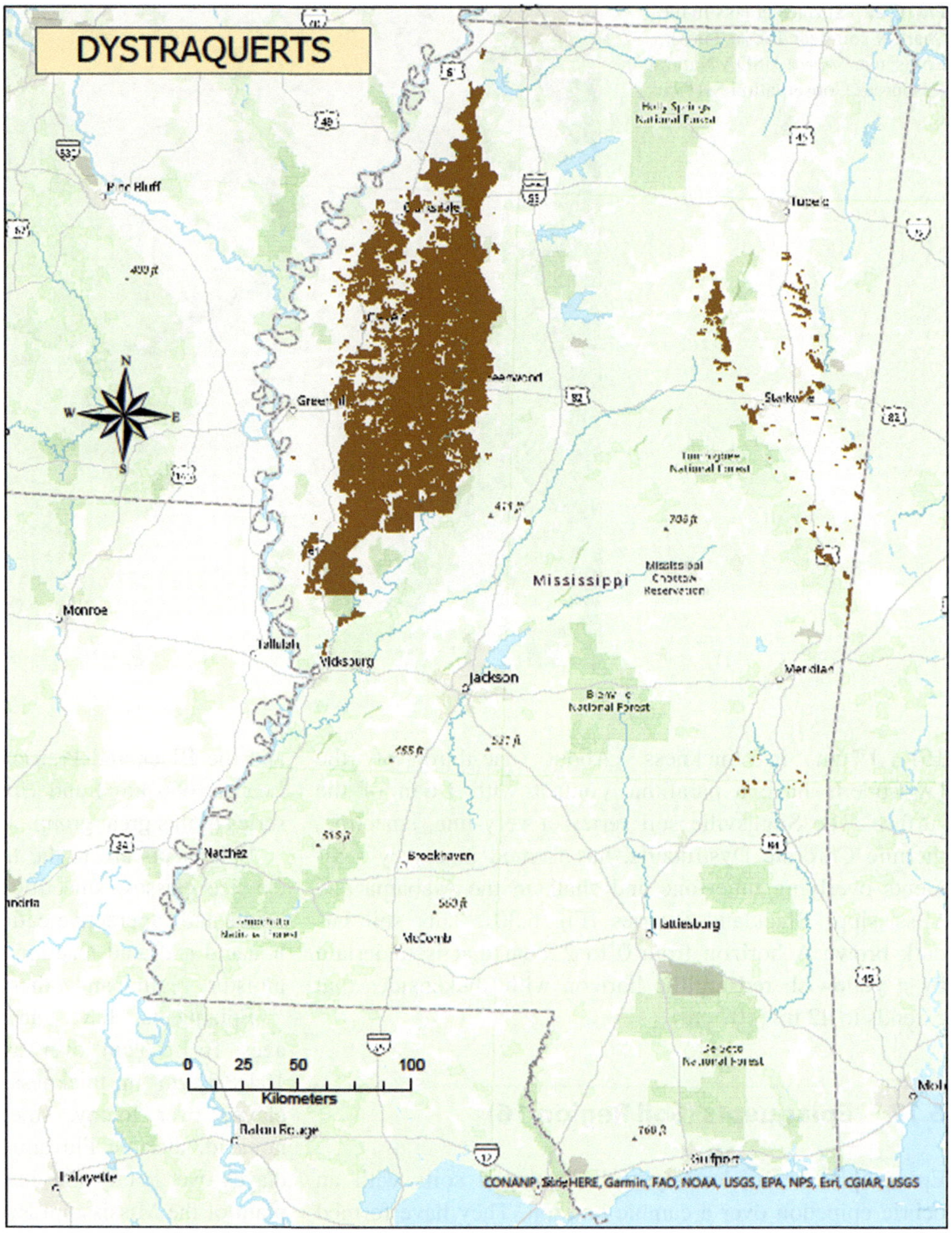

Fig. 6.23 Distribution of Dystraquerts in Mississippi. *Source* USDA Natural Resources Conservation Service, Mississippi

on alluvial terraces and uplands. Paleaquults have maximum slopes of 3.5 ± 3.0%, are very deep, and range from somewhat poorly drained to poorly drained. The native vegetation is southern pine forest, pines and hardwoods, or hardwoods and loblolly pine. The mean annual air temperature is 18 ± 1.2 ℃, and the mean annual precipitation is 1400 ± 167 mm.

There are eight soil series in the Paleaquults great group in Mississippi that cover 1,800 km^2, primarily in the Gulf Coastal Plain (MLRA 133C) and the East Gulf Coast Flatwoods (MLRA 152A) (Fig. 6.33). The Smithton (643 km^2) and Trebloc (610 km^2) are the most extensive soil series in his great group.

Fig. 6.24 Alligator soil series, a very-fine, smectitic, thermic Chromic Dystraquerts, has formed in clayey alluvium in backswamps and meander scrolls, sloughs, and flood plains of the Mississippi River. The scale is in decimeters. *Source* USDA Natural Resources Conservation Service

Paleaquults are mainly in the coarse-loamy and fine-silty particle-size classes, the siliceous mineralogy class, and the semiactive, active, and subactive cation-exchange activity classes. They have an aquic soil moisture regime and a thermic soil temperature regime.

Paleaquults have an ochric epipedon that averages 46 ± 39 cm over an argillic horizon that averages 133 ± 30 cm in thickness. An albic subsurface horizon overlays the argillic horizon in 38% of the Paleaquults. The Grady soil series, a fine, kaolinitic, thermic Typic

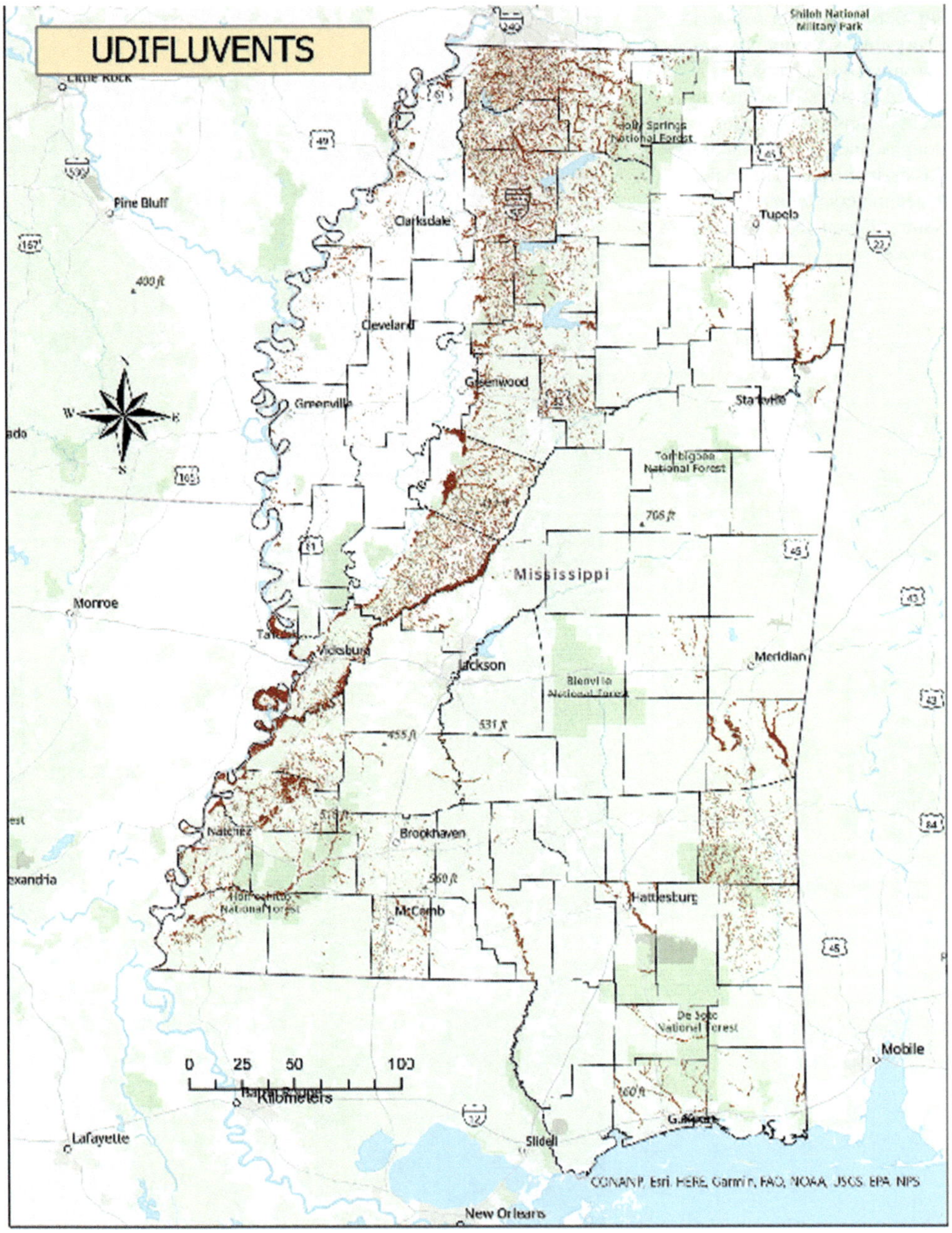

Fig. 6.25 Distribution of Udifluvents in Mississippi. *Source* USDA Natural Resources Conservation Service, Mississippi

Paleaquults, is poorly drained, and is derived from thick beds of clayey marine sediments in the Southern Coast Plain (Fig. 6.34). The Grady soil has a dark gray ochric epipedon to 12 cm and a gray argillic horizon to 150 cm.

6.19 Fraglossudalfs (Soil Region 18)

Fraglossudalfs are well-developed soils with an ochric epipedon over an albic, argillic, fragipan, and glossic horizon. They have formed in silty loess or loess over loamy

Fig. 6.26 Collins soils series, a coarse-silty, mixed, active, acid Aquic Udifluvents, has formed in silty alluvium on flood plains of streams in the Southern Mississippi Valley Alluvium (MLRA 134). The scale is in meters. *Source* USDA Natural Resources Conservation Service

alluvium on alluvial terraces and in uplands. Fraglossudalfs have maximum slopes of $7.2 \pm 4.6\%$, are very deep, and range from moderately well drained to somewhat poorly drained. The native vegetation is hardwoods. The mean annual air temperature is 17 ± 1.7 °C, and the mean annual precipitation is 1325 ± 58 mm.

The Fraglossudalfs great group includes five soil series which cover 1000 km^2 (Fig. 6.35). The Oaklimeter soil

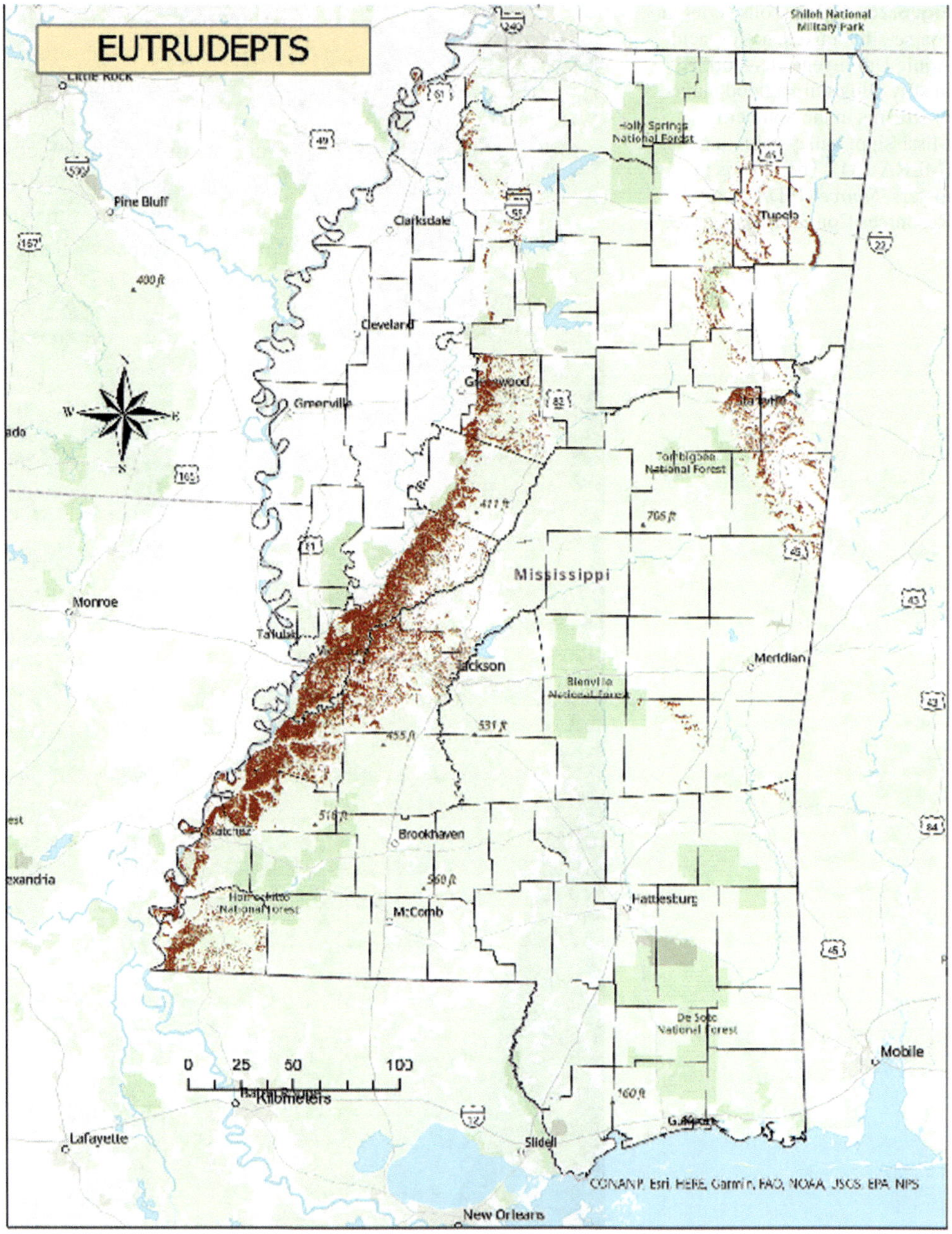

Fig. 6.27 Distribution of Eutrudepts in Mississippi. *Source* USDA Natural Resources Conservation Service, Mississippi

series accounts for 84% of the area. Fraglossudalfs primarily occur in the Southern Mississippi Valley Loess region (MLRA 134).

Fraglossudalfs are in the fine-silty particle-size class, the mixed or siliceous mineralogy class, and the active cation-exchange activity class. They have a udic soil moisture regime and a thermic soil temperature regime (Table 6.1).

Fraglossudalfs have an ochric epipedon that averages 21 ± 7.9 cm in thickness, an albic horizon of 19 ± 19 cm (80% of soil series), an argillic horizon of 119 ± 29 cm, a fragipan of 71 ± 16 cm, and a glossic horizon averaging 32 ± 24 cm in thickness (Table 6.2).

The Grenada soil series, a fine-silty, mixed, active, thermic Oxyaquic Fraglossudalfs, is very deep, moderately well drained, and has formed in loess (Fig. 6.36). The Grenada

Fig. 6.28 Bruin soil series, a coarse-silty, mixed, superactive, thermic Oxyaquic Eutrudepts, has formed in silty alluvium on natural levees in the alluvial plain of the Mississippi River. The scale is in decimeters. *Source* USDA Natural Resources Conservation Service

soil has a dark grayish brown ochric epipedon to 18 cm, a yellowish brown cambic horizon to 65 cm, a light gray albic horizon to 70 cm, a yellowish brown fragipan-argillic horizon (Btx/E) to 125 cm, and a yellowish brown fragipan-argillic horizon (Btx) to 140 cm.

6.20 Hapluderts (Soil Region 19)

Hapluderts are weakly developed soils with a mollic or ochric epipedon over a cambic horizon. They have formed in clayey and clayey over chalk alluvium and residuum in flood plains and on uplands. Hapluderts have slopes averaging 8.6 ± 8.0%, range from shallow to very deep, and are moderately well drained to well drained. The native vegetation is eastern red cedar, hackberry, and grasses. The mean annual air temperature is 18 ± 0.74 ℃, and the mean annual precipitation is 1265 ± 55 mm.

There are seven soil series in the Hapluderts great group in Mississippi that cover 700 km^2, primarily in the Alabama and Mississippi Blackland Prairie (MLRA 135A) (Fig. 6.37). The Brooksville, Okolona, and Griffith have the greatest areas within the Hapluderts of Mississippi.

Hapluderts are in the fine, very-fine, and clayey particle-size classes and the smectitic soil mineralogy class. They have a udic soil moisture regime and a thermic soil temperature regime (Table 6.1).

Hapluderts have either a mollic epipedon averaging 70 ± cm (57% of soil series) or an ochric epipedon averaging 19 ± 19 cm (43% of soil series) over a cambic horizon that averages 87 ± 42 cm. However, 19% of the Hapluderts lack a diagnostic subsurface horizon.

The Okolona soil series, a fine, smectitic, thermic Oxyaquic Hapluderts, is a deep, well drained soil in uplands that is formed in calcareous clayey material underlain by marly clay and chalk (Fig. 6.38). The soil has a very dark grayish brown mollic epipedon to 46 cm (18 in.), an olive cambic horizon to 90 cm (3 ft), and a dark grayish brown C horizon to 180 cm (6 ft) that are underlain by platy chalk.

6.21 Summary

The soils of Mississippi are divided into 19 dominant great groups, including the Fragiudalfs, Paleudults, Hapludults, Hapludalfs, Fragiudults, Endoaquepts, Endoaqualfs,

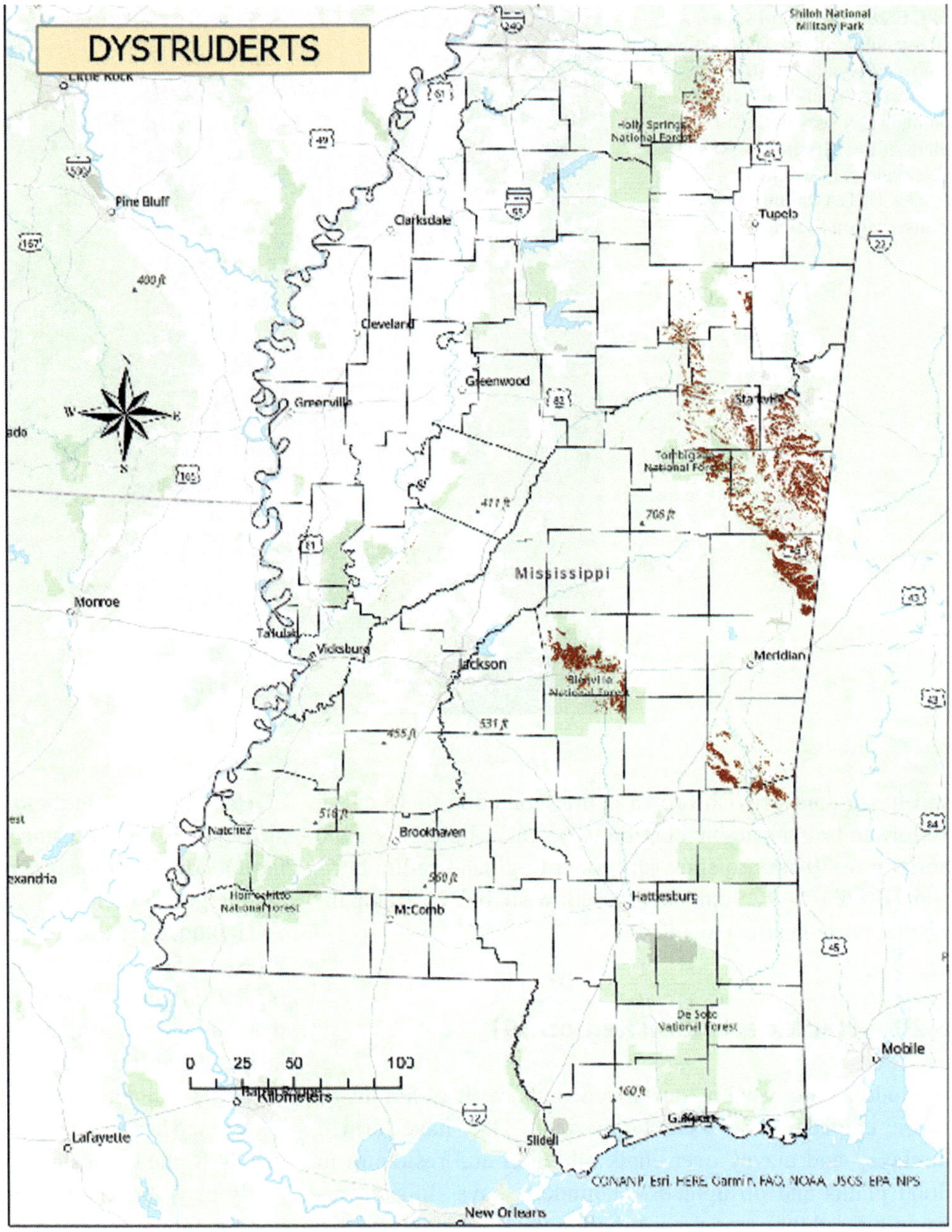

Fig. 6.29 Distribution of Dystruderts in Mississippi. *Source* USDA Natural Resources Conservation Service, Mississippi

Fig. 6.30 Suggsville soil series, a very-fine, smectitic, thermic Chromic Dystruderts, has formed in clayey seidments overlying limestone and chalk in the Alabama and Mississippi Blackland Prairies. The scale is in inches. *Source* USDA Natural Resources Conservation Service

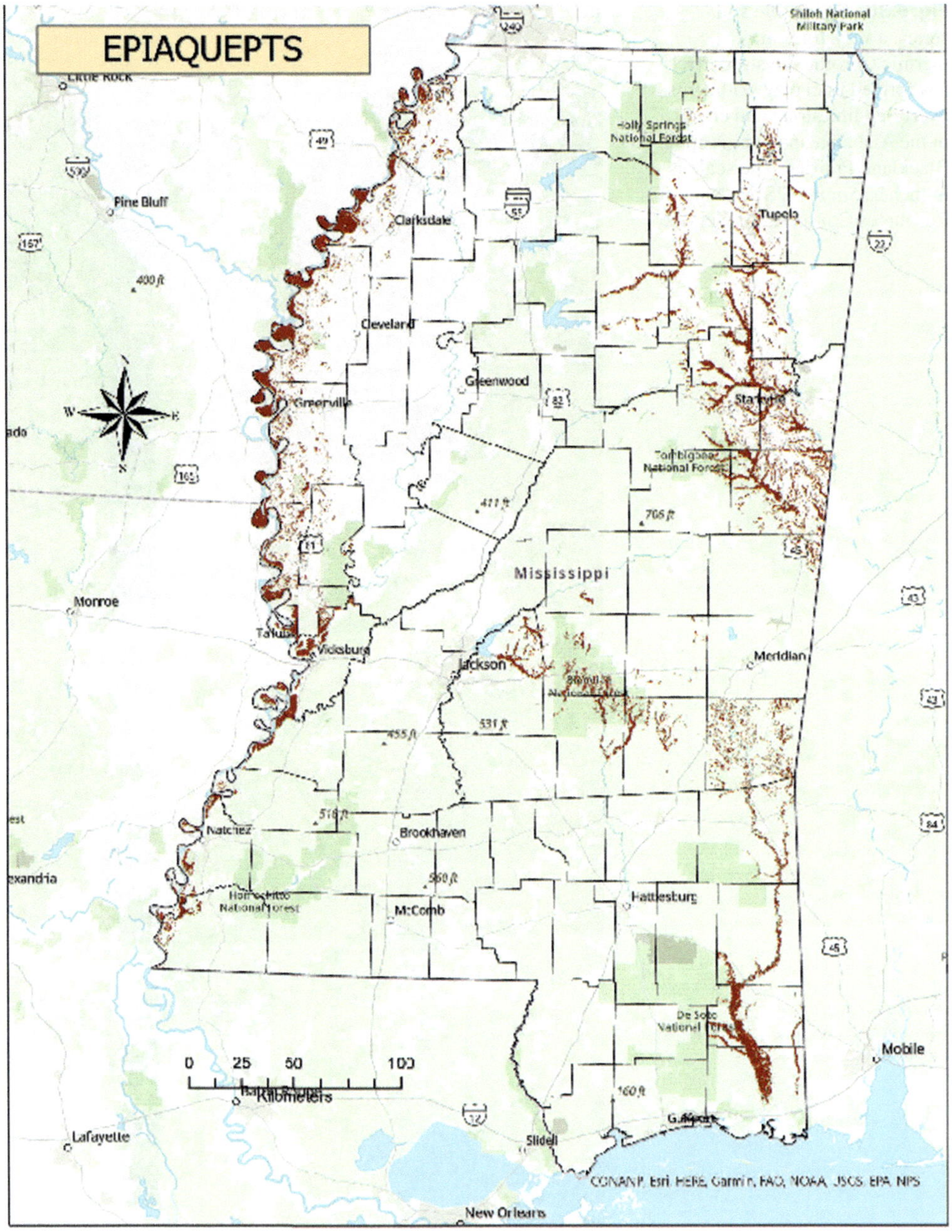

Fig. 6.31 Distribution of Epiaquepts in Mississippi. *Source* USDA Natural Resources Conservation Service, Mississippi

Fig. 6.32 Newellton soil series, a clayey over loamy, smectitic over mixed, superactive, nonacid, thermic Fluvaquentic Epiaquepts, has formed in clayey over loamy alluvium on natural levees in the alluvial plain of the Mississippi River and its tributaries. The scale is in inches. *Source* USDA Natural Resources Conservation Service

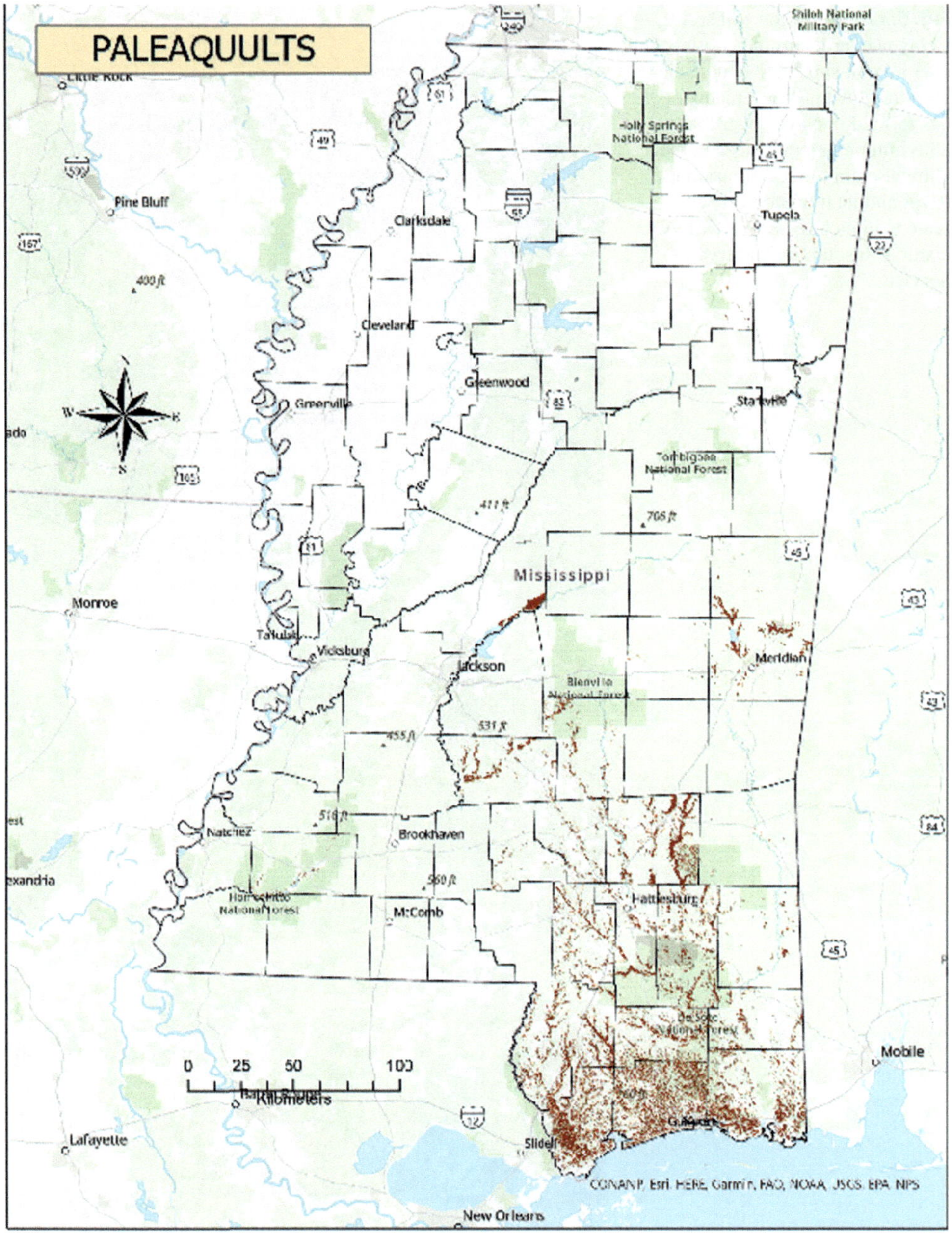

Fig. 6.33 Distribution of Paleaquults in Mississippi. *Source* USDA Natural Resources Conservation Service, Mississippi

Fig. 6.34 Grady soil series, a fine, kaolinitic, thermic Typic Paleaquults, is poorly drained, and is derived from thick beds of clayey marine sediments in the Southern Coast Plain. The Grady soil has a dark gray ochric epipedon to 12 cm and a gray argillic horizon to 150 cm. The scale is in meters. *Source* USDA Natural Resources Conservation Service

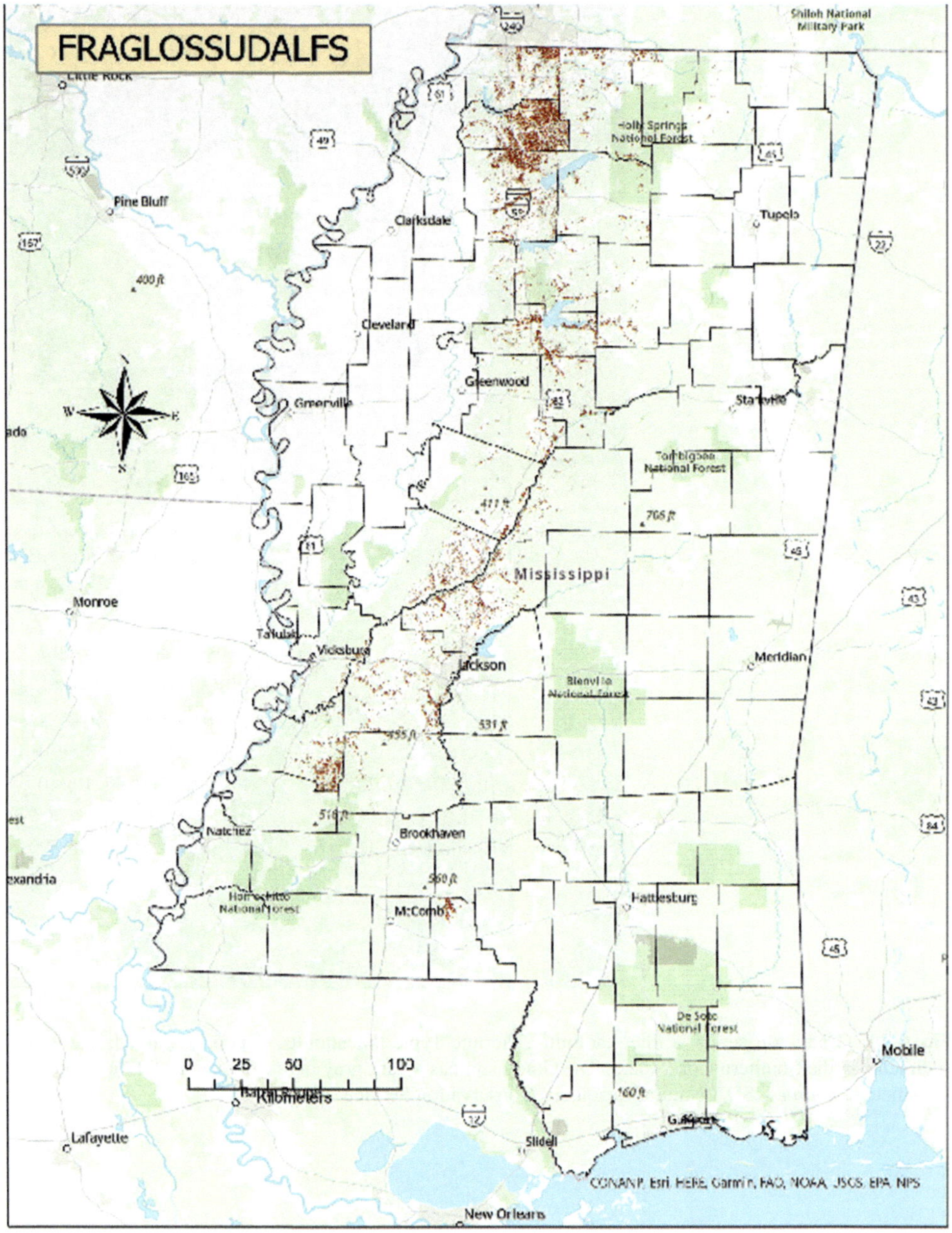

Fig. 6.35 Distribution of Fraglossudalfs in Mississippi. *Source* USDA Natural Resources Conservation Service, Mississippi

Fig. 6.36 Grenada soil series, a fine-silty, mixed, active, thermic Oxyaquic Fraglossudalfs, is very deep, moderately well drained, and has formed in loess. The Grenada soil has a dark grayish brown ochric epipedon to 18 cm, a yellowish brown cambic horizon to 65 cm, a light gray albic horizon to 70 cm, a yellowish brown fragipan-argillic horizon (Btx/E) to 125 cm, and a yellowish brown fragipan-argillic horizon (Btx) to 140 cm. The scale is in meters. *Source* USDA Natural Resources Conservation Service

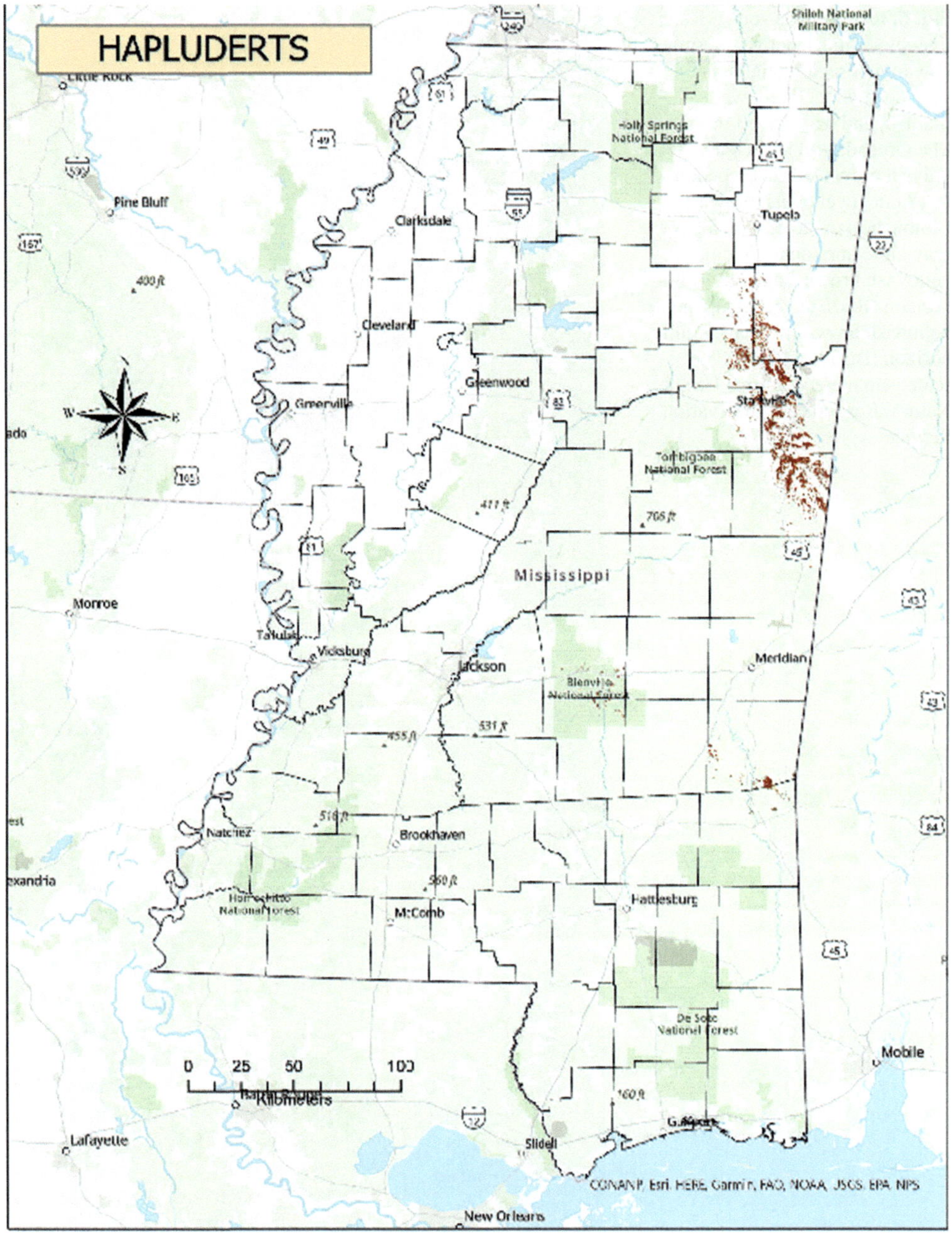

Fig. 6.37 Distribution of Hapluderts in Mississippi. *Source* USDA Natural Resources Conservation Service, Mississippi

Fig. 6.38 Okolona soil series, a fine, smectitic, thermic Oxyaquic Hapluderts, is a deep, well drained soil in uplands that is formed in calcareous clayey material underlain by marly clay and chalk. The soil has a very dark grayish brown mollic epipedon to 46 cm (18 in.), an olive cambic horizon to 90 cm (3 ft), and a dark grayish brown C horizon to 180 cm (6 ft) that are underlain by platy chalk. The photo is from Hale County, Alabama, and the scale is in feet. *Source* USDA Natural Resources Conservation Service

Paleudalfs, Dystrudepts, Fluvaquents, Epiaquerts, Dstraquerts, Udifluvents, Eutrudepts, Dystruderts, Epiaquepts, Paleaquults, Fraglossudalfs, and Hapluderts. Each group is discussed in terms of soil-forming factors, numbers of soil series and area, occurrence by Major Land Resource Area, family classes (particle size, mineralogy, cation-exchange activity, reaction, soil moisture, and soil temperature), thicknesses of diagnostic horizons, and representative soil series.

7 Ultisols

7.1 Distribution

Ultisols account for 34% of the land area and 34% of the soil series in Mississippi (Table 5.2). Ultisols occur in three of the six MLRAs in the state, primarily in the Gulf Coastal Plain (MLRA 133C), but also in the East Gulf Coast Flatwoods (MLRA 152A). The dominant suborders are Udults (82%) and Aquults (18%). Four great groups account for 90% of the Ultisols on an area basis, including the Hapludults (34%), Paleudults (34%), Paleaquults (12%), and Fragiudults (10%) (Table 5.2). The most extensive Ultisols include the Smithdale (3180 km^2), Ora (3066 km^2), Savannah (2563 km^2), Sweatman (2416 km^2), Ruston (2018 km^2), and McLaurin (1767 km^2) soil series.

Ultisols in Mississippi have a mean annual air temperature averaging 18 ± 1.3 °C and receive 1420 ± 135 mm of precipitation per year. The native vegetation on Ultisols ranges from pure pines to pines and hardwoods, to pure hardwoods. Maximum slopes average $13 \pm 12\%$, and elevations range from sea level to 75 m. Ultisols are formed predominantly in marine deposits and alluvium. They seldom occur in loess. Parent materials range from clayey to sandy, usually are deep to very deep, and are most commonly moderately well drained to well drained. Ultisols occur on alluvial terraces in lowlands and on dissected marine terraces in uplands. Ultisols in Mississippi generally are of middle to early Pleistocene age or older.

Vanderford and Shaffer (1966) found similar modes of genesis of silt-rich bisequal soils with a fragipan in the Loess Hills (Loring, Granada, and Calloway soils series; Fragiudalfs and Fraglossudalfs) and in the Southern Coastal Plain (Ora, Paden, and Pheba soil series; Fragiudults). Some of the earliest research on Ultisols was conducted on what were previously known as red-yellow podzolic soils in Mississippi (Templin et al. 1951; Dyal et al. 1951). Focusing on the Cahaba (Typic Hapludults), Leaf (Typic Albaquults), and Ruston (Typic Paleudults) series in the Upper Coastal Plain, they noted high amounts of clay in the B horizon, high exchangeable acidity, an abundance of kaolinite, quartz, and gibbsite in the clay-size fraction. Differences in soil properties were attributed to differences in parent material composition rather than in climate. Some of the definitive work on plinthite formation was done by Aide et al. (2004) on the Escambia (Plinthaquic Paleudults) and Irvington (Plintic Fragiudults) soil series in the Southern Coastal Plain of Mississippi. Plinthite formation occurred in conjunction with Btx horizon degradation and in a zone of iron-accumulation having a firm to very firm consistency. Plinthite was viewed as a redoximorphic feature of highly weathered soils and was considered to be an iron-rich and humus-poor pedogenic product that hardens irreversibly on exposure to sunlight. Ufnar (2007) reported that clay coatings (argillans) in Ultisols in the Eastern Gulf Coast Flatwoods (MLRA 152A) were polygenetic and comparable to those of mid-Cretaceous paleosols of North America. Nash (1979) studied the clay mineralogy of a drainage sequence of Udults and Aqualfs on Pliocene–Pleistocene terraces of the Tombigbee River in northeastern Mississippi. Smectite was the dominant clay mineral, especially in the zone of maximum clay accumulation. In deeper horizons, especially in wetter soils, the aluminum interlayers in hydroxyl Al-interlayered 2:1 clay were replaced by Al polymers.

7.2 Properties and Processes

Ultisols contain an ochric epipedon that averages 32 ± 30 cm over an argillic horizon that averages 122 ± 45 cm in thickness. About 15% of the Ultisols in Mississippi have a fragipan that averages 97 ± 42 cm; 13% have plinthite averaging 97 ± 38 cm, 8% have an albic subsurface horizon averaging 34 ± 38 cm, and 3% have a kandic horizon that averages 104 ± 18 cm in thickness.

About 67% of the Ultisols are in three particle-size classes, including the fine-loamy (31%), coarse-loamy

D. Johnson et al., *The Soils of Mississippi*, World Soils Book Series,
https://doi.org/10.1007/978-3-031-36235-4_7

(24%), and loamy (12%). About 71% are in the siliceous and 25% are in the mixed mineralogy classes. Nearly three-quarters (94%) of the Ultisols are in the three cation-exchange activity classes: semiactive (56%), subactive (23%), and active (15%). Ultisols have a thermic soil temperature regime and either a udic (84%) or aquic (16%) soil moisture regime.

Photographs of major great groups of Ultisols are given in Chap. 6, including a Paleudults (McLaurin soil series; Fig. 6.4), a Hapludults (Cuthbert soil series; Fig. 6.6), a Fragiudults (Ora soil series; Fig. 6.10), and a Paleaquults (Grady oil series; Fig. 6.34). Figure 7.1 is the Lucy soil series, a loamy, kaolinitic, thermic Arenic Kandiudults, formed in sandy and loamy marine and alluvial sediments in the lower Gulf Coastal Plain (MLRA 133C). This soil has a dark grayish brown plow layer (Ap horizon) to 14 inches (35 cm) over a strong brown albic horizon to 28 in. (70 cm) and a deep yellowish red to red argillic horizon.

Fig. 7.1 Lucy soil series, a loamy, kaolinitic, thermic Arenic Kandiudults, has formed in sandy and loamy marine and alluvial sediments in the lower Gulf Coastal Plain (MLRA 133C). This photo was taken in Houston County, Alabama. The scale is in feet. *Source* USDA Natural Resources Conservation Service

Table 7.1 summarizes properties of six dominant Ultisol great groups in Mississippi. The argillic horizons range from 63 to 158 cm (average = 118 cm). All of the soils have a base saturation averaging less than 35% 125 cm below the upper boundary of the argillic horizon. Below the A horizon, the aluminum saturation ranges from 70 to 96%, reflecting the abundance of exchangeable acidity in Ultisols. The dominant soil-forming processes in Ultisols are argilluviation, ferruginization, Fragification and plinthization are important in some Ultisols.

7.3 Use and Management

Ultisols in Mississippi are used primarily for commercial timber production, including the extensive Smithdale, Ruston, and Ora soil series. Some Ultisols have been cleared and are cropped, primarily the Ora, Savannah, Sweatman, McLaurin, Benndale, and Prentiss soil series. The Ruston, Trebloc, and Cuthbert soil series are used for pasture. Other uses are wildlife, recreation, watershed protection, and development.

7.4 Summary

Ultisols are the most abundant soil order in Mississippi in terms of land area and number of soil series. Ultisols are most common in the eastern portion of the state. Udults and Aquults are the primary suborders. Four great groups account for 90% of the Ultisols on an area basis, including the Hapludults, Paleudults, Paleaquults, and Fragiudults. Ultisols have a mean annual air temperature averaging 18 ± 1.3 °C and receive 1420 ± 135 mm of precipitation per year. The vegetation on Ultisols is predominantly southern pines with varying amounts hardwoods. Ultisols form from alluvium and marine deposits in uplands. Ultisols in Mississippi contain an ochric epipedon over an argillic horizon that averages 122 ± 45 cm in thickness. Some Ultisols in Mississippi contain a fragipan or plinthite. Ultisols are mainly in the loamy particle-size classes, the siliceous mineralogy class, and the semiactive, subactive, or active cation-exchange activity class. Ultisols have a thermic soil temperature regime and a udic or aquic soil moisture regime.

Table 7.1 Analytical properties of some Ultisols found in Mississippi

Horizon[a]	Depth (cm)	Clay (%)	Silt (%)	Sand (%)	SOC (%)	CEC7 ($cmol_c$/kg)	Base sat (%)	pH H_2O	Al sat (%)	1.5 mPa H_2O/ clay	Cole
Atmore; coarse-loamy, siliceous, semiactive, thermic Plinthic Paleaquults; Harrison, MS;											
Pedon no. 40A5220											
A	0–13	5.5	58.9	35.6	1.42	4.2	2	3.9	88	0.75	0.005
Eg1	13–23	6.6	59.1	34.3	0.72	2.9	2	3.8	92	0.58	0.005
Eg2	23–70	8.5	60.7	30.8	0.4	2.9	2	3.8	95	0.54	0.006
Btvg1	70–99	11.9	58.9	29.2	0.31	4	3	3.8	93	0.50	0.009
Btvg2	99–130	21.5	52.1	26.4	0.13	5.8	7	4.0	86	0.47	0.017
Btvg3	130–150	41.9	25.4	32.7	0.06	12.1	10	4.2	80	0.43	0.036
Btvg4	150–198	34.0	22.7	43.3	0.1	10.1	14	4.2	76	0.40	0.022
C1	198–229	23.8	18.6	57.6	0.1	7.5	17	3.9		0.43	
C2	229–284	30.6	36.0	33.4	0.1	12.6	28	3.7	60	0.41	
Benndale; coarse-loamy, siliceous, semiactive, thermic Typic Paludults, Harrison, MS;											
Pedon no. 40A5222											
Ap	0–15	6.8	20.1	73.1	1.16	5.9	39	5.4	4	0.69	0.004
E1	15–28	11.7	24.1	64.2	0.22	4.3	9	4.8	80	0.42	0.006
E2	28–61	11.8	24.2	64.0	0.05	4.3	7	4.9	87	0.42	0.004
E3	61–89	10.7	19.6	69.7	0.77	3.1	6	4.9	88	0.41	0.008
E4	89–109	10.8	16.7	72.5	0.02	3.9	8	5.3	82	0.42	0.004
Bt1	109–132	13.9	15.0	71.1		3.1	3	4.9	95	0.42	0.004
Bt2	132–150	16.8	11.8	71.4		4.4	2	4.9	96	0.43	0.004
Bt3	150–185	17.2	9.8	73.2	0.02	4.2	2	4.9	96	0.44	0.004
Bt4	185–213	24.1	7.3	68.6	0.01	5.7	5	4.8	92	0.45	
Bt5	213–257	27.4	7.4	65.2	0.03	5.6	7	4.8	91	0.42	
Cabaha; fine-loamy, siliceous, semiactive, thermic Typic Hapludults; Tangipahoa Parish, LA;											
Pedon no. 40A3991											
A	0–13	6.4	34.7	58.9	1.48	8.8	48	5.7			
E	13–28	7.6	35.2	57.2	0.52	5.6	21	4.9			
Bt1	28–41	16.0	36.0	48.0	0.13	6.3	17	4.5			
Bt2	41–56	23.6	37.3	39.1	0.14	9.3	14	4.5			
Bt3	56–79	21.1	37.4	41.5	0.08	8.5	15	4.7			
Bt4	79–91	14.5	30.7	54.8	0.02	5.9	15	4.6			
2C1	91–100	8.4	17.8	73.8	0.02	3.5	17	4.6			
2C2	100–122	3.6	7.1	89.3		1.9	37	4.7			
Ora; fine-loamy, siliceous, semiactive, thermic Typic Fragiudults; Covington, MS; Pedon no											
40A4815											
O1/O2											
A1	0–8	5.6	46.6	47.8	2.92	9.2	23	4.9		0.68	
E	8–13	5.4	45.5	49.1	0.82	3.8	29	5.2		0.46	
A2	13–25	5.9	44.7	49.4	0.22	2.0	20	5.2		0.34	
B	25–33	11.9	47.4	40.7	0.17	3.4	38	5.1		0.34	
Bt1	33–61	23.0	37.0	40.0	0.18	8.3	35	5.0		0.38	

(continued)

Table 7.1 (continued)

Horizon[a]	Depth (cm)	Clay (%)	Silt (%)	Sand (%)	SOC (%)	CEC7 ($cmol_c/kg$)	Base sat (%)	pH H_2O	Al sat (%)	1.5 mPa H_2O/ clay	Cole
Bt2	61–81	9.8	27.9	62.3	0.08	3.0	17	5.0		0.37	
Btx1	81–104	7.5	20.8	71.7	0.04	2.3	17	5.0		0.39	
Btx2	104–142	18.5	19.8	61.7	0.06	4.1	34	5.0		0.36	
Ruston; fine-loamy, siliceous, semiactive, thermic Typic Paleudults; Jackson, MS; pedon no											
40A4825											
A	0–13	7.2	34.6	58.2	1.14	6.4	20	5.6			
E	13–25	13.3	34.1	52.6	0.37	5.4	20	6.0			
Bt1	25–43	19.9	31.8	48.3	0.15	6.0	18	6.0			
Bt2	43–132	20.6	29.0	50.4	0.06	5.5	11	5.9			
Bt3	132–183	16.2	27.1	56.7	0.02	4.1	10	5.5			
Smithdale; fine-loamy, siliceous, subactive, thermic Typic Hapludults; Wayne, MS; pedon no											
14N0296											
A	0–5	3.7	20.6	75.7	1.8	3.8	18	5.1	53	1.24	
E	5–28	4.2	22.2	73.6	0.8	1.8	17	5.0	70	0.52	
Bt1	28–99	40.4	11.5	48.1	0.2	7.6	13	5.2		0.40	
Bt2	99–127	42.2	8.1	49.7	0.2	8.1	10	5.2	86	0.38	
BC1	127–160	30.3	6.6	63.1	0.1	5.8	9	5.1	89	0.38	
BC2	160–211	23.2	6.8	70.0	0.1	4.6	4	5.0	94	0.39	

[a]Diagnostic horizons in bold-face are diagnostic: *Bt* argillic; *Bv* plinthite; *Bx* fragipan
Red type indicates SOC was estimated at a pH of 8

References

Aide M, Pavich Z, Lilly ME, Thornton R, Kingery W (2004) Plinthite formation in the coastal plain region of Mississippi. Soil Sci 169:613–623

Dyal RS, Martin IL, Templin EH (1951) Red-yellow poodzolic soils of the southeastern United States: II. Character of the clay fractions of Ruston, Stephenville, Boswell, Windthorst, Cahaba, Leaf, and Axtell. Agron J 43:482–487

Nash VE (1979) Mineralogy of soils developed on Pliocene-Pleistocene terraces of the Tombigbee River in Mississippi. Soil Sci Soc Am J 43:616–623

Templin EH, Martin IL, Dyal RS (1951) Red-yellow podzolic soils of the southeastern United States: I. Morphology of the Ruston, Stephenville, Boswell, Windthorst, Cahaba, Leaf, and Axtell series. Agron J 43:482–487

Ufnar DF (2007) Clay coatings from a modern soil chronosequence: a tool for estimating the relative age of well-drained paleosols. Geoderma 141:181–200

8 Alfisols

Abstract

Alfisols account for 32% of the soil area and 25% of the soil series in Mississippi (Table 5.2). Alfisols occur in all of the six MLRAs but are most common Southern Mississippi Valley Loess (MLRA 134), followed by the Gulf Coastal Plain (MLRA 133C), and the Southern Mississippi River Alluvium (MLRA 131A). Udalfs are the most common suborder, composing 81% of the Alfisols, followed by Aqualfs (19%). Four great groups account for 92% of the Alfisols on an area basis, including the Fragiudalfs (35%), Hapludalfs (27%), Endoaqualfs (15%), and Paleudalfs (15%) (Table 5.2).

8.1 Distribution

Alfisols account for 32% of the soil area and 25% of the soil series in Mississippi (Table 5.2). Alfisols occur in all of the six MLRAs but are most common Southern Mississippi Valley Loess (MLRA 134), followed by the Gulf Coastal Plain (MLRA 133C), and the Southern Mississippi River Alluvium (MLRA 131A). Udalfs are the most common suborder, composing 81% of the Alfisols, followed by Aqualfs (19%). Four great groups account for 92% of the Alfisols on an area basis, including the Fragiudalfs (35%), Hapludalfs (27%), Endoaqualfs (15%), and Paleudalfs (15%) (Table 5.2).

The most extensive Alfisols include the Providence (4482 km^2), Loring (4349 km^2), Memphis (4276 km^2), Forestdale (2560 km^2), Dundee (1508 km^2), Lorman (1080 km^2), and Kipling (1020 km^2) soil series.

Vanderford and Shaffer (1966) found similar modes of genesis of silt-rich bisequal soils with a fragipan in the Loess Hills (Loring, Granada, and Calloway soils series; Fragiudalfs and Fraglossudalfs) and in the Southern Coastal Plain (Ora, Paden, and Pheba soil series; Fragiudults). Bartelli (1973) studied the nature and origin of albic materials in loess-derived soils of Mississippi, including the Grenada and Calloway soil series (Fraglossudalfs), the Henry soil series (Fragiaqualfs), and Calhoun soil series (Glossaqualfs). The albic materials were deduced to form by ferrolysis, which involves destruction of clay caused by cation-exchange reactions involved iron in repetitive reduction–oxidation cycles taking place under conditions of alternating wetness and dryness.

Buntley et al. (1977) and Lindbo et al. (1995, 1997, 2000) contributed greatly to our understanding of fragipan formation by examining Alfisols in Mississippi with and without a fragipan, including the Memphis (Typic Hapludalfs; no fragipan), Loring (Oxyaquic Fragiudalfs), and Grenada (Oxyaquic Fraglossudalfs) soil series. The fragipan typically forms within 100 cm of the surface and has vertical gray seams and associated redoximorphic features.

Lindbo et al. (2000) provided some of the early information on formation of a glossic horizon in a Fraglossudalf (Grenada soil series) in Mississippi. The glossic horizon occurred on top of a fragipan and contained bleached coatings (albic material) along primary ped faces. Iron-manganese nodules were also present in the glossic horizon. Field and laboratory studies suggested that the upper part of the fragipan was undergoing degradation.

Wilson and others (2015, 2017a, 2017b) studied soil pipe networks in Fragiudalfs and Fraglossudalfs at the Goodwin Creek Experimental Watershed (GCEW) in Panola County, Mississippi. Piping was attributed to the fragipan horizon holding up water, resulting in subsurface water flow and collapse of the soil following conversion of forest land to cropland (Fig. 8.1).

8.2 Properties and Processes

Alfisols contain an ochric epipedon averaging 21 ± 15 cm over an argillic horizon averaging 122 ± 51 cm (91% of soil series). Other diagnostic subsurface horizons in Alfisols

D. Johnson et al., *The Soils of Mississippi*, World Soils Book Series,
https://doi.org/10.1007/978-3-031-36235-4_8

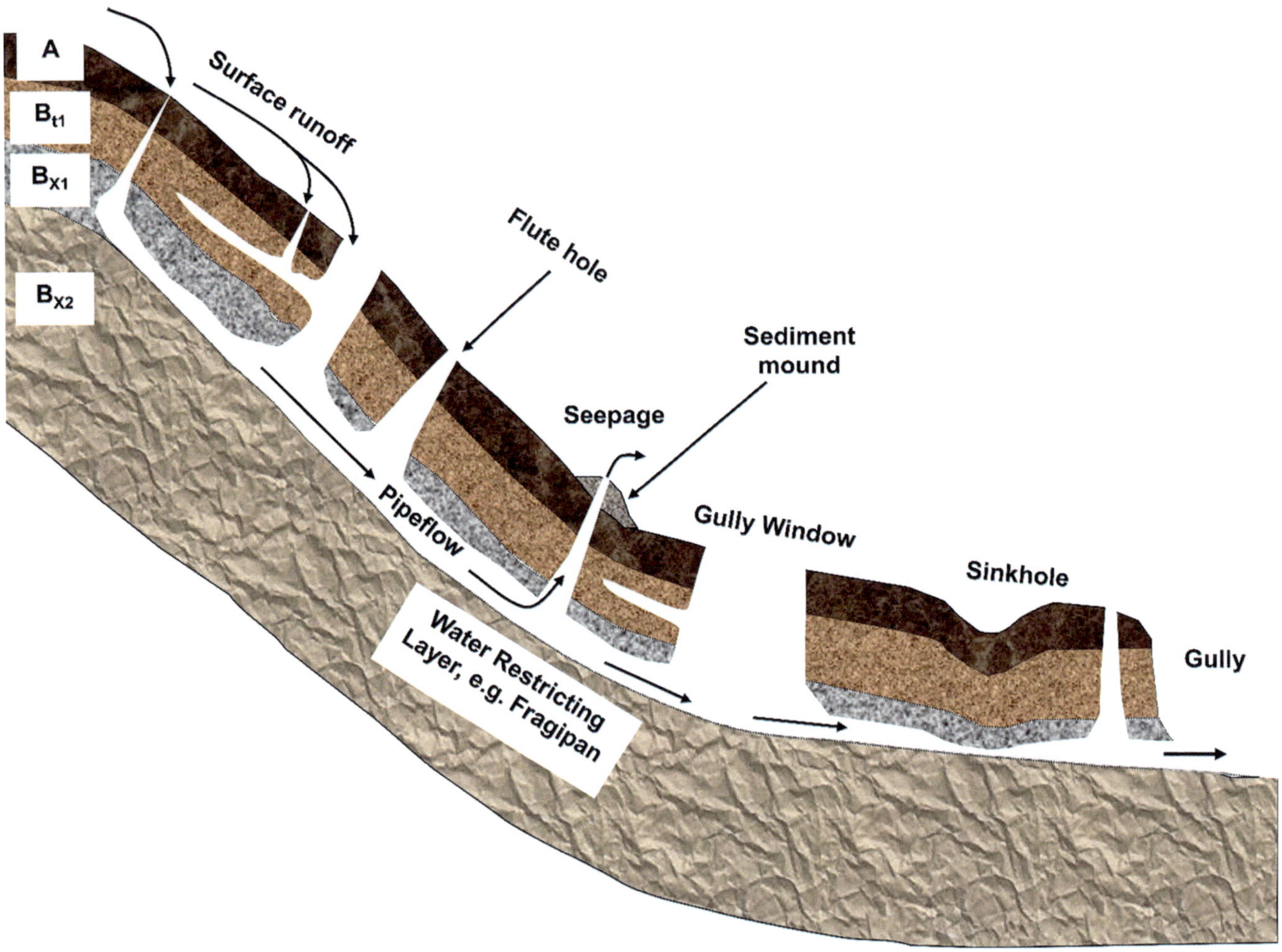

Fig. 8.1 Pipeflow and soil collapse due to the presence of a fragipan in Fragiudalfs and Fraglossudalfs at the Goodwin Creek Experimental Watershed in Panola County, Mississippi (Wilson et al. 2015) *Source* by J. Bockheim

of Mississippi include the glossic with a thickness of 71 ± 49 cm (35%), the albic with a thickness of 28 ± 20 cm (26%), the fragipan with a thickness of 62 ± 23 cm (16%), and the natric horizon with a thickness of 101 ± 27 cm (12% of soil series).

Alfisols in Mississippi have a mean annual air temperature averaging 18 ± 1.4 °C and receive an average of 1320 ± 86 mm of precipitation per year. Alfisols have an average maximum slope of 17 ± 18%. Most Alfisols are very deep and well drained to somewhat poorly drained. The vegetation on Alfisols is predominantly hardwoods and pines or hardwoods. In Mississippi, Alfisols are most commonly derived from loess, but they may originate from alluvium, marine deposits, or residuum. Dominant textures are silty and/or clayey. Dominant landforms are uplands, but Alfisols may also occur on alluvial terraces, marine terraces, or hillslopes. Alfisols in Mississippi commonly are of late Pleistocene age.

In Mississippi, Alfisols generally are in the fine-silty (51%), fine-loamy (19%), or fine (18%) particle-size class; the mixed (60%) or siliceous (30%) mineralogy class; and the active (70%) cation-exchange activity class. Alfisols in Mississippi have a thermic soil temperature regime and either a udic (70%) or aquic (30%) soil moisture regime.

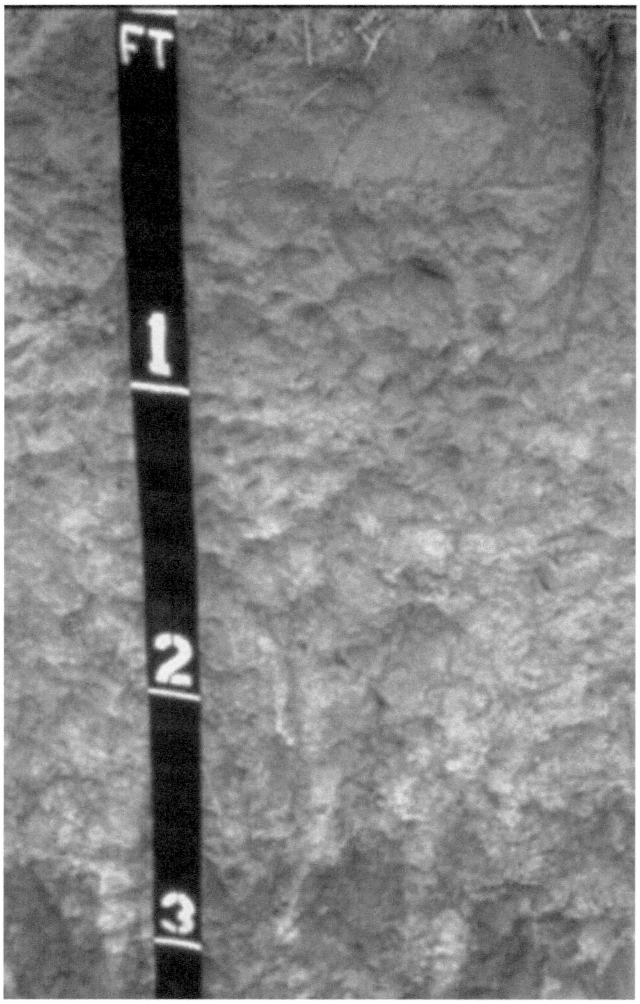

Fig. 8.2 Guyton soil series, a fine-silty, siliceous, active, thermic Typic Glossaqualfs, has formed in thick loamy coastal plain sediments of Pleistocene age. The scale is in feet. *Source* USDA Natural Resources Conservation Service

Photographs of soil series representing major great groups of Alfisols are given in Chap. 6, including a Fragiudalfs (Providence soil series; Fig. 6.1), a Hapludalfs (Dubbs soil series; Fig. 6.7), an Endoaqualfs (Forestdale soil series; Fig. 6.13), a Paleudalfs (Susquehanna soil series; Fig. 6.15), and a Fraglossudalfs (Grenada soil series, Fig. 6.36). The Guyton soil series, a fine-silty, siliceous, active, thermic Typic Glossaqualfs, has formed in thick loamy coastal plain sediments of Pleistocene age (Fig. 8.2). This soil contains a grayish brown A horizon to 6 in. (15 cm), a gleyed light brownish gray albic horizon to 23 in. (58 cm), and a grayish brown to gray argillic horizon to 70 in. (178 cm).

Table 8.1 contains data from six Alfisols representing four great groups in Mississippi. The soils have an ochric epipedon ranging from 11 to 35 cm (average = 20 cm) in thickness and an argillic horizon ranging from 101 to 197 cm (average = 139 cm). The soils all have a base saturation in excess of 35%. The dominant soil-forming processes in Alfisols are argilluviation, base-cation accumulation, gleization, glossification, and fragification.

8.3 Use and Management

Alfisols in Mississippi are used primarily for cropland and grazing and to a lesser extent for forestry. Cotton, corn, soybeans, and small grains are the most common crops, particularly on the Providence, Loring, Memphis, Forestdale, Dundee, and Dubbs soil series. Major Alfisol forest soils include the Lorman, Providence, Memhis, Dundee, Kipling, Susquehanna, and Guyton soil series.

8.4 Summary

Alfisols account for 32% of the soil area and 25% of the soil series in Mississippi. They are most common in in the Southern Mississippi Valley Loess (MLRA 134). The most common suborders are Udalfs (81%) and Aqualfs (19%). Four great groups account for 92% of the Alfisols on an area basis, including the Fragiudalfs, Hapludalfs, Endoaqualfs, and Paleudalfs. The most extensive Alfisols include the Providence, Loring, Memphis, Forestdale, and Dundee soil series. Alfisols contain an ochric epipedon over a deep over an argillic horizon. Some Alfisols have an albic, glossic, fragipan or natric horizon. Alfisols have a mean annual air temperature 18 ± 1.4 °C and receive an average of 1320 ± 86 mm of precipitation per annum. Alfisols have an average maximum slope of 17 ± 18%. Most Alfisols are very deep and well drained to somewhat poorly drained. The vegetation on Alfisols is predominantly hardwoods and pines or hardwoods. Parent materials are loess, alluvium, marine, or residuum of silty, loamy, or clayey composition on alluvial in uplands, alluvial terraces, marine terraces, or hillslopes. Mississippi Alfisols generally are in fine-silty, fine-loamy, or fine particle-size class; the mixed or siliceous mineralogy class; and the active cation-exchange activity class. Alfisols have a thermic soil temperature regime and either a udic or aquic soil moisture regime. Alfisols are important to the agricultural economy of Mississippi but are also used for pasture and forestry.

Table 8.1 Analytical properties of some Alfisols found in Mississippi

Horizon[a]	Depth (cm)	Clay (%)	Silt (%)	Sand (%)	SOC (%)	CEC7 ($cmol_c/kg$)	Base sat (%)	pH H_2O	Al sat (%)	1.5 mPa H_2O/ clay	Cole
Dubbs; fine-silty, mixed, active, thermic Typic Hapludalfs; Tallahatchie, MS; pedon no											
10N0425											
A	0–11	19.9	45.5	34.6	0.6	12.9	54	5.2	12	0.41	0.013
A/Bt	11–16	27.8	41.7	30.5	0.2	17.3	38	4.8	41	0.40	0.033
Bt1	16–23	28.7	41.4	29.9	0.2	18.1	33	4.8	48	0.41	0.031
Bt2	23–51	26.3	60.7	13.0	0.1	18	31	4.7	51	0.43	0.020
Bt3	51–74	21.6	61.3	17.1	0.1	15.5	37	4.8	42	0.46	0.018
Bt4	74–90	20.2	65.6	14.2	0.1	15.5	43	4.8	36	0.47	0.020
Bt5	90–122	20.4	66.4	13.2	0.1	16.0	48	4.9	23	0.49	0.035
2Bt	122–145	29.1	66.6	4.3	0.1	20.9	67	5.3	4	0.51	0.073
3Bt1	145–163	20.3	61.8	17.9	0.1	15.5	75	5.7	2	0.53	0.050
3Bt2	163–178	23.9	62.3	13.8	0.1	18.4	79	5.8		0.51	0.034
3Bt3	178–208	23.0	62.1	14.9	0.1	17.4	61	5.0	12	0.48	0.031
C	208–220	13.9	52.4	33.7	0.1	12.4	83	6.6		0.52	0.023
Forestdale; fine, smectitic, thermic Typic Endoaqualfs, Sunflower, MS; pedon no. 10N0428											
Ap1	0–14	16.4	81.8	1.8	0.6	11.1	93	6.6		0.50	0.015
Ap2	14–35	25.0	73.7	1.3	0.4	14.4	72	5.2	9	0.43	0.022
Bt1	35–51	35.6	63.4	1.0	0.2	23.8	54	4.6	34	0.47	0.043
Bt2	51–76	41.7	56.4	1.9	0.2	30.7	54	4.6	30	0.46	0.064
Bt3	76–101	40.5	57.6	1.9	0.2	27.7	71	4.6	18	0.46	0.078
Bt4	101–127	59.1	39.0	1.9	0.2	36.8	87	4.8	5	0.42	0.101
Bt5	127–148	59.3	39.7	1.0	0.2	37.6	97	6.3		0.39	0.149
Bt6	148–164	51.8	47.2	1.0	0.2	32.9	100	6.8		0.41	0.112
Ab	164–176	49.0	49.6	1.4	0.2	32.4	100	7.0		0.43	0.091
Bssg	176–200	60.0	39.2	0.8	0.2	37.1	100	7.4		0.41	0.124

(continued)

Table 8.1 (continued)

Horizon[a]	Depth (cm)	Clay (%)	Silt (%)	Sand (%)	SOC (%)	CEC7 ($cmol_c/kg$)	Base sat (%)	pH H_2O	Al sat (%)	1.5 mPa H_2O/ clay	Cole
Kipling; fine, smectitic, thermic Vertic Paleudalfs; Noxubee, MS; pedon no											
91P1061											
Ap1	0–14	22.4	56.3	21.3	1.44	17.9	60	5.0	7	0.46	0.027
Ap2	14–23	26.2	49.9	23.9	0.36	16.8	46	4.8	29	0.42	0.020
Bt1	23–37	31.0	47.5	21.5	0.23	19.3	45	5.0	34	0.41	0.033
Bt2	37–58	45.6	38.9	15.5	0.23	29.1	37	4.9	46	0.40	0.083
Bt3	58–83	51.4	35.0	13.6	0.21	31.7	37	4.9	54	0.37	0.098
Btss1	83–115	57.4	30.6	12.0	0.19	35.4	46	4.8	44	0.36	0.100
Btss2	115–142	54.5	33.6	11.9	0.18	33.6	63	4.8	25	0.37	0.097
Btss3	142–152	52.0	35.9	12.1	0.18	33.4	75	5.2	12	0.36	0.087
2Bss1	152–190	47.8	38.6	13.6	0.15	34.9	100	7.0		0.39	0.092
2Bss2	190–240	46.9	39.9	13.2	0.07	35.0	97	7.6		0.41	0.101
2Bss3	240–275	51.2	35.2	13.6	0.08	37.2	100	7.7		0.42	
3Cr	275–290	41.9	38.6	19.5	0.06	8.5	100	8.0		0.28	0.016
Loring; fine-silty, mixed, active, thermic Oxyaquic Fragiudalfs; Hinds, MS; pedon no. 81P0290											
Ap	0–13	16.9	80.7	2.4	2.18	14.4	94	6.3		0.64	0.019
Bt	13–36	22.7	75.9	1.4	0.49	12.0	100	7.1		0.46	0.028
Bt/E	36–69	20.5	77.8	1.7	0.16	10.6	54	4.6	28	0.47	
Btx1	84–99	18.4	79.6	2.0	0.12	10.2	51	4.5	32	0.47	0.007
Btx2	107–122	17.6	81.6	0.8	0.07	11.4	67	4.8	14	0.52	0.015
Btx3	122–155	18.4	81.0	0.6	0.06	12.1	82	5.3	5	0.53	0.015

(continued)

Table 8.1 (continued)

Horizon[a]	Depth (cm)	Clay (%)	Silt (%)	Sand (%)	SOC (%)	CEC7 ($cmol_c/kg$)	Base sat (%)	pH H_2O	Al sat (%)	1.5 mPa H_2O/ clay	Cole
Memphis; fine-silty, mixed, active, thermic Typic Hapludalfs; Warren, MS; pedon no. 40A4812											
Ap	0–10	9.7	89.3	1.0	1.00	6.3	76	6.0		0.46	
E	10–23	19.7	79.9	0.4	0.33	8.5	76	5.8		0.42	
Bt11	23–33	26.5	73.0	0.5	0.24	12.0	83	5.7		0.42	
Bt12	33–53	28.7	70.7	0.6	0.14	13.8	62	5.4		0.42	
Bt13	53–76	26.2	72.8	1.0	0.11	13.6	54	5.4		0.41	
Bt14	76–99	23.5	75.5	1.0	0.10	13.3	54	5.2		0.43	
Bt2	99–124	21.2	78.0	0.8	0.08	12.9	55	5.5		0.44	
C1	124–155	18.2	81.2	0.6	0.08	12.3	59	5.3		0.47	
C2	155–193	19.4	80.1	0.5	0.08	12.6	65	5.2		0.47	
Providence; fine-silty, mixed, active, thermic Oxyaquic Fragiudalfs; Lincoln, MS; pedon no											
40A4821											
A	0–8	9.2	73.0	17.8	3.92	14.2	44	5.3		0.73	
E	8–18	11.1	74.2	14.7	0.84	6	15	5.0		0.38	
Bt1	18–23	14.5	74.3	11.2	0.4	6.3	30	5.1		0.37	
Bt2	23–36	20.9	70.2	8.9	0.23	7.7	47	5.1		0.39	
Bt3	36–58	29.6	62.8	7.6	0.14	11.4	40	5.2		0.42	
Bt4	58–76	24.7	65.3	10.0	0.1	10.4	44	5.3		0.43	
2Btx/Ex	76–97	15.5	56.1	28.4	0.06	6.7	61	5.5		0.43	
2Btx1	97–135	14.6	34.5	50.9	0.02	5.1	45	5.4		0.38	
2Btx2	135–152	16.0	26.8	57.2	0.03	5.0	56	5.4		0.36	

[a] Diagnostic horizons in bold-face are diagnostic: *Bt*, *Btss*, *Btx* argillic

References

Bartelli LJ (1973) Soil development in loess in the southern Mississippi Valley. Soil Sci 118:254–260

Buntley GJ, Daniels RB, Gamble EE, Brown WT (1977) Fragipan horizons in soils of the Memphis-Loring-Grenada sequence in west Tennessee. Soil Sci Soc Am J 41:401–407

Dyal RS, Martin IL, Templin EH (1951) Red-yellow podzolic soils of the southeastern United States: II. Character of the clay fractions of Ruston, Stephenville, Boswell, Windthorst, Cahaba, Leaf, and Axtell. Agron J 43:482–487

Lindbo DL, Rhoton FE, Bigham JM, Jones FS, Smeck NE, Hudnall WH, Tyler DD (1995) Loess toposequences in the lower Mississippi River Valley I: Fragipan morphology and identification. Soil Sci Soc Am J 59:487–500

Lindbo DL, Rhoton FE, Hudnall WH, Smeck NE, Bigham JM (1997) Loess stratigraphy and fragipan occurrence in the lower Mississippi Valley. Soil Sci Soc Am J 61:195–210

Lindbo DL, Rhoton FE, Hudnall WH, Smeck NE, Bigham TDD (2000) Fragipan degradation and nodule formation in glossic Fragiudalfs of the lower Mississippi River Valley. Soil Sci Soc Am 64:1713–1722

Templin EH, Martin IL, Dyal RS (1951) Red-yellow podzolic soils of the southeastern United States: I. Morphology of the Ruston, Stephenville, Boswell, Windthorst, Cahaba, Leaf, and Axtell series. Agron J 43:482–487

Vanderford HB, Shaffer ME (1966) Comparison of fragipan and bisequal soils of the Gulf Coastal Plain with soils of Southern Loess Belt. Soil Sci Soc Am Proc 30:494–498

Wilson GV, Rigby JR, Dabney SM (2015) Soil pipe collapses in a loess pasture of Goodwin Creek watershed, Mississippi: role of soil properties and past land use. Earth Surf Proc Landf 40:1448–1463

Wilson GV, Wells R, Kuhnle R, Fox G, Nieber J (2017a) Sediment detachment and transport processes associated with internal erosion of soil pipes. Earth Surf Proc Landf 43:45–63

Wilson GV, Nieber JL, Fox GA, Dabney SM, Ursic M, Rigby JR (2017b) Hydrologic connectivity and threshold behavior of hillslopes with fragipans and soil pipe networks. Hydrol Proc 31:2477–2496

9 Inceptisols

9.1 Distribution

Inceptisols compose 16% of the soil series and 17% of the land area of Mississippi (Table 5.2). Inceptisols are most common in the Gulf Coastal Plain (MLRA 133C), the Southern Mississippi Valley Loess (MLRA 134), the Southern Mississippi River Alluvium (MLRA 131A), and the Alabama and Mississippi Blackland Prairie (MLRA 135A). Inceptisols in Mississippi are limited to the major river valleys, particularly the Pearl, Big Black, Yalobusha, Yokona, Tallahatchie, and Tombigbee. About 55% of the Inceptisol area is Aquepts and 45% is Udepts. The most extensive great groups include the Endoaquepts (41% of the Inceptisol area), followed by Dystrudepts (27%), Eutrudepts (17%), and Epiaquepts (13%).

The most extensive Inceptisols in Mississippi are the Mantachie (1845 km^2), Dowling (1694 km^2), Oaklimeter (1237 km^2), and Arkabutla (942 km^2) soil series.

9.2 Properties and Processes

The key concept of Inceptisols is that they are more strongly developed than Entisols but lack the characteristics of more strongly developed soil orders such as the Ultisols or Alfisols. In Mississippi, Inceptisols have an ochric epipedon averaging 19 ± 9.0 cm over a cambic horizon averaging 108 ± 50 cm in thickness. About 8% of the Inceptisols in the state lack a diagnostic subsurface horizon. Inceptisols in Mississippi have an average annual air temperature of 18 ± 1.2 °C and a mean annual precipitation of 1340 ± 100 mm. Inceptisols are deep to very deep and moderately well drained to somewhat poorly drained. Maximum slopes are 6 ± 12%. Inceptisols most commonly support bottomland hardwoods or mixed hardwood vegetation, although some Inceptisols have hardwoods and pines. The parent materials are alluvium of loamy, silty, or clayey composition that occur on flood plains.

More than two-thirds (69%) of the Inceptisols are in the coarse-silty (30%), fine-silty (22%), or fine (17%) particle-size classes; 58% have a mixed and 25% a siliceous mineralogy; 61% are in the active and 22% in the superactive cation-exchange activity classes; and 28% are in the non-acid and 25% in the acid reaction classes. Inceptisols have a thermic soil temperature regime, and 53% of the soil series have an aquic and 47% a udic soil moisture regime.

Photographs of major great groups of Inceptisols are given in Chap. 6, including an Endoaquepts (Commerce; Fig. 6.12), a Eutrudepts Bruin; Fig. 6.28, and an Epiaquepts (Newellton; Fig. 6.31).

The representative five Inceptisol soil series in Table 9.1 are more strongly developed than Entisols (see Chap. 11) but do not meet the requirements for other soil orders. The soils have cambic horizons ranging from 15 to 101 cm and averaging 72 cm in thickness. The Adler, Leeper, Ariel, and Rosebloom soil series are base-enriched, but the Mantachie is base-depleted below 53 cm. All of the Inceptisols in Table 9.1 and derived from alluvium. Cambisolization and gleization are the dominant soil-forming processes in Inceptisols of Mississippi.

9.3 Use and Management

Inceptisols are intensively cropped and are also used in timber production. The Mantachie, Oaklimeter, Arkabutla, and Ariel soil series are used for cotton, corn, and soybean production. The Dowling, Natchez, Jena, and Leeper are used in forestry operations primarily for hardwood species.

9.4 Summary

Inceptisols compose 16% of the soil series and 17% of the land area of Mississippi and occur in primarily in the Southern Coastal Plain, the Southern Mississippi Valley

D. Johnson et al., *The Soils of Mississippi*, World Soils Book Series,
https://doi.org/10.1007/978-3-031-36235-4_9

Table 9.1 Analytical properties of some Inceptisols found in Mississippi

Horizon[a]	Depth (cm)	Clay (%)	Silt (%)	Sand (%)	SOC (%)	CEC7 ($cmol_c$/kg)	Base sat (%)	pH H_2O	Al sat (%)	1.5 mPa H_2O/clay	Cole
Adler; coarse-silty, mixed, superactive, thermic Fluvaquentic Eutrudepts; Tallahatchie, MS;											
pedon no. 10N0427											
Ap	0–10	6.9	90.5	2.6	1.1	6.1	100	6.9		0.68	0.010
Bw1	10–25	8.0	89.3	2.7	0.2	5.3	74	5.8	3	0.50	0.004
Bw2	25–43	13.3	82.3	4.4	0.1	6.9	96	6.3		0.49	0.006
Bw3	43–62	15.7	80.3	4.0	0.1	8	100	7.0		0.50	0.015
Bw4	62–105	17.5	78.8	3.7	0.1	9.3	100	7.3		0.51	0.013
BC	105–150	13.5	73.3	13.2	0.1	7.8	100	8.1		0.52	0.007
C1	150–188	11.4	85.5	3.1	0.1	6.9	100	8.2		0.50	0.009
C2	188–200	10.6	78.1	11.3	0.1	6.5	100	8.1		0.52	0.007
Ariel; coarse-silty, mixed, active, thermic Fluvaquentic Dystrudepts; Panola, MS; pedon no											
81P0841											
Ap2	0–18	10.3	89.2	0.5	0.77	7.2	61	4.6	15	0.53	0.009
Bw1	18–41	9.4	89.4	1.2	0.59	6.9	81	5.8		0.56	
Bw2	41–51	9.3	89.1	1.6	0.4	5.7	77	5.8		0.53	
Bw3	51–94	12.6	85.5	1.9	0.29	6.2	56	4.9	17	0.48	0.012
Bw4	94–119	10.5	86.6	2.9	0.21	5.6	54	4.8	19	0.51	
B/Eb	119–147	10.5	87.5	2.0	0.14	5.6	34	4.6	46	0.51	0.002
Btx/B1	147–168	13.4	84.5	2.1	0.14	6.4	27	4.7	56	0.45	
Btx/B2	168–191	15.1	81.0	3.9	0.14	7.2	28	4.9	51	0.42	0.006
Leeper; fine, smectitic, nonacid, thermic Vertic Epiaquepts; Lee, MS; pedon no. 81P0293											
91P1061											
Ap	0–13	35.7	55.2	9.1	1.60	25.4	100	7.6		0.39	0.055
A	13–25	42.5	51.0	6.5	1.29	27.5	100	7.7		0.39	0.055
Bw1	25–61	34.0	47.3	18.7	0.91	22.9	100	7.6		0.42	0.060
Bw2	61–102	36.1	40.0	23.9	0.60	23.1	100	7.9		0.43	
C	102–152	39.4	39.9	20.7	0.56	25.4	100	8.0		0.43	
Mantachie; fine-loamy, siliceous, active, acid, thermic Fluventic Endoaquepts; Lee, MS; pedon no											
07N0110											
Ap1	0–13	10.8	21.7	67.5	1.0	8.1	77	5.7	5	0.51	0.015
Ap2	13–25	13.1	25.6	61.3	0.5	8.6	87	5.8	4	0.47	0.014
Bw1	25–36	14.2	25.6	60.2	0.3	8.4	56	5.2	35	0.44	0.017
Bg1	36–53	15.5	25.4	59.1	0.2	8.7	36	4.9	57	0.40	0.015
Bg2	53–94	16.3	24.3	59.4	0.1	9.4	23	4.8	72	0.40	0.020
Bg3	94–137	19.4	24.8	55.8	0.1	11.1	20	4.8	76	0.41	0.021
Bg4	137–173	18.7	26.2	55.1	0.1	11.5	24	4.9	70	0.48	0.025
Bg5	173–218	20.2	25.9	53.9	0.1	12.7	43	5.0	49	0.50	0.028

(continued)

Table 9.1 (continued)

Horizon[a]	Depth (cm)	Clay (%)	Silt (%)	Sand (%)	SOC (%)	CEC7 ($cmol_c/kg$)	Base sat (%)	pH H_2O	Al sat (%)	1.5 mPa H_2O/clay	Cole
Rosebloom; fine-silty, mixed, active, acid, thermic Fluvaquentic Endoaquepts; Oktibbeha, MS;											
pedon no. 15N0129											
A	0–18	30.3	53.9	15.8	2.6	18.3	57	4.2	15	0.49	
Bg1	18–36	17.5	46.8	35.7	0.7	8.4	44	4.7	34	0.46	0.010
Bg2	36–65	14.9	43.4	41.7	0.3	6.6	39	4.9	45	0.45	0.010
Bg3	65–90	12.7	32.7	54.6	0.2	5.7	40	4.8	47	0.43	
Cg1	90–137	20.7	53.6	25.7	0.3	9.7	39	4.8	49	0.45	
Cg2	137–200	25.0	48.1	26.9	0.3	10.9	33	4.8	54	0.41	

[a] Diagnostic horizons in bold-face are diagnostic: *Bw*, *Bg* cambic

Loess, the Southern Mississippi River Alluvium, and the Blackland Prairies. They form along major rivers in the state. The dominant suborders is Aquepts and Udepts, and the most extensive great groups are Endoaquepts, Dystrudepts, Eutrudepts, and Epiaquepts. Inceptisols generally are deep to very deep and moderately well drained to somewhat poorly drained. The native vegetation commonly is bottomland hardwoods and hardwoods. Inceptisols are commonly formed in alluvium from loamy, silty, or clayey materials on floodplains. Inceptisols have an ochric epipedon over a cambic horizon. Inceptisols are in coarse-silty and finer particle-size classes, the mixed or siliceous mineralogy class, and the active cation-exchange activity class. They have a thermic soil temperature regime and an aquic or udic soil moisture regime. Inceptisols are used primarily for crop or forest production.

10 Vertisols

10.1 Distribution

Vertisols compose 8% of the soil series and 10% of the land area of Mississippi (Table 5.2). Vertisols occur in three of the six MLRAs in Mississippi, including the Southern Mississippi River Alluvium (MLRA 131A), Alabama and Mississippi Blackland Prairie (MLRA (135A), and the Gulf Coastal Plain (MLRA 133C). About 69% of the Vertisols are Aquerts and 31% are Uderts (Table 6.2). Epiaquerts compose 36% of the Vertisol great groups, followed by Dystraquerts (33%), Dystruderts (23%), and Hapluderts (8%). The most extensive Vertisols are the Sharkey (3080 km^2), Alligator (2857 km^2), and Vaiden (884 km^2) soil series.

Pettry (1993) described the types, distribution, and extent of clayey, expansive soils in Mississippi and presented physical, chemical, and mineralogical data of representative soils. About 18% of the total state area was composed of expansive soils. Nearly three-quarters (72%) of the expansive soils were in the Southern Mississippi River Alluvium (MLRA 131A) and the Blackland Prairie (MLRA 135A). Nearly one-half (42%) of the soil area was in what are now classified as Vertisols. The Coefficient of Linear Extensibility (COLE) of selected Vertisol horizons ranged between 0.14 and 0.22 and averaged 0.18. Cheng and Pettry (1993) measured horizontal and vertical movements of two expansive Vertisols in the Blackland Prairie, including the Okolona (Oxyaquic Hapluderts) and Vaiden (Aquic Dystruderts) soil series. Vertical movement ranged between 24 and 27 cm and horizontal movement from 20 to 36 mm. Pettry and Switzer (1996) characterized the Sharkey soil in Mississippi, recommending that it be reclassified as a Vertisol. Their study led to a reexamination of expansive soils in the state and several extensive Alfisols and Inceptisols were reclassified as Vertisols, including the Alligator, Vaiden, Wilcox, and Oktibbeha soil series. Ewing (2011) described foundation repairs on the Eudora Welty historical house near Jackson, MS, due to the presence of expansive soils.

10.2 Properties and Processes

Vertisols have an ochric epipedon averaging 12 ± 6 cm (79% of soil series) or a mollic epipedon averaging 70 ± 31 cm in thickness (21%) that overlays a cambic horizon averaging 122 ± 48 cm (53%) and/or an argillic horizon averaging 76 ± 43 cm in thickness (37%). About 16% of the Vertisols in Mississippi lack a diagnostic subsurface horizon.

The mean annual temperature for Vertisols in Mississippi is 18 ± 0.98 °C, and the mean annual precipitation is $1,325 \pm 77$ mm. Vertisols are deep to very deep and often are somewhat poorly drained to poorly drained. The maximum slope for Vertisols is 9 ± 10%. The native vegetation on Vertisols ranges from bottomland hardwoods to southern pines, often with a grassy understory. Parent materials are alluvium, residuum, and marine deposits, often of clayey and chalky composition, on uplands and in flood plains.

All of the Vertisols are in the fine, very-fine, and clayey particle-size classes, all have a smectitic mineralogy, a thermic soil temperature regime, and a udic (65%) or aquic (35%) soil moisture regime. Photographs of major great groups of Vertisols are given in Chap. 6, including an Epiaquerts (Sharkey soil series; Figs. 6.21 and 6.22), a Dystraquerts (Alligator soil series; Fig. 6.24), a Dystruderts (Suggsville soil series; Fig. 6.30), and a Hapluderts (Okolona soil series; Fig. 6.38).

Laboratory data are provided for four extensive Vertisols in Mississippi, including the Alligator, Brooksville, Sharkey, and Vaiden soil series (Table 10.1). Clay concentrations in the Bss horizon range from 51 to 77%; all of the soils are smectitic; and COLE values in slickenside horizons range between 0.11 and 0.21. The soils tend to have a high base saturation. The dominant soil forming processes in Mississippi Vertisols are vertization and gleization. Information about soil forming processes is given in Chap. 13.

D. Johnson et al., *The Soils of Mississippi*, World Soils Book Series,
https://doi.org/10.1007/978-3-031-36235-4_10

Table 10.1 Analytical properties of some Vertisols found in Mississippi

Horizon[a]	Depth (cm)	Clay (%)	Silt (%)	Sand (%)	SOC (%)	CEC7 ($cmol_c/kg$)	Base sat (%)	pH H_2O	Al sat (%)	1.5 mPa H_2O/clay	Cole
Alligator; very-fine, smectitic, thermic Chromic Dystraquerts; Leflore, MS; pedon no. 40A4759											
Ap	0–10	73.7	33.4	2.9	1.58	*48.6*	*69*	4.8			
Bg	10–25	77.3	21.8	0.9	0.50	*55.2*	*72*	4.6			
Bssg1	25–69	73.4	26.0	0.6	0.36	*53.3*	*77*	4.6			
Bssg2	69–102	70.2	28.9	0.9	0.36	*50.7*	*82*	4.8			
Bssg3	102	65.6	33.4	1.0	0.36	*48.9*	*92*	6.4			
Brooksville; fine, smectitic, thermic Aquic Hapluderts; Chickasaw, MS; pedon no. 40A3963											
81P0841											
Ap	0–15	39.8	56.3	3.9	1.71	32.6	80	5.8		0.40	0.030
A1	15–41	45.6	51.1	3.3	1.03	34.1	82	5.8		0.40	0.076
A2	41–66	48.7	48.3	3.0	0.63	34.7	84	6.4		0.39	0.093
Bw1	66–94	51.0	46.5	2.5	0.37	36.4	89	7.0		0.38	0.095
Bw2	94–140	53.7	44.2	2.1	0.29	37.9	94	7.4		0.38	0.106
Cg	140–180	53.2	42.4	4.6	0.11	38.1	100	7.7		0.37	0.084
Sharkey; very-fine, smectitic, thermic Chromic Epiaquerts; Tunica, MS; pedon no. 98P0096											
Ap	0–9	71.1	26.8	2.1	2.10	48.0	100	6.9		0.37	
A1	9–26	72.2	25.6	2.2	1.60	48.5	92	7.1		0.37	0.174
A2	26–56	74.9	23.6	1.5	1.10	47.5	84	5.9		0.37	0.206
Bssg1	56–99	76.9	21.6	1.5	0.70	49.3	82	5.2	4	0.37	0.201
Bssg2	99–160	75.1	23.9	1.0	0.5	49.8	95	6.4		0.37	0.207
Bssyg	160–200	64.9	32.1	3.0	0.5	45.6	100	7.2		0.40	0.177
Vaiden; very-fine, smectitic, thermic Aquic Dystruderts; Newton, MS; pedon no. 81P0560											
Ap	0–13	39.4	54.3	6.3	2.53	35.3	100	5.7		0.57	0.101
Btss1	13–56	66.0	30.8	3.2	0.57	48.5	91	5.5	1	0.44	0.176
Btss2	56–84	63.1	33.5	3.4	0.27	49.2	100	6.6		0.45	0.141
2C	84–152	19.1	47.1	33.8	0.17	13.7	100	8.1		0.52	0.017

[a]Horizons in bold-face are diagnostic: *Bw*, *Bg* cambic
Italic type means CEC8 instead of CEC7 and base saturation by summation

10.3 Use and Management

Several extensive Vertisols are important for Mississippi's agriculture, including the Sharkey, Alligator, and Vaiden soil series. The Louin, Eutaw, and Smithdale are important forest soils in Mississippi.

10.4 Conclusions

Vertisols compose 8% of the soil series and 10% of the land area of Mississippi. They most commonly occur in the Alabama and Mississippi Blackland Prairie, the Southern Coastal Plain, and the Southern Mississippi River Alluvium. About 69% of the Vertisols are Aquerts and 31% are Uderts. Epiaquerts are the most extensive Vertisol great group, followed by Dystraquerts, Dystruderts, and Hapluderts. Vertisols most commonly have an ochric epipedon and less commonly a mollic epipedon that overlays a cambic and/or an argillic horizon. About 16% of the Vertisols in Mississippi lack a diagnostic subsurface horizon. The mean annual temperature for Vertisols in Mississippi is 18 ± 0.98 °C, and the mean annual precipitation is 1325 ± 77 mm. Vertisols are deep to very deep and often are somewhat poorly drained to poorly drained. The maximum slope for Vertisols is 9 ± 10%. The native vegetation on Vertisols ranges from bottomland hardwoods to southern pines, often with a grassy understory. Parent materials are alluvium, residuum, and marine deposits, often of clayey and chalky composition, on uplands and in flood plains. All of the Vertisols are in the fine, very-fine, and clayey particle-size classes and all have a smectitic mineralogy, a thermic soil temperature regime, and a udic (65%) or aquic (35%) soil moisture regime. The dominant soil forming processes in Mississippi Vertisols are vertization and gleization.

References

Cheng Y, Pettry DE (1993) Horizontal and vertical movements of two expansive soils in Mississippi. Soil Sci Soc Am J 57:1542–1547

Ewing RC (2011) Foundation repairs due to expansive soils: Eudora Welty House, Jackson, Mississippi. J Perform Constr Facil 25:50–55

Pettry DE (1993) Expansive soils in Mississippi. Miss Agric For Exp Sta Bull 986:32 pp

Pettry DE, Switzer RE (1996) Sharkey soils in Mississippi. Miss Agric For Exp Sta Bull 1057:48 pp

11 Entisols

11.1 Distribution

Entisols account for 12% of the soil series and 8.4% of the land area of Mississippi (Table 5.2). Entisols occur in all six Major Land Resource Areas in Mississippi, but are most common in the Gulf Coastal Plain (MLRA 133C), the East Gulf Coast Flatwoods (MLRA 152A), and the Southern Mississippi River Alluvium (MLRA 131A). About one-half (51%) of the Entisols on an area basis are in the Aquents suborder and 40% in the Fluvents (Table 5.2). About 90% of the Entisol soil series are in two great groups, including the Fluvaquents (50%) and Udifluvents (40%). The most extensive Entisols are the Falaya (1574 km^2), Collins (1500 km^2), Bibb (1020 km^2), and Gillsburg (912 km^2) soil series.

11.2 Properties and Processes

Entisols are the least developed soils in Mississippi and usually contain only an ochric epipedon, with a mean thickness of 19 ± 15 cm. About 8% of the soil series have a cambic horizon that averages 47 ± 38 cm in thickness.

Entisols in Mississippi have a mean annual air temperature of 18 ± 1.5 °C and receive $1{,}360 \pm 90$ mm of precipitation. They occur on maximum slopes of $6 \pm 9\%$. They are deep to very deep and range from excessively drained to very poorly drained. The native vegetation is dominantly bottomland hardwoods or hardwoods mixed with southern pines. Entisols have formed predominantly in alluvium (68% of soil series) but also occur in eolian and marine deposits. They are often sandy or silty. Entisols form most commonly in floodplains.

More than three-quarters (80%) of the Entisols in Mississippi are in the sandy (42%), coarse-loamy (19%), or coarse-silty (19%) particle size classes. Entisols commonly have a mixed (46%) or siliceous (31%) mineralogy, an active (38%) or superactive (19%) cation-exchange activity, and an acid (32%) or nonacid (30%) reaction. Entisols have a thermic soil temperature regime and a udic (62%) or aquic (38%) soil moisture regime.

Photographs of key Entisol great groups are given in Chap. 6, including Fluvaquents (Falaya soil series; Fig. 6.19) and Udifluvents (Collins soil series; Fig. 6.26). Two additional photographs are provided here for less common Entisols in Mississippi. The Demopolis soil series, a loamy, carbonatic, thermic, shallow Typic Udorthents, has formed in soft limestone residuum in the Blackland Prairie (MLRA 135A) (Fig. 11.1). This soil has a dark grayish brown ochric horizon to 6 in. (15 cm), with a paralithic contact with soft limestone at 13 in. (33 cm). The Lakeland soil series, a thermic, coated Typic Quartzipsamments, has formed in thick beds of eolian or marine and/or alluvial sands in the Gulf Coastal Plain (MLRA 133C) (Fig. 11.2). This soil has a very dark grayish brown ochric horizon to 3 in. (7.5 cm) followed by yellowish brown sand lacking a diagnostic subsurface horizon.

Table 11.1 provides data for two Entisols in Mississippi. Both of these extensive Entisols are formed in recent alluvium and neither contains a diagnostic subsurface horizon. The dominant soil-forming processes in Entisols are gleization and weak forms of cambisolization and other processes. These are discussed in Chap. 15.

11.3 Use and Management

Many extensive Entisols are used for irrigated crops, including corn, cotton, soybeans, and small grains. These include the Falaya, Collins, and Iuka soil series. The Deerford and Crevasse soil series are used for grazing. The Bibb, Gillsburg, and Bruno soil series are used for forest products.

D. Johnson et al., *The Soils of Mississippi*, World Soils Book Series,
https://doi.org/10.1007/978-3-031-36235-4_11

Fig. 11.1 Demopolis soil series, a loamy, carbonatic, thermic, shallow Typic Udorthents, has formed in soft limestone residuum in the Blackland Prairie (MLRA 135A). The scale is in inches. *Source* USDA Natural Resources Conservation Service

Fig. 11.2 Lakeland soil series, a thermic, coated Typic Quartzipsamments, has formed in thick beds of eolian or marine and/or alluvial sands in the Gulf Coastal Plain (MLRA 133C). The scale is in feet. *Source* USDA Natural Resources Conservation Service

11.4 Summary

Entisols account for 12% of the soil series and 8.4% of the land area of Mississippi. Entisols occur in all six Major Land Resource Areas in Mississippi, but are most common in the Southern Coastal Plain, the East Gulf Coast Flatwoods, and the Southern Mississippi River Alluvium. Entisols are roughly equally divided into the Aquents and Fluvents suborders and in the Fluvaquents and Udifluvents great groups. Entisols are the least developed soils in Mississippi and usually contain only an ochric epipedon, with a mean thickness of 19 ± 15 cm. Entisols in Mississippi have a mean annual air temperature of 18 ± 1.5 °C and receive $1,360 \pm 90$ mm of precipitation.

Table 11.1 Analytical properties of some Entisols found in Mississippi

Horizon[a]	Depth (cm)	Clay (%)	Silt (%)	Sand (%)	SOC (%)	CEC7 ($cmol_c$/kg)	Base sat (%)	pH H_2O	Al sat (%)	1.5 mPa H_2O/ clay	Cole
Collins; coarse-silty, mixed, active, acid, thermic Aquic Udifluvents; Panola, MS; pedon no											
81P0842											
Ap	0–18	7.3	90.8	4.5	0.54	5.3	64	4.9	8	0.58	0.009
C1	18–48	9.8	88.3	6.9	0.26	6.2	65	5.0	11	0.52	0.012
C2	48–91	11.4	87.6	6.5	0.44	6.3	51	5.0	18	0.49	
C3	91–117	11.8	77.4	7.4	0.40	6.1	46	4.8	28	0.48	
C4	117–152	17.1	73.9	9.0	0.15	7.9	37	4.8	37	0.46	0.015
Cg1	152–193	18.8	76.1	7.8	0.13	9.2	46	4.7	30	0.46	
Cg2	193–254	21.4	33.4	10.7	0.12	14.5	84	5.1	4	0.5	
Falaya; coarse-silty, mixed, active, acid, thermic Aeric Fluvaquents; Tate, MS; pedon no											
40A3971											
Ap	0–20	19.0	80.0	1.0	1.17	12.2	79	5.6		0.64	
A	20–41	25.9	72.9	1.2	0.8	14.8	60	5.2		0.57	
Bw	41–76	21.3	77.2	1.5	0.55	12.0	64	5.4		0.56	
Cg1	76–130	21.6	75.9	2.5	0.87	11.6	34	4.9		0.54	
Cg2	130–180	19.3	77.6	3.1	0.16	9.5	33	4.9		0.49	

They are deep to very deep and range from excessively drained to very poorly drained. The native vegetation is dominantly bottomland hardwoods or hardwoods mixed with southern pines. Entisols have formed predominantly in sandy or silty alluvium on floodplains.

More than three-quarters (80%) of the Entisols in Mississippi are in the sandy, coarse-loamy, or coarse-silty particle size classes, the mixed or siliceous mineralogy class, the active or superactive cation-exchange activity class, and the acid or nonacid reaction class. Entisols have a thermic soil temperature regime and a udic or aquic soil moisture regime. The dominant soil-forming processes in Entisols are gleization and weak forms of cambisolization. Entisols are used for cropping, pastures, and forestry operations.

12 Histosols, Mollisols, and Spodosol

12.1 Distribution

Histosols account for a very small number of soil series (5) and area (527 km^2) in Mississippi. There are four Mollisols that cover 413 km^2 and one Spodosol that occupies 1 km^2. Histosols occur in the Gulf Coastal Plain (MLRA 133C) and the Eastern Gulf Coast Flatwoods (MLRA 152A) in southernmost Mississippi. The Mollisols occur mainly in the Alabama and Mississippi Blackland Prairie (MLRA 135A), but also in Southern Mississippi Alluvium (MLRA 131A) and the Gulf Coastal Plain (MLRA 133C). The Spodosol occurs in the Eastern Gulf Coastal Flatwoods (MLRA 152A). The most extensive Histosols are the Dorovan (177 km^2) and Croatan (126 km^2) soil series, and the most extensive Mollisol is the Bowdre (104 km^2) soil series.

The Histosols are very deep and very poorly drained. They have a mean annual air temperature of 19 °C and receive 1400 mm of precipitation per year. The maximum slope averages 1%. The vegetation is bottomland hardwoods, pond pine, or cordgrass and marsh. The parent materials are organic sediments that have accumulated in basins.

The Mollisols range from shallow to very deep and are either well drained or somewhat poorly drained. They have a mean annual air temperature averaging 17 °C and receive 1330 mm of precipitation per year. Mollisols occur on maximum slopes averaging 17%. The vegetation is either bottomland hardwoods or redcedar-osage orange. The parent materials are either residuum on dissected uplands or alluvium in flood plains. The soils are underlain by clayey chalk or limestone.

The Spodosol is very deep and very poorly drained, has a mean annual air temperature averaging 20 °C, and receives 1650 mm of precipitation per year (Pettry et al. 1994). The Spodosol occurs on a maximum slope of 5%. The vegetation is slash and longleaf pines. The parent material is sandy marine deposits in upland flats.

12.2 Properties and Processes

Histosols are organic soils with a histic epipedon averaging 117 cm in thickness. About 80% of the Histosol area contains sapric materials, the most highly decomposed stage of organic matter, and 20% contains hemic materials, which is intermediate in decomposition stage. About 60% of the Histosols are dysic ($pH<4.5$) and 40% are euic ($pH>4.5$). The Maurepas Histosol is the only hyperthermic soil in Mississippi.

The Mollisols have a mollic epipedon averaging 32 cm in thickness and, with one exception, lack a diagnostic subsurface horizon. Half of the Mollisols are Udolls/Hapludolls (suborder/great group) and half are Rendolls/Haprendolls. Three-quarters of the Mollisols are in the clayey or fine particle-size class; 50% are in the carbonatic, and 50% are in the smectitic mineralogy class; and all have a thermic soil temperature regime and a udic soil moisture regime.

The Leon soil series, a sandy, siliceous, thermic Aeric Alaquods, is formed in sandy marine sediments of the Eastern Gulf Coast Flatwoods (MLRA 152A) (Fig. 12.1). This soil has a 38-cm thick ochric epipedon over a spodic horizon that is 107 cm thick.

The dominant processes are paludization and gleization in Histosols, humification and cambisolization in Mollisols, and podsolization and gleization in the Spodosol.

There are limited laboratory data for the Histosols, Mollisols, and Spodosol in Mississippi. However, data are provided for the Catalpa (Hapludolls), Leon (Alaqods), and Maurepas (Haplosaprists) soil series (Table 12.1).

12.3 Use and Management

The Histosols are used primarily for wildlife protection and watershed management but some have been drained and cropped and others are managed for forest products. The

D. Johnson et al., *The Soils of Mississippi*, World Soils Book Series,
https://doi.org/10.1007/978-3-031-36235-4_12

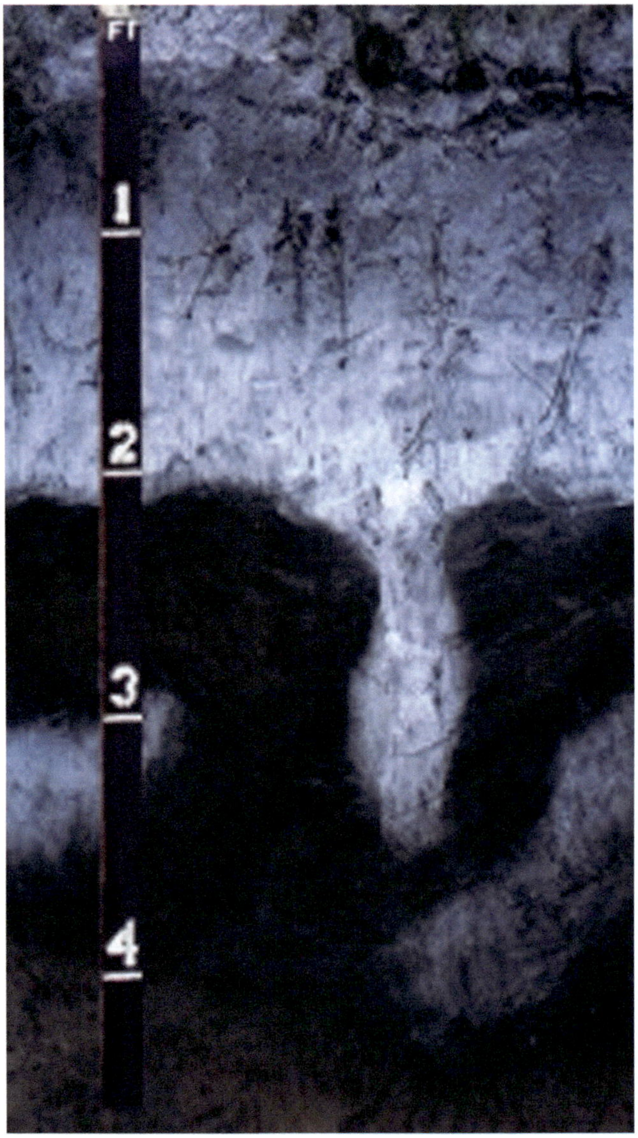

Fig. 12.1 Leon soil series, a sandy, siliceous, thermic Aeric Alaquods, has formed in sandy marine sediments of the Eastern Gulf Coast Flatwoods. This soil has a deep, gray albic horizon with an irregular boundary over a dark brown spodic (Bh) horizon. The scale is in feet. *Source* USDA Natural Resources Conservation Service

Mollisols are cropped for soybeans, small grain, cotton, and corn, used for grazing, or managed for forest products. The Leon Spodosol is in suburban Biloxi and is used for development.

12.4 Summary

Five Histosols, four Mollisols, and one Spodosol account for 941 km^2 in Mississippi. Whereas Mollisols are most common in the Blackland Prairies, Histosols and the Spodosol are most common in the Eastern Gulf Coast Flatwoods. The Histosols are very deep and very poorly drained. The Mollisols range from shallow to very deep and are either well drained or somewhat poorly drained. The Spodosol is very deep and very poorly drained. Histosols are organic soils with a histic epipedon averaging 117 cm in thickness. The Mollisols have a mollic epipedon averaging 32 cm in thickness and, with one exception, lack a diagnostic subsurface horizon. The Leon soil series is a Spodosol with an ochric epipedon over a spodic horizon that is 107 cm thick.

Table 12.1 Analytical properties of a Mollisol, Spodosol, and Histosol found in Mississippi

Horizon[a]	Depth (cm)	Clay (%)	Silt (%)	Sand (%)	SOC (%)	CEC7 ($cmol_c$/kg)	Base sat (%)	pH H_2O	Fe_p (%)	Al_p (%)	1.5 mPa H_2O/ clay	Cole
Catalpa; fine, smectitic, thermic Fluvaquentic Hapludolls; Lee, MS: pedon no. 81P0294												
Ap1	0–13	30.4	44.4	25.2	2.15	27.7	100	7.4			0.44	
Ap2	13–25	30.2	43.5	26.3	1.74	27.2	100	7.7			0.43	0.057
Bt1	25–51	39.6	44.2	16.2	1.19	31.5	100	7.8			0.46	0.071
Bt2	51–107	44.7	44.5	10.8	0.71	30.4	100	7.8			0.40	0.070
Leon; sandy, siliceous, thermic Aeric Alaquods; Santa Rosa, FL; pedon no. S57_046												
A	0–5	2.0	3.1	92.1	1.95	**8.6**	**4**	3.8			0.64	
E/Eg	5–40	0.6	3.5	95.9	0.31	**1.6**	**5**	4.6			0.57	
Bh1	40–53	5.3	7.9	86.8	1.66	**11.0**	**1**	4.4		0.3	0.56	
Bh2	53–64	4.8	5.9	89.3	1.70	**10.0**	**1**	4.8	tr	0.3	0.54	
Bw1	64–81	3.2	5.2	91.6	0.36	**3.8**	**2**	4.9			0.49	
Bw2	81–112	1.2	2.7	96.2	0.40	**1.0**	**3**	4.8				
Eg1	112–142											
Eg2	142–203											
Maurepas; euic, hyperthermic Typic Haplosaprists; Iberia Parish, LA; pedon no. 40A4089												
Oa1	0–15				32.9	103.0		**5.7**				
Oe	15–30				31.8	84.6		**5.6**				
Oa2	30–51				44.9	98.4		**5.9**				
Oa3	51–94				41.1	90.5		**6.1**				
Oa4	94–122				40.8	92.2		**6.1**				
Oa5	122–155				41.1			**6.2**				
Oa6	155–244											
2Ag	244–335											
2Cg	335–366											

[a] Horizons in bold face are diagnostic, including mollic (Ap), spodic (Bh), and histic (O)
Numbers in bold are reported for CEC8 rather than CEC7, base saturation by sum of cations, and pH in $CaCl_2$ rather than H_2O

Reference

Pettry DE, Switzer RE, Hinton RB, Johnson DB (1994) Spodosols: a new soil order recognized in Mississippi. Soil Horizons 35(1):1–6

13 Soil-Forming Processes in Mississippi

13.1 Introduction

Specific soil processes are determined by the soil-forming factors and are expressed in diagnostic horizons, properties, and materials, which are then used to classify soils: soil-forming factors → soil-forming processes → diagnostic horizons, properties, materials → soil taxonomic system Bockheim and Gennadiyev (2000) identified 13 generalized soil-forming processes, of which 12 occur in Mississippi soils. An additional process, cambisolization, is added here and will be defined forthwith. The dominant soil-forming processes in Mississippi are argilluviation, gleization, base cycling, cambisolization, vertization, glossification, fragification, ferralitization (plinthite formation), and humification (Table 13.1). Solodization, paludization, sulfurization, and podsolization occur to a limited extent.

13.2 Argilluviation

Argilluviation (lessivage) refers to the movement and accumulation of clays in the solum. Argilluviation is a dominant process in Alfisols, Ultisols. The evidence for argilluviation in Mississippi soils is the presence of argillans (i.e., clay skins) and abrupt increases in the clay content from the eluvial (A horizon) to the Bt horizon. This process is favored by long duration precipitation, parent materials enriched in carbonate-free clays, stable landscape positions, backslopes rather than eroding shoulders, and a time interval of more than 2000 years (Bockheim and Hartemink 2013). Argilluviation is ubiquitous in Mississippi and occurs in soils of all MLRAs but to a lesser extent in the Blackland Prairies (MLRA 135A).

13.3 Gleization

Gleization (hydromorphism) refers to the presence of aquic conditions as evidenced by reductimorphic or redoximorphic features such as mottles and gleying. In Mississippi, gleization occurs dominantly in Aqu-suborders and aquic and Oxyaquic subgroups of Ultisols, Alfisols, Entisols, Vertisols, and Inceptisols. Gleization in Mississippi is favored by depressions in the landscape, proximity to water and by parent materials that restrict drainage by virtue of texture, a layer that restricts moisture movement. Gleization is most common in soils in of the Southern Mississippi Alluvium (MLRA 131A).

13.4 Base Cycling

Base cycling refers to the ability of the vegetation in cycling base cations such as calcium (Ca), magnesium (Mg), and potassium (K) and maintains moderately acid (pH 5.6–6.0) or slightly acid (pH 6.1–6.5) conditions in soils. Some plant species are notorious base cyclers, such as magnolia, dogwood, and pecan (hickory), yellow poplar, green ash, and baldcypress (Blinn and Buckner 1989; Self and Londo 2019). Others, including all of the southern pines and many of the oaks, initiate and/or perpetuate acidic conditions. Soils with high base conditions include all of the Alfisols (except possibly Ultic Hapludalfs) and Eutric great groups of Inceptisols. Soils of the Southern Mississippi Loess region (MLRA 134) generally have high base cycling rates.

D. Johnson et al., *The Soils of Mississippi*, World Soils Book Series,
https://doi.org/10.1007/978-3-031-36235-4_13

Table 13.1 Quantification of soil-forming processes in Mississippi

Process	Taxa	Order	Sub-order	Great group	Sub-group	Other	Total
Argilluviation	Alfisols, Ultisols	132	0	0	0	0	132
Gleization	Aqu-suborders; Aquic, Oxyaquic subgroups in Alfisols, Entisols, Inceptisols, Ultisols, and Vertisols	0	66	0	53	0	119
Base cycling	Mollisols and Alfisols; Eutr-great groups in Inceptisols	61	0	7	0	0	68
Cambisolization	Cambic horizon in Inceptisols, Mollisols, Vertisols	0	0	0	0	47	47
None	Entisols; lack of diagnostic subsurface horizon in Mollisols and Vertisols	26	0	0	0	9	35
Vertization	Vertisols; Vertic subgroups in Alfisols and Entisols	20	0	0	11	0	31
Glossification	Gloss- and Fragloss-great groups and Glossic subgroups in Alfisols	0	0	12	14	0	26
Fragification	Fragi- and Fragloss-great groups in Alfisols and Ultisols and Frag-subgroups in Ultisols	0	0	18	3	0	21
Ferralitization (plinthite)	Plinthic and Plinthaquic subgroups in Ultisols	0	0	0	9	0	9
Humification	Mollisols; Hum- great group in Inceptisols, Umbr- great group in Ultisols; Cumulic, Rendollic subgroups of Inceptisols	4	0	2	2	0	8
Solodization (natric)	Natr-great groups in Alfisols	0	0	6	0	0	6
Paludization	Histosols	5	0	0	0	0	5
Sulfurization (sulfuric)	Sulf- great groups in Entisols and Histosols	0	0	3	0	0	3
Podzolization	Spodosol	1	0	0	0	0	1
	Total	249	66	48	92	56	455

13.5 Cambisolization

Cambisolization refers to a collection of weak soil-forming processes that leads to the formation of a cambic horizon, which is present largely in Inceptisols, including Dystrudepts, Endoaquepts, Epiaquepts, Eutrudepts, but also in Hapluderts and Dystruderts. Cambic horizons in Mississippi feature processes that do not lead to the significant accumulation of clay or salts. A cambic horizon may form in less than 1000 years on gravelly parent materials (Ciolkosz and Waltman 1995). Cambisolization is important in the Southern Mississippi River Alluvium (MLRA 131A), Gulf Coastal Plain (MLRA 133C), Southern Mississippi Valley Loess (MLRA 134), and Blackland Prairies (MLRA 135A).

13.6 Vertization

Vertization represents a collection of subprocesses occurring in soils with very high amounts of smectitic clay, which enables soils to undergo shrinking and swelling that leads to cracking on the soil surface, tilted, wedge-shaped aggregates, and slickensides on aggregate faces. In Mississippi, vertization is most common in Vertisols but also in Vertic subgroups of Inceptisols and Alfisols. Soil series featuring vertization commonly are in fine or very-fine particle-size classes and the smectitic mineralogy class. Vertization is most common in soils of the Southern Mississippi Valley Alluvium (MLRA 131A) and Blackland Prairies (MLRA 135A).

13.7 Glossification

Glossification refers to conditions whereby an argillic, kandic, or natric horizon is degraded by an advancing albic horizon. This process occurs in Gloss- and Fragloss-great groups and Gloss-subgroups in Alfisols, primarily in the Gulf Coastal Plain (MLRA 133C) and the Southern Mississippi Valley Loess (MLRA 134).

13.8 Fragification

Fragification is the process whereby a reversibly cemented (in water) fragipan develops in the soil, primarily Alfisols and Ultisols. This process occurs mainly in the Gulf Coastal Plain (MLRA 133C) and the Southern Mississippi Valley Loess (MLRA 134).

13.9 Ferralitization

Soils of the inter-tropical regions undergo a series of processes in which in aluminum (Al) and iron (Fe) are concentrated and Si is lost in the profile as a result primary and secondary mineral weathering Duchaufour (1982) envisioned this process as containing three phases, including fersiallitization, ferrallitization, and ferrugination. These three phases are characterized by an increasing degree of weathering of primary minerals, an increasing loss of silicon (Si), and an increased dominance of secondary clays from incongruent dissolution. Ferrallitization occurs in Ultisols of Mississippi and is partially well displayed in those containing plinthite. The Malbis, Atmore, and Saucier soil series are the most extensive soils containing plinthite in the state and occur mainly in the Gulf Coastal Plain (MLRA 133C) and the Eastern Gulf Coast Flatwoods (MLRA 152A).

13.10 Humification

Humification refers to the accumulation of well-humified organic compounds in the upper mineral soil. Soils reflecting humification include the Mollisols, the Hum-great groups in Inceptisols, the Umbr-great group in Ultisols, and Cumulic and Rendollic subgroups of Inceptisols. In Mississippi, humification is favored by grassland vegetation and base-rich parent materials. Common native, warm-season grasses that enhance humification in Mississippi include big bluestem (*Andropogon gerandii*), broomsedge (*Andropogon virginicus*), eastern gamagrass (*Tripsacum dactyloides*), switchgrass (*Panicum virgatum*), little bluestem (*Schizachyrium scoparium*), and indiangrass (*Sorgastrum nutans*) (Hamrick et al. 2022). Humification is important in the Gulf Coastal Plain (MLRA 133C), the Blackland Prairies (MLRA 135A), and the Eastern Gulf Coast Flatwoods (MLRA 152A).

13.11 Solodization

This process involves argilluviation of the dispersed colloids as manifested by the development of an acid A horizon with very little colloidal material over a clay-enriched Btn horizon. Solodization is evident in natric great groups of Alfisols in Mississippi.

13.12 Paludization

This term pertains primarily to the deep (> 40 cm) accumulation of organic matter (histic materials) on the landscape usually in marshy areas. All five soils featuring paludization in Mississippi are in the Histosol order, but soils containing histic materials can occur in other orders. Ripening is a subprocess of paludization and refers to chemical, physical, and biological changes following drainage and aeration of organic materials.

13.13 Sulfurization

Sulfurization is a process whereby pyrite or other iron sulfide minerals undergo oxidation and hydrolysis of the Fe usually following drainage, yielding organic materials that have a pH of 3.5 or less and high concentrations of jarosite, schwertmannie, or other iron and/or aluminum sulfates or hydroxysulfate minerals.

13.14 Podzolization

Podzolization is a complex collection of processes that includes eluviation of base cations, weathering transformation of Fe and Al compounds, mobilization of Fe and Al in surface horizons, and transport of these compounds to the spodic (Bs) horizon as Fe and Al complexes with fulvic acids and other complex polyaromatic compounds. Weaker degrees of podzolization occur in spodic subgroups of Entisols and Andisols. Podzolization is distinguished from ferrallitization in that the Fe and Al complexed with organic acids is transported in the solum, whereas with ferrallitization, the Fe and Al are residual and are accompanied by strong desilication.

13.15 Soils with Minimal Soil-Forming Processes

Approximately 17% of the soil series in Mississippi lack a diagnostic subsurface horizon and evidence of key soil-forming processes. These soils include nearly all Entisols and Histosols and in some Inceptisols, Mollisols, and Vertisols.

13.16 Summary

The dominant soil-forming processes in Mississippi are argilluviation, the transfer of clay into the subsoil; gleization, reducing conditions from restricted drainage; base cycling, the cycling of base cations such as Ca, Mg, and K by the vegetation; cambisolization, the development of B horizons with weak color and structure; vertization, the development of cracking and slickensides in parent materials enriched in smectitic clays, calcification; glossification; fragification; ferralitization; and humification, the accumulation of well-humified materials in the upper mineral soil (Table 13.2).

Table 13.2 Concentration of foliar calcium and soil pH range for major tree species occurring in Mississippi

Species	% Foliar Ca[a]	Soil pH range[b]
S magnolia	2.38	5.0–6.0
Am elm	1.80	5.5–8.0
Dogwood	1.60	5.0–8.0
Pecan	1.58	4.8–7.5
Yellow-poplar	1.50	4.5–7.5
Green ash	1.46	3.6–7.5
Baldcypress	1.37	4.6–7.5
Black cherry	1.36	5.0–6.5
Cherrybark oak	1.20	4.5–6.2
White oak	1.18	4.5–6.2
E cottonwood	1.11	3.6–7.5
Am sycamore	1.10	4.4–7.5
Hackberry	1.06	5.0–7.5
Sweetgum	1.00	3.6–7.5
Am holly	0.97	3.5–6.0
Black tupelo	0.96	4.6–7.0
Nuttall oak	0.93	3.6–6.8
Red maple	0.75	4.4–7.5
Am beech	0.62	5.0–7.5
Longleaf pine	0.33	4.5–7.0
Shortleaf pine	0.23	4.5–7.0
Loblolly pine	0.12	4.5–7.0
Slash pine	0.10	4.5–7.0
Southern live oak		6.0–7.5
Shumard's oak		4.4–6.2
S red oak		5.0–7.0
Water oak		3.6–6.3
Willow oak		3.6–6.3

[a] Blinn and Buckner (1989)
[b] Self and Londo (2019)

References

Blinn CR, Buckner ER (1989) Normal foliar nutrient levels in North American forest trees. Minnesota Gric Exp Sta, Univ of Minnesota, St. Paul MN

Bockheim JG, Gennadiyev AN (2000) The role of soil-forming processes in the definition of taxa in soil taxonomy and the world reference base. Geoderma 95:53–72

Bockheim JG, Hartemink AE (2013) Distribution and classification of soils with clay-enriched horizons in the USA. Geoderma 209–210:133–160

Ciolkosz EJ, Waltman WJ (1995) Cambic horizons in Pennsylvania soils. Pennsylvania State Univ. Agron Ser 133:1–26

Duchaufour P (1982) Pedology: pedogenesis and classification. Allen and Unwin, Boston MA

Hamrick R, Burger LW, Jones J, Strickland B (2022) Native warm-season grass restoration in Mississippi. Mississippi State Univ. Exten Publ 2435

Self B, Londo AJ (2019) Soil pH and tree species suitability in Mississippi. Mississippi State Univ. Exten Publ 2311

Benchmark, Endemic, Rare, and Endangered Soils in Mississippi

14

14.1 Introduction

Benchmark soils are those that (i) have a large extent within one or more MLRAs, (ii) hold a key position in the *Soil Taxonomy*, (iii) have a large amount of data, (iv) have special importance to one or more significant land uses, or (v) are of significant ecological importance. About 6.3% of the soil series in the United States have been designated benchmark soils (Table 14.1).

Endemic soils are defined as the only soil in a family (Bockheim 2005). The proportion of soil series identified in the USA that are endemic is 31%. Rare soils are those with an area less than 10,000 ha (Ditzler 2003). About 62% of the soil series in the USA occupy less than 10,000 ha (100 km^2).

Endangered soils are those that are endemic and rare. The proportion of soil series in the United States that is endangered is 22%. Moreover, there are certainly endemic soils with an area exceeding 10,000 ha that, depending on land use, have become endangered. A list of all benchmark, endemic, rare, and endangered soil series in Mississippi is given in Appendix E.

14.2 Benchmark, Endemic, Rare, and Endangered Soils

About 19% of the soil series in which Mississippi is the lead state have been designated as benchmark soils, which is three times that for the US (Table 14.1). About 34% of the soil series recognized in Mississippi are the only soil in the family and, therefore, may be considered endemic (Table 14.1). This is comparable with the 31% value for all soil series in the US. About 10% of the soil series in Mississippi occur only within the confines of the state, which is substantially lower than for most states. About 46% of the soil series recognized in Mississippi occupy less than 100 km^2 (10,000 ha) each and, therefore, may be considered rare (Table 14.1). This is less than the 62% value for the nation as a whole. About 16% of the soil series in Mississippi are rare and endemic to the state, i.e., are considered "endangered" (Table 14.1). Nearly, one-quarter (22%) of the soil series in the USA are endangered.

14.3 Highly Represented Soil Taxa

About 34% of the soil series in the US with a xeric soil moisture regime occur in Mississippi (Table 14.1). Xer-soil suborders that are dominant in Mississippi include the Xererts (85% all US soil series), Xeralfs (70%), Xerults (69%), Xerepts (48%), Xerands (30%), and Xerolls (24%).

14.4 Summary

About 19% of the soil series in Mississippi have been designated as benchmark soils, because of their large extent, position in Soil Taxonomy, large database, significant land uses, or ecological significance. About one-third (34%) of the soil series recognized in the state are the only soil in the family and, therefore, may be considered endemic. About 10% of the soil series recognized in the state only occur in Mississippi, which is substantially lower than for most states, particularly in the western US. About 46% of the soil series mapped in the state occupy less than 100 km^2 and, therefore, may be considered rare. About 16% of the soil series in the state are rare and endemic and, therefore, may be considered endangered.

D. Johnson et al., *The Soils of Mississippi*, World Soils Book Series,
https://doi.org/10.1007/978-3-031-36235-4_14

Table 14.1 Percentage of benchmark, endemic, rare, and endangered soils in Mississippi and the US

Soil	MS	US
Benchmark	19	6.3
Endemic	34	31
Rare	46	62
Endangered	16	22

References

Bockheim JG (2005) Soil endemism and its relation to soil formation theory. Geoderma 129:109–124

Ditzler C (2003) Endangered soils. National Coop Soil Surv Newsletter No. 25, Nov 2003, pp 1–2

Land Use in Mississippi

15

15.1 Introduction

About 89% of the land in Mississippi is privately owned, with 5% in federal care and 6% managed by state or other public agencies. About 62% of the state is forested (Fig. 15.1), and 89% of the forest area is managed privately. Nearly one-quarter (22%) of the state is in cropland and 12% in pasture. Land use of Mississippi soil series is given in Appendix F. Key environmental issues in Mississippi are soil erosion and water quality, climate warming, and loss of biodiversity.

15.2 Forest Land

Forests cover 77,760 km^2 or 62% of Mississippi (Oswalt, 2013; Mississippi Forestry Commission, 2020). The dominant forest types are loblolly-shortleaf pine (38%), oak-hickory (26%), oak-gum-cypress (13%), oak-pine (10%), longleaf-slash pine (4.4%), and elm-ash-cottonwood (4.0%). The direct economic output from Mississippi's forest products industry was about $3 billion in 2019, including wood furniture (38%), solid wood products (29%), pulp and paper (18%), and logging (13%) (Tanger and Measells, 2020). Wood furniture includes wood windows, doors, kitchen cabinets, countertops, and household and office furniture. Solid wood products include sawmills, wood preservation, veneer and plywood manufacturing, engineered wood member and truss manufacturing, and other products. Logging refers to the economy of cutting and transporting timber and producing wood chips in the field.

Loblolly pine grows throughout the state except for the Mississippi River bottoms and, because of its abundance and wide range of uses, is the principal commercial tree species (Fig. 15.2). The southern or yellow pines, including loblolly pine, shortleaf pine, slash pine, and longleaf pine, are important for saw timber and pulp and paper production. The furniture industry relies primarily on hardwood species, including oaks, sweetgum, yellow poplar, and pecan.

About 34% of the soil series in the state are used mainly for forestry; another 44% are used partially for forestry but also for agriculture and grazing. The latter soils occur primarily in the Southern Mississippi Valley Alluvium (MLRA 131A) and the Alabama and Mississippi Blackland Prairie (MLRA 135A). Mississippi forest soils are primarily Ultisols and Alfisols.

15.3 Cropland

About 22% of the land in Mississippi is used for cropland (Fig. 15.1). The key crops are soybeans, corn, and cotton, followed by hay, sweet potatoes, horticultural crops, rice, wheat, peanuts, oats, pecans, and sorghum. In 2022, Mississippi agriculture attained a $9.7 billion value overall (MSU Extension). The Mississippi Delta (MLRA 131A) is the most important agricultural area in Mississippi (Dabney et al., 2001). Key agricultural soils in the state are the Providence, Loring, Memphis, Sharkey, Ora, Alligator, Forestdale, Savannah, Mantachie, McLaurin, Falaya, Dundee, Collins, Oaklimeter, and Kipling soil series. The three main agricultural crops in Mississippi in terms of economic significance and area are soybeans, corn, and cotton (Table 15.1).

Soybeans are a $1.8 billion crop in Mississippi in 2022. During that same year, over 120 million bushels of soybeans valued at were produced on 3087 farms in Mississippi (Fig. 15.3). The majority of the soybeans in Mississippi are grown in the Delta, especially Bolivar, Sunflower, Washington, Tallahatchie, and Coahoma Counties. The main soils for growing soybeans in Mississippi are the Providence, Loring, Memphis, Sharkey, Ora, Mantachie, Dowling, Falaya, Dundee, and Collins soil series.

D. Johnson et al., *The Soils of Mississippi*, World Soils Book Series,
https://doi.org/10.1007/978-3-031-36235-4_15

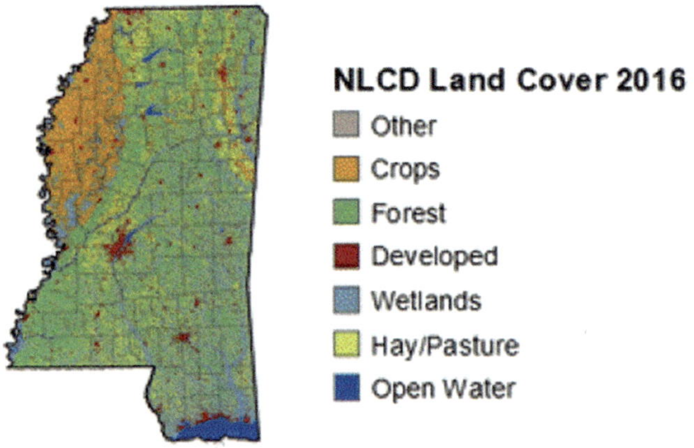

Fig. 15.1 Land use in Mississippi. *Source* USDA Forest Service

In 2022, corn generated $630 million in revenue in Mississippi, with 130 million bushels produced in 1427 farms. Most of the state's corn is grown in the Mississippi Delta, with Washington and Leflore Counties as the leading corn producers (Fig. 15.4). The main soils for growing corn in Mississippi are the Loring, Ruston, Falaya, Dundee, Collins, Oaklimeter, and Bibb soil series.

Cotton was first grown in what is now Mississippi in 1795 in the Spanish-ruled Natchez area. In the middle of the nineteenth century, Mississippi was the leading cotton-producing state in the country. Cotton is inextricably linked with racial injustice in Mississippi and the US. In 2021, cotton generated $558 million, with over 1 billion bales produced on 780 farms. Mississippi ranks third in the US for cotton production. Today, cotton is grown mainly in the Delta region (Fig. 15.5). The main soils for growing cotton in Mississippi are the Providence, Loring, Memphis, Sharkey, Ora, Ruston, Mantachie, McLaurin, Falaya, Dundee, and Collins soil series.

In addition to cotton, soybeans, and corn, other major agricultural crops in Mississippi are hay ($164 million), sweet potatoes ($112 million), horticultural crops ($108 million), rice ($97 million), wheat ($36 million), peanuts ($13 million), oats, pecans ($3.2 million), and sorghum. Sweet potatoes are grown primarily in the Gulf Coastal Plain (MLRA 133C) on the Providence, Falkner, Bude, Dulac, Adaton, Longview, and Mayhew soil series. Rice is grown primarily in the Delta (MLRA 131A) on the Sharkey, Alligator, Forestdale, Dowling, Falaya, Dundee, and Collins soil series. Wheat is grown throughout the state, primarily on the Loring, Ora, Alligator, Savannah, Ruston, Dundee, and Arkabutla soil series.

Mississippi ranks tenth in the country in peanut production, producing 69,700 million pounds annually (Fig. 15.6). The majority of the peanuts are grown in the Gulf Coastal Plain (MLRA 133C) on Udults with a siliceous mineralogy class. The runner variety of peanuts requires soils with a sandy loam texture, primarily for lifting purposes. The main soils for peanut production in Mississippi are the McLaurin, Benndale, Prentiss, Stough, Jena, Heidel, Iuka, Ochlockonee, and Latonia soil series. These soils have a coarse-loamy particle-size class or are in Arenic

Fig. 15.2 Hauling loblolly pine logs near Tishomingo. Photo by J. Bockheim

Table 15.1 Mississippi's top commodities based on cash receipts (Miss. State Univ. Ext. 2021)

Commodity	Cash receipts (million $)
Broilers/eggs	2650
Soybeans	1490
Forest products	1290
Corn	748
Cotton	558
Cattle	400
Catfish	232
Hay	160
Sweet potatoes	110
Specialty crops	108

(50–100 cm of sandy textured surface) or Grossarenic (more than 100 cm of sandy textured surface) subgroups.

The most important soil series for growing oats are the Providence, Memphis, Ora, Savannah, Ruston, Mantachie, Falaya, Dundee, and Collins. Sorghum is grown to a limited extent in Mississippi, mainly in the Delta on the Sharkey, Alligator, Forestdale, Dowling, Dundee, Collins, Dubbs, and Commerce soil series. Pecans (Fig. 15.7) occur naturally in oak-hickory forests of Mississippi but can be grown throughout the state, particularly in the Delta and near the Gulf Coast, on well drained, fertile sites, including the Askew, Bassfield, Bruin, Iuka, Lexington, Longview, Ochlockonee, Ora, Tippah, and Vaiden soil series.

Fig. 15.3 Mississippi soybean harvesting. *Source* Miss. State Univ. Ext. Serv

Fig. 15.4 Corn production in Mississippi. *Source* Miss. State Univ. Ext. Serv

Fig. 15.5 Irrigated cotton production in Leflore County. *Source* USDA Natural Resources Conservation Service

Fig. 15.6 Peanut production in Mississippi. *Source* Miss. Dep. of Agric. & Commerce; mdac.ms.gov

15.4 Pasture Land

About 12% of the land in Mississippi is used for pasture. Pasture land is used mainly for cattle, but also for hogs and free-range chickens. Key pasture soil series in the state are the Loring, Memphis, Ora, Savannah, Ruston, Mantachie, McLaurin, Falaya, Dundee, Collins, and Oaklimeter. In 2022, poultry and egg production is a $3.84 billion industry in Mississippi. Approximately, 737 million broilers were produced in 2021 on 1,237 broiler farms. In 2018, only 16% of the poultry operations were cage-free (free-range) in Mississippi. In 2022, livestock generated $456 million in revenue in Mississippi, with 920,000 head of cattle and calves produced on 15,890 farms.

15.5 Developed Land and Wildlife, Watershed, and Recreation Lands

About 2% of Mississippi's area in 2016 had been developed (Fig. 15.1). Developed land includes soils in urban and suburban areas. Soils that are not used extensively for crops, grazing, forestry, or development are used for wildlife habitat, watershed protection, and recreation. Only 4.7% of the soil series of Mississippi fall into this category.

15.6 Key Environmental Issues

It can be debated which are the most important environmental issues facing Mississippi residents and land managers. However, soil erosion and water quality, climate change, and loss of biodiversity appear to be key concerns in the state.

Fig. 15.7 Pecan orchard near Starkville. Photo by Kat Lawrence Miss. State Univ

15.6.1 Soil Erosion and Water Quality

The Great Mississippi Flood of 1927

The great Mississippi flood of 1927 was the most destructive river flood in the history of the United States and significantly affected the soils of the lower Mississippi River Valley. Approximately 70,000 km^2 (17.3 million acres) were inundated at depths of up to 9 m (30 ft). The uninflated cost of damage is estimated to be between $246 million and $1 billion (Watkins 1997).

About 500 people died and over 630,000 people were directly affected. More than 200,000 African Americans were displaced from their homes along the Lower Mississippi River and had to live for lengthy periods in relief camps. As a result of this disruption, many joined the Great Migration from the South to the industrial cities of the North and Midwest (Hornbeck and Naidu 2014).

Several musicians mentioned the 1927 flood in their music. Among these are "Backwater Blues" by Bessie Smith, "When the Levee Breaks" by Memphis Minnie and covered by Led Zeppelin, "Down in the Flood" by Bob Dylan and "High Water Everywhere" by Charley Patton.

Accelerated soil erosion and degradation of water quality occur throughout Mississippi, particularly in the Delta region (Mississippi Department of Environmental Quality 2022a, b). The issue of erosion in Mississippi from historical and cultural perspectives has been addressed by Day and Erdman (2017) and Mandelman (2020). In the US, the average rate of sheet and rill erosion cropland declined from 7.1 to 4.6 t/ac (16–10 T/ha) over the period 1992 to 2017, a 35% reduction (US Department of Agriculture 2020). In 2017, the average rate of erosion on cultivated cropland in the Mississippi Delta region was 4.5 t/ac (10 T/ha), compared with 3.0 t/ac (6.7 T/ha) for the US as a whole (Table 15.2). The greater water erosion rate in Mississippi than the US is due to the silty nature of many of the soils which

Table 15.2 Average annual sheet and rill erosion on non-federal rural land in Mississippi and the US in 2017 (tons/acre/year)

Treatment	Mississippi	US
Cultivated crops	4.52	3.00
Non-cultivated crops	0.97	0.69
Total crops	4.25	2.67
Pasture	1.27	0.61

Source National Resources Inventory Summary Report, 2017

affects its erodibility, high intensity and duration of rainfall, moderately steep slopes in many regions, crop type and tillage practice, and supporting management practices.

Accelerated soil erosion results in (i) a loss in arable land and crop productivity, (ii) a decline in water quality, (iii) increased flooding, and (iv) damage to dams and levees. Sediment, nutrients, and pesticides from soil erosion have impacted surface water quality in Mississippi, particularly in the Mississippi Delta (Locke 2004), the northern part of the southern Mississippi loess region, and along the Gulf Coast. Rapid subsidence and collapse have resulted in wetland loss in the Mississippi Delta Plain (Morton et al. 2005).

15.6.2 Climate Change

According to the US Environmental Protection Agency (2016), unlike most of the nation, Mississippi has not become warmer during the last 50–100 years. However, soils have become drier, annual rainfall has increased, and more rain arrives in heavy downpours. There is an increase risk in droughts and fires in the state. The changing climate likely will increase the frequency of severe tropical cyclones, such as Hurricanes Katrina and Rita in 2005 (Becker et al. 2014). Based on the historical relation between climate and growth and General Circulation Models (GCMs), a 7.2 °F (4 °C) increase in air temperature and reduced precipitation would result in a 9% reduction in cotton yield in the Mississippi Delta region (Reddy et al. 2002). Global warming is associated with a 35-cm rise in sea level (EPA 2016) and a loss of 1.2 million acres (4800 km^2) of wetlands (Day et al. 2005) along the Gulf Coast during the past century. Rising sea level along the Mississippi coast has caused submerging of wetlands and dry land, eroded beaches and barrier islands, threatened levees, and poses a threat to coastal communities along Mississippi Sound.

15.6.3 Loss of Biodiversity

A loss in biodiversity of plants in terrestrial ecosystems is of concern in Mississippi. These losses have resulted from land use and climate changes. Some examples include the loss of wetlands in the Delta (Day et al. 2019), bottomland hardwood forests, especially in the Delta (Forest Stewards Guild 2016; Stanturf et al. 2000), prairie belts to agriculture (Barone 2005), pine savannas from fire and grazing (Fowler and Beckage 2020), and Gulf Coast ecosystems from development, oil spills, and land subsidence.

15.7 Summary

About 89% of the land in Mississippi is privately owned. The land cover in Mississippi is 62% forest, 22% cropland, and 12% pasture. The dominant forest type is loblolly-shortleaf pine, followed by oak-hickory, oak-gum-cypress, oak-pine, longleaf-slash pine, and elm-ash-cottonwood. The major forest product is wood furniture, followed by solid wood products (29%) and pulp and paper (18%). Key agricultural crops are soybeans, corn, and cotton, followed by hay, sweet potatoes, horticultural crops, rice, wheat, peanuts, oats, pecans, and sorghum. Pasture land is used for cattle and calves, hogs, and free-range poultry. Major environmental issues in Mississippi include soil erosion and water quality, climate change, and loss of biodiversity.

References

Barone JA (2005) Historical presence and distribution of prairies in the Black Belt of Mississippi and Alabama. Castanea 70:170–183

Becker A, Matson P, Fischer M, Mastrandrea M (2014) Towards seaport resilience for climate change adaptation: stakeholder perceptions of hurricane impacts in Gulfport (MS) and Providence (RI). Progress in Planning. Available at https://doi.org/10.1016/j.progress.2013.11.002.

Dabney SM, Rebich RA, Pote JW (2001) The Mississippi Delta MSEA program, pp 1094–1110. In: Stott DE, Mohtar RH, Steinhardt GC (eds) Sustaining the global farm. Selected papers 10th Internat. Soil Conserv Org Meeting 24–29 May, 1999 Purdue Univ

Day JW, Barras J, Clairain, E, Johnston J, 8 co-authors (2005) Implications of global climatic change energy cost and availability for the restoration of the Mississippi delta. Ecol Engineer 24:253–265

Day JW, Erdman JA (2017) Mississippi River restoration: pathways to a sustainable future. Springer Nature, Switzerland

Day JW, Shaffer GP, Cahoon DR, DeLaune RD (2019) Canals, backfilling and wetland loss in the Mississippi Delta. Estuar Coast Shelf Sci 227:106325

EPA (2016) What climate change means for Mississippi. U.S. Envir. Prot. Agency. EPA 430-F-16-026

Fowler NL, Beckage B (2020) Savannas of North America, pp 123–150. In: Scogings PF, Sankaran M (eds) Woody plants and large herbivores. Wiley

Hornbeck R, Naidu S (2014) When the levee breaks, Black migration and economic development in the American South. Am Econ Rev 104:964–990

Locke MA (2004) Mississippi Delta management systems evaluation area: overview of water quality issues on a watershed scale, pp. 1–15. In: Nett et al (eds) Water Quality Assessments in the Mississippi Delta. ACS Symp. Ser. Am. Chem. Society. Washington, D.C.

Mandelman A (2020) The place with no edge: an intimate history of people, technology, and the Mississippi River Delta. Louisiana State Univ Press

Mississippi Forestry Commission (2020) Mississippi's Forest Action Plan 2020. Mississippi Forestry Commission, Jackson, MS

Mississippi Department of Agriculture and Commerce (2021) Annual Report. (http://mdac.ms.gov/agency-info/mississippi-agriculture)

Mississippi Department of Environmental Quality (2022a) State of Mississippi water quality assessment 2022a section 305 (b) report. (https://www.mdeq.ms.gov/water/field-services/water-quality-assessment)

Mississippi Department of Environment Quality (2022b) Mississippi 2022b list of impaired water bodies. (https://www.mdeq.ms.gov/)

Morton RA, Bernier JC, Barras JA, Ferina NF (2005) Historical subsidence and wetland loess in the Mississippi Delta plain. Gulf Coast Assoc Geol Soc Trans 55:555–571

Oswalt SN (2013) Mississippi's Forests, 2013. US For Serv, SO Res Sta, Resour Bull SRS-204

Reddy KR, Doma PR, Mearns LO, Boone MYL, Hodges HF, Richardson AG, Kakani VG (2002) Simulating the impacts of climate change on cotton production in the Mississippi Delta. Climate Res 22:271–281

Stanturf JA, Gardiner ES, Hamel PB, Devall MS, Leininger TD, Warren Jr ME (2000) Bottomland hardwood ecosystems in the lower Mississippi alluvial valley. J For 98:10–16

Tanger SM, Measells MK (2020) The economic contributions of forestry and forest products. Mississippi Miss State Univ Ext

U.S. Department of Agriculture (2020) Summary report: 2017 national resources inventory, Natural Resources Conservation Service, Washington, DC, and Center for Survey Statistics and Methodology, Iowa State University, Ames, Iowa. https://www.nrcs.usda.gov/wps/portal/nrcs/main/national/technical/nra/nri/results/

Watkins TH (1997) Boiling over. The New York Times April 13, p 34

16 Yield Potential of Mississippi Soils

16.1 Introduction

This chapter focuses on the yield potential of Mississippi soils for agricultural crops, pastures, and forests. For croplands and pastures, yield potential is based on yields reported in county soil survey reports and Land Capability Classes (LCC) reported in county soil survey reports and the Web Soil Survey. The LCCs are grouped on the basis of their capability to produce common cultivated crops and pasture plants without deteriorating over time, with or without irrigation. Forest site quality is based on site index data of key indicator tree species that are provided in archived soil survey reports and in Web Soil Survey. Since many of the crop and forest yield data are outdated and considerable soil and site changes have been made over time, these data may not always be reliable.

16.2 Crop Yield Potential

As pointed out in Chap. 15, there are a variety of crops produced in Mississippi. The most common crops are soybeans, corn, and cotton; however, oats, wheat, rice, sweet potatoes, and sorghum grain are also commonly produced in the state. In archived county soil survey reports, yields for specific crops are given in bushels per acre (soybeans, corn, rice, sweet potatoes, sorghum, oats, and wheat) or for cotton lint in lbs/acre. Archived county soil survey reports and Web Soil Survey provide the Land Capability Class (LCC) for soil map units.

The land capability classification (LCC) is a system of grouping soils primarily on the basis of their capability to produce common cultivated crops and pasture plants without deteriorating over a long period of time (Quandt 2020). The classification has two categories: capability class and capability subclass. Capability class is the broadest category in the land capability classification system. Class codes 1, 2, 3, 4, 5, 6, 7, and 8 represent Land Capability Classes. Class 1 soils have few limitations restricting their use. Class 2 soils have moderate limitations that reduce the choice of plants or require moderate conservation practices. Class 3 soils have severe limitations that reduce the choice of plants or require special conservation practices, or both. Class 4 soils have very severe limitations that restrict the choice of plants or require very careful management, or both. Class 5 soils have little or no hazard of erosion but have other limitations, impractical to remove that limit their use mainly to pasture, range, forestland, or wildlife food and cover. Class 6 soils have severe limitations that make them generally unsuited to cultivation and that limit their use mainly to pasture, range, forestland, or wildlife food and cover. Class 7 soils have very severe limitations that make them unsuited to cultivation and that restrict their use mainly to grazing, forestland, or wildlife. Class 8 includes miscellaneous areas that have limitations precluding their use for commercial plant production and limiting their use for recreation, wildlife, or water supply.

Capability subclass is the second category in the land capability classification system. The subclass represents the dominant limitation that determines the capability class. Class codes e, w, s, and c are used for land capability subclasses. Within a capability class, where the kinds of limitations are essentially equal, the subclasses have the following priority: e, w, s, and c. Subclass e is composed of soils for which the susceptibility to erosion is the dominant problem or hazard affecting their use. Erosion susceptibility and past erosion damage are the major soil factors that affect soils in this subclass. Subclass w is composed of soils for which excess water is the dominant hazard or limitation affecting their use. Poor soil drainage, wetness, a high water table, and overflow are the factors that affect soils in this subclass. Subclass s is comprised of soils that have soil limitations within the rooting zone, such as shallowness of the rooting zone, stones, low moisture-holding capacity, low fertility that is difficult to correct, and salinity or sodium content. Subclass c contains soils for which the climate

D. Johnson et al., *The Soils of Mississippi*, World Soils Book Series,
https://doi.org/10.1007/978-3-031-36235-4_16

(the temperature or lack of moisture) is the major hazard or limitation affecting their use. Subclasses are not assigned to soils or miscellaneous areas in capability classes 1 and 8.

Land Capability Classes and subclasses are reported for detailed soil map units, both managed and unmanaged. Therefore, a range of LCCs may be reported for a soil series occurring on a variety of slopes and having undergone a variety of management histories.

Land use capabilities of all Mississippi soil series are provided in Appendix F. Nearly, one-half (45%) of Mississippi's soil series, accounting for 32% of the soil area, are "high-quality soils," i.e., are in Land Capability Classes 1 and 2 (Fig. 16.1). For the US as a whole, 23% of the soil map units are in LCCs 1 and 2 (Natural Resources Conservation Service 1997). Mississippi is ranked 15th in the nation in proportion of high-quality soils. These soils are located primarily in the Southern Mississippi Valley Alluvium (MLRA 131A) and the Alabama and Mississippi Blackland Prairie (MLRA 135A) (Fig. 16.2). These areas have been designated the US Department of Agriculture as "prime farmland," because they contain the best combination of physical and chemical characteristics for producing food, feed, forage, and fiber, and oilseed crops.

Of the 232 soil series in Mississippi, 173 (75%) are used to some extent for agricultural crops (Appendix F). The most productive extensive soil series for agricultural crops are in the Udifluvents, Paleudults, Hapludalfs, Endoaquepts, Eutrudepts, and Dystrudepts great groups (Table 16.1). These soils tend to be deep to very deep, well drained to somewhat poorly drained and in the fine-silty, fine-loamy, or the fine particle-size class, the mixed or siliceous mineralogy class, and the active or semiactive cation-exchange activity class.

Table 16.2 shows yields for common crops grown in Mississippi for the best LCC specified for a soil map unit within a particular soil series without irrigation. The most productive soils for soybeans are the Dubbs, Collins, Commerce, Morganfield, Sharkey, and Vaiden series. The Morganfield, Commerce, Dubbs, Collins, Ariel, Cascilla, Iuka, Mantachie, and Vicksburg soil series are the top cotton-producing soil series. For corn, the top soils are the Morganfield, Malbis, Vicksburg, Collins, Ariel, Cascilla, Iuka, Adler, Jena, Ochlockonee, and Bibb series. Rice is restricted to 11 soil series, with the Sharkey, Forestdale, and Alligator being most productive. Sweet potatoes are grown mainly on seven soil series, including the Falkner, Providence, Bude, Dulac, Longview, Adaton, and Mayhew. Sorghum is most productive on the Dundee, Dubbs, Commerce, Collins, Sharkey, Alligator, Forestdale soil series. The best-yielding soils for oats are the Providence, Iuka, Ochlockonee, Cahaba, Ruston, Ora, Prentiss, Savannah, Stough, and Shubuta series. High-yield soils for wheat include the Cahaba, Ora, Dubbs, Adler, Morganfield, Arkabutla, Commerce, and Tensas series.

Oldham (2012) provided comprehensive management guidelines for agronomic crops grown in Mississippi.

16.3 Grazing Land Quality

Of the 232 soil series in Mississippi, 186 (80%) are used to some extent for grazing (Appendix F). Cattle and poultry are important agricultural commodities in Mississippi. Major forage crops for cattle include barley, alfalfa, clover, and a number of warm-season grasses, including sudangrass, millet, sorghum, bahaiagrass, dallisgrass, and bermudagrass. Oldham (2012) gives a comprehensive treatment management guidelines for pasture crops grown in

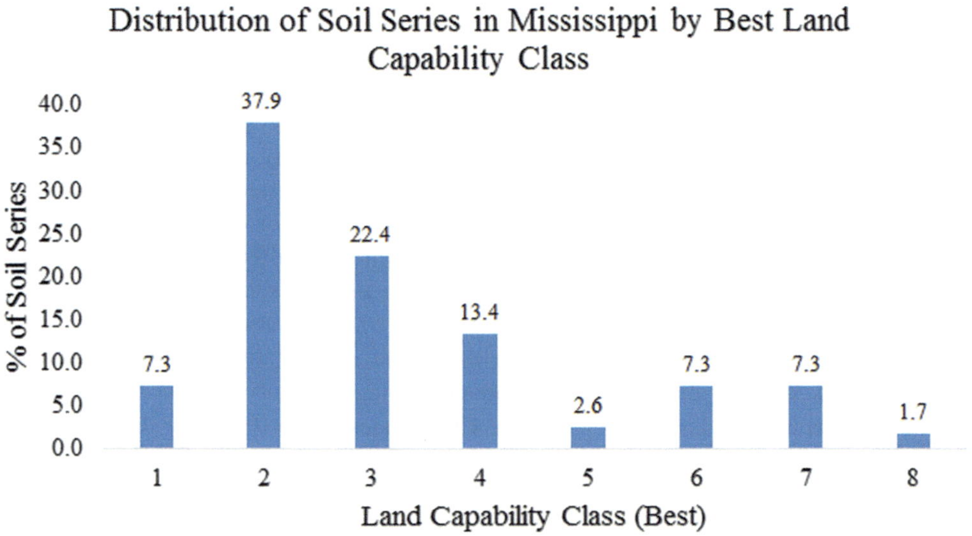

Fig. 16.1 Distribution of soil series in Mississippi by best land capability class

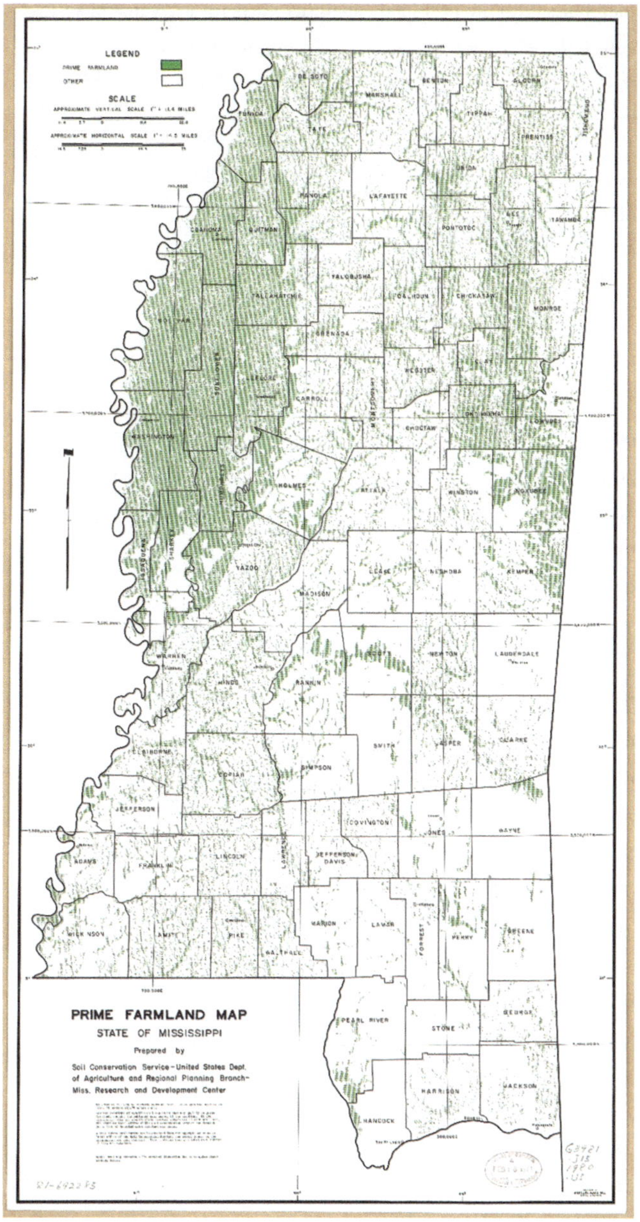

Fig. 16.2 Prime farmland map of Mississippi. *Source* USDA Soil Conservation Service 1980

Mississippi. Most of the broiler operations in Mississippi involve caged chickens in houses. These operations use specialized feeds that rarely are generated onsite. Less than 16% of the broiler farms in Mississippi are free-range operations.

Table 16.1 Average maximum land capability class for soil great groups in Mississippi

Great group	Land capability class (best)
Udifluvents	2.0 ± 0.71
Glossudalfs	2.0 ± 0.82
Fragiudults	2.1 ± 0.35
Dystrudepts	2.1 ± 0.74
Fragiudalfs	2.2 ± 0.45
Fraglossudalfs	2.4 ± 0.89
Eutrudepts	2.4 ± 1.7
Endoaqualfs	2.5 ± 0.58
Epiaquepts	2.6 ± 0.53
Fluvaquents	2.6 ± 1.3
Paleudults	2.6 ± 1.4
Paleudalfs	2.8 ± 1.4
Endoaquepts	2.9 ± 1.2
Hapluderts	2.9 ± 1.6
Epiaquerts	3.0 ± 1.0
Dystruderts	3.2 ± 0.41
Dystraquerts	3.2 ± 0.50
Hapludalfs	3.2 ± 2.1
Glossaqualfs	3.4 ± 0.55
Natraqualfs	3.5 ± 0.55
Paleaquults	3.9 ± 0.99
Hapludults	4.3 ± 2.1
Quartzipsamments	5.0 ± 2.1
Haplosaprists	7.2 ± 0.50

16.4 Forest Site Quality

There are 210 soil series in Mississippi that are partially or wholly managed for forestry. There is a strong correlation between forest type and soil taxon in Mississippi. In loblolly pine-shortleaf pine, which is the dominant forest type in Mississippi, the primary great groups are Hapludults, Paleudults, and Paleudalfs. In oak-hickory forests, Hapludalfs are most common. Oak-gum-cypress forests contain mainly Fluventic and Fluvaquentic Dystrudepts, Endoaquepts, Fluvaquents, and Haplosaprists. Oak-pine forests are commonly underlain by Hapludalfs, Hapludults, and Fraglossudalfs. Longleaf pine-slash pine

Table 16.2 Some yields[a] for common crops grown in Mississippi on soil map units with the best land capability class for a given soil series

Soil series	MLRA	Land Capability Class	Soybeans (bu/ac)	Cotton lint (lb/ac)	Corn (bu/ac)	Rice (bu/ac)	Sweet potatoes (bu/ac)	Sorghum (bu/ac)	Oats (bu/ac)	Wheat (bu/ac)
Adaton	133C	3w	30	550	70		225		50	
Adler	134	2w	35	825	110				80	50
Alaga	133C, 152A	3s	20	450	60				45	30
Alligator	131A	3e, 3w	35	600	60	130		50	60	45
Almo	133C, 134	3w	27		50		225			
Angie	133C	2e	25		50				50	
Annamaine	133C	2w	40	800	100					40
Ariel	134	2w	40	850	110					40
Arkabutla	134	2w	40	750	100				75	50
Askew	131A	2e	35	700	80					40
Atmore	133C, 152A	4w	20	400	40					30
Atwood	133C, 134	2e	40	775	95				80	
Bama	133C	1		800	90					
Basin	133C, 152A	2w	35	450	75				65	
Bassfield	133C	2s, 2e	30	650	75					35
Baxterville	133C	2e	30	715	75				65	
Bayou	133C, 152A	4w			40					
Beauregard	133C, 152A	2w	25		65					
Belden	133C, 135A	2w	35	650	80					
Benndale	133C	2e	30	750	80				60	40
Beulah	131A	2s	35	600	75				60	45
Bibb	133C, 152A	5w	35	650	110					40
Bigbee	133C	5w	25	450	55					30
Bonn	133C, 134	4s	15		20					
Boswell	133C	3e	25	450	50				60	
Bowdre	131A	2w	35	625	45				50	
Brewton	133C	3w	35	650	60					
Brooksville	135A	2e	35	700	60				80	45
Bruin	131A	1	35	825	85				60	
Bruno	131A	3s	25	400	50				30	30
Bude	133C, 134	2w	25	625	85		250		60	35
Byram	134	2ee	30	700	85					
Cahaba	133C	2e	40	800	100				100	50
Caledonia	133C	1	40	750	100					
Calhoun	131A, 134	3w	25	475	50				50	30
Calloway	134	2e	40	750	95				65	35
Cascilla	134	1	40	850	110				90	38
Catalpa	135A	2w	40	750	90				80	45
Chastain	133C	3w	27	500	50				60	
Chenneby	133C, 134	2w	35	700	100					30
Collins	131A, 133C, 134	2w	45	850	110	55		60	80	40
Columbus	133C	2w	30	650	90					35

(continued)

Table 16.2 (continued)

Soil series	MLRA	Land Capability Class	Soybeans (bu/ac)	Cotton lint (lb/ac)	Corn (bu/ac)	Rice (bu/ac)	Sweet potatoes (bu/ac)	Sorghum (bu/ac)	Oats (bu/ac)	Wheat (bu/ac)
Commerce	131A	2w, 2e	45	900	100			65	70	50
Convent	131A	3w	35	700	90				70	
Crevasse	131A	5w	40	300	20				20	
Daleville	133C	5w	25							
Darco	133C	3s	15		45				45	
Deerford	134	3w	30	475						
Dexter	133C, 134	1	35	825	100				65	
Dogue	133C	2e	45		125					
Dowling	131A	4w	40	350	40	90		25	45	5
Dubbs	131A	1	55	850	105			65	80	50
Dulac	133C	3e	35	700	75		250		80	40
Dundee	131A	2e, 2w	40	800	100	55		65	80	45
Escambia	133C, 152A	2w	40	600	100					
Eustis	133C	7e, 7s	25	350	60				50	
Falaya	131A, 133C, 134	2w	40	750	100	70			80	35
Falkner	133C, 134	2w	35	625	75		300		60	35
Forestdale	131A	3w	35	675	80	130		45	60	40
Freeland	133C	4e	25	700	75		250		80	
Freest	133C, 135A	2e	35	650	80					35
Freestone	133C	2w	22	450	45				50	
Frizzell	133C	2e	30	430	55					
Frost	134	4w			40				25	
Gillsburg	133C, 134	2w	40	750	100					35
Greenville	133C	1	30	800	65				20	
Grenada	134	2e	40	700	95				80	40
Griffith	135A	2w	40	750	85				80	35
Guyton	133C, 152A	4w	23	400	50					
Harleston	133C	2w	40	750	100				65	35
Hatchie	133C	2w	25	600	65		300		70	
Heidel	133C	6e	20		60					
Henry	134	3w	35	625	70				50	30
Houlka	133C, 135A	2w	40	725	80				50	40
Houston	135A	2e, 2s	35	700	75				80	
Ichusa	133C, 135A	3e	25	550						
Irvington	133C, 152A	2e	35	650	80					45
Iuka	133C, 152A	2w	40	850	110				100	
Jena	133C	2w	40	750	110				55	40
Johns	133C	2w		650	120					
Johnston	133C	4w	40		80				90	
Kipling	135A	3w	35	550	70				55	35
Kirkville	133C	2w	40	700	100					40
Kolin	133C	4e	30	650	75					40

(continued)

Table 16.2 (continued)

Soil series	MLRA	Land Capability Class	Soybeans (bu/ac)	Cotton lint (lb/ac)	Corn (bu/ac)	Rice (bu/ac)	Sweet potatoes (bu/ac)	Sorghum (bu/ac)	Oats (bu/ac)	Wheat (bu/ac)
Latonia	133C, 152A	2w, 2s	25	750	80				65	35
Lax	133C, 134	2e	30	500	70					
Leaf	133C, 152A	3w	35							
Leeper	133C, 135A	2w	40	750	90				75	40
Lenoir	133C	3w		525	100					
Leverett	134	1	40	800	90				75	40
Lexington	133C, 134	3e	35	750	90					28
Longview	133C	2e	40	650	85		225		55	35
Loring	134	2e	40	750	100				72	45
Lorman	133C, 134	6e	20							
Lucedale	133C	1	40	750	80				64	50
Lucy	133C	2e	33	650	80					
Luverne	133C	4e	30	600	95				60	
Maben	133C	7e	25	550	40				45	30
Malbis	133C	1	40	800	120					35
Mantachie	133C	2w	40	850	90				80	40
Marietta	133C, 135A	2w	40	750	100				70	40
Mashulaville	133C	3w	20							
Mathiston	133C	2w	35	700	95					
Mayhew	133C, 135A	3e	30	500	70		225			30
Maytag	133C, 135A	4e	25		75					
McLaurin	133C	2e	30	750	85				65	35
McRaven	134	2w	35	700	100					
Memphis	134	2e	40	825	95				80	40
Mhoon	131A	2w	45	500	80	75			60	
Mooreville	133C, 135A	2w	35	750	90					35
Morganfield	134	2w	45	950	125				75	50
Neshoba	133C	2e	35	650	75					
Newellton	131A	3e	30	600	45				50	
Oaklimeter	131A, 133C, 134	2w	40	750	105					40
Ocilla	133C	4w	35		70					
Ochlockonee	133C, 152A	2w	40	750	110				100	40
Okolona	135A	2e	35	700	60				80	45
Oktibbeha	133C, 135A	3e	35	600	60				55	35
Olivier	134	2e, 2w	22	375	60				50	
Olla	133C	7e, 7s	25	600	70					
Ora	133C	2e	40	750	85				100	50
Orangeburg	133C	7e		650	80				100	
Ozan	133C	3w	20							
Paden	133C	2e, 2w	35	700	80				65	
Pelahatchie	135A	2e, 2w	35	700						40
Pheba	133C	3w	30	575	75				80	

(continued)

Table 16.2 (continued)

Soil series	MLRA	Land Capability Class	Soybeans (bu/ac)	Cotton lint (lb/ac)	Corn (bu/ac)	Rice (bu/ac)	Sweet potatoes (bu/ac)	Sorghum (bu/ac)	Oats (bu/ac)	Wheat (bu/ac)
Poarch	133C, 152A	1	30	650	90					
Pooleville	133C, 134	2w	30	625	80					
Prentiss	133C	2e, 2w	35	750	85				100	40
Providence	133C, 134	2e	40	800	90		250		100	40
Quitman	133C	2e, 2w	30	650	80				55	35
Richland	134	3e	25	500	50				60	
Riedtown	134	1	40	750	100					
Robinsonville	131A, 135A	1	42	825	115				75	
Rosebloom	134	3w	30	550	60	65			55	25
Rosella	133C	3w	30	500	70					
Ruston	133C	2e	30	800	100				100	45
Saffell	133C, 134	7e	30	500	60				60	
Saucier	133C, 152A	3e	35	650	90					
Savannah	133C	2ee	35	700	85				100	45
Sessum	135A	4w	25						45	25
Sharkey	131A	3e, 3w	45	800	85	130		50	65	32
Shubuta	133C	2e	30	600	70				100	
Silverdale	131A	3s	25	400	135					
Smithdale	133C	4e	30	600	65					30
Smithton	133C	4w	24							
Steens	133C	2w	30		75					
Stough	133C, 152A	2w	25	775	80				100	35
Suffolk	133C	1		800	115					
Sumter	133C, 135A	3e	30	450	50				70	
Susquehanna	133C	6e	20	400	40					30
Sweatman	133C	6e	20	400	50					30
Talla	133C, 134	2w	25	650	70					20
Tensas	131A	3w	40	600						50
Tippah	133C, 134	2e	35	650	80				70	35
Tippo	131A	2w	30	625	80				55	40
Trebloc	133C	5w	25	550	60					32
Trinity	133C	2w	30	750	90				80	
Troup	133C	6s	22	450	60					
Tunica	131A	3w	40	650	70	70			60	30
Tuscumbia	135A	3w	30		70					
Tutwiler	131A	1, 2e	35	850	140				55	35
Una	133C, 135A	3w	25							
Urbo	133C, 135A	2w	35	750	95				80	40
Vaiden	133C, 135A	3e	45	600	50				60	30
Vancleave	133C, 152A	2w		600	80					
Velda	133C, 134	2w	40	750	100					
Verdun	134	4s	20							
Vicksburg	131A, 134	1	40	850	115				80	40

(continued)

Table 16.2 (continued)

Soil series	MLRA	Land Capability Class	Soybeans (bu/ac)	Cotton lint (lb/ac)	Corn (bu/ac)	Rice (bu/ac)	Sweet potatoes (bu/ac)	Sorghum (bu/ac)	Oats (bu/ac)	Wheat (bu/ac)
Vimville	133C	3w	30							25
Wadley	133C, 152A	3s		500	60					
Waverly	134	3w	30	475	55	70			60	
Wilcox	133C, 135A	3e, 3w	25	400	40				35	30

Source Pettiet (1974), Pettry and Koos (1980); county soil survey reports for Mississippi
[a] The yields for common crops are taken from archived soil survey reports, the average and medium of which is 46 years. Additionally, the data do not reflect positive and negative changes that have been made in the soil and site

Table 16.3 Mississippi's most productive forest soil great groups based on site index (base age = 50 years)

Species	High site index	Major soil great groups
Green ash	≥ 90	Epiaquepts, Endoaquepts, Eutrudepts
Sweetgum	≥ 100	Dystrudepts, Eutrudepts, Udifluvents, Endoaquepts
Yellow poplar	≥ 100	Dystrudepts, Eutrudepts, Udifluvents
Cherrybark oak	≥ 90	Dystrudepts, Endoaquepts, Paleudalfs, Hapludalfs, Epiaquepts
	≥ 95	Endoaquepts, Dystrudepts, Eutrudepts
Water oak	≥ 90	Hapludalfs, Paleudalfs, Epiaquerts
Shumard's oak	≥ 95	Endoaquepts, Epiaquepts
Nuttall oak	≥ 90	Endoaquepts, Epiaquepts, Hapludalfs, Eutrudepts, Dystrudepts
Willow oak	≥ 75	Fragiudults, Fluvaquents, Hapludalfs, Dystruderts
Southern red oak	≥ 110	Endoaquepts, Udifluvents, Eutrudepts
Eastern cottonwood	≥ 105	Eutrudepts
American sycamore	≥ 75	Paleudults, Fragiudults
Longleaf pine	≥ 80	Paleudalfs, Paleudults, Hapludults, Glossaqualfs
Shortleaf pine	≥ 90	Paleudults, Fragiudults, Hapludults, Udifluvents, Glossaqualfs
Slash pine	≥ 90	Paleudults, Dystrudepts, Paleudalfs, Hapludults, Paleaquults, Udifluvents, Dystruderts
Loblolly pine	> 90	Paleudults, Dystrudepts, Paleudalfs, Hapludults, Paleaquults, Udifluvents, Dystruderts

Sources Broadfoot (1969), Clatterbuck (1987), Carmean et al. (1989)

forests most commonly occur on Paleudults, Hapludults, and Quartzipsamments. Elm-ash-cottonwood lowlands generally feature Endoaquepts, Aquic Udifluvents, and Epiaquepts.

Based on site index, the most productive soil series for loblolly, longleaf, shortleaf, and slash pine are in the Paleudults, Fragiudults, Paleudalfs, and Hapludults great groups (Table 16.3). For bottomland hardwoods (green ash, sweetgum, cherrybark oak, and yellow poplar), the most productive soil series are in the Eutrudepts, Endoaquepts, Dystrudepts, Epiaquepts, and Udifluvents great groups.

The most productive soil series occupying 50 km^2 or more for loblolly pine-shortleaf pine are the Freest, Harleston, Heidel, Louin, Neshoba, Petal, and Quitman soil series (Table 16.4). For oak-hickory, productive soil series include the Brantley, Bruin, Calhoun, Ichusa, Iuka, Kinston, Newellton, Ochlockonee, and Urbo soil series. The most productive soil series for oak-gum-cypress are the Ariel, Arkabutla, Bowdre, Cascilla, Chenneby, Dundee, Forestdale, Gillsburg, Guyton, Johnston, Mathiston, Ochlockonee, and Tensas. The most productive oak-pine forests occur on the Bassfield, Bruno, Bude, Columbus, Hatchie, Jena, Kipling, Kirkville, Leverett, Mayhew, Memphis, Natchez, Nugent, Oaklimiter, Okeelala, and Prentiss. High-quality longleaf pine-slash pine soil series include the Atmore, Basin, Benndale, Escambia, Latonia,

Table 16.4 Forest of Mississippi and their productivity

Soil series	Great group	Forest type	Tree indicator species (site index)
Adler	Eutrudepts	Elm-ash-cottonwood	Green ash (95), sweetgum (100), Am sycamore (115), E cottonwood (120), water oak (100), willow oak (100)
Alaga	Quartzipsamments	Longleaf pine-slash pine	Slash pine (80), loblolly pine (80), longleaf pine (70)
Alligator	Dystraquerts	Elm-ash-cottonwood	Sugarberry (90), green ash (70), honeylocust (80), sweetgum (90), water oak (90), willow oak (95), Nuttall oak (90), cedar elm (90)
Almo	Fragiaqualfs	Oak-hickory	Sweetgum (80), loblolly pine (80), water oak (80), willow oak (80)
Angie	Paleudults	Loblolly pine-shortleaf pine	loblolly pine (92)
Annemaine	Hapludults	Loblolly pine-shortleaf pine	Sweetgum (80), yellow poplar (90), shortleaf pine (70), slash pine (80), loblolly pine (80), Am sycamore (90)
Arat	Hydraquents	Oak-gum-cypress	Water tupelo (50), baldcypress (50)
Ariel	Dystrudepts	Oak-gum-cypress	Sweetgum (100), yellow poplar (110, loblolly pine (95), E cottonwood (115), water oak (105), cherrybark oak (110)
Arkabutla	Endoaquepts	Oak-gum-cypress	Green ash (95), sweetgum (100), loblolly pine (100), E cottonwood (110), water oak (100), cherrybark oak (105), willow oak (100), Nuttall oak (110)
Arundel	Hapludults	Loblolly pine-shortleaf pine	Shortleaf pine (60), loblolly pine (70)
Askew	Hapludalfs	Oak-hickory	Sweetgum (90), E cottonwood (100), water oak (90), cherrybark oak (90), willow oak (90), Nuttall oak (90)
Atmore	Paleaquults	Longleaf pine-slash pine	Slash pine (90), longleaf pine (72), loblolly pine (75)
Atwood	Paleudalfs	Oak-hickory	Sweetgum (85), loblolly pine (90), cherrybark oak (90)
Bama	Paleudults	Loblolly pine-shortleaf pine	Longleaf pine (80), loblolly (90)
Basin	Paleudults	Longleaf pine-slash pine	Slash pine (90), loblolly pine (90)
Bassfield	Hapludults	Oak-pine	Sweetgum (90), shortleaf pine (80), loblolly pine (90), cherrybark oak (90)
Baxterville	Paleudults	Longleaf pine-slash pine	Slash pine (86), longleaf pine (70), loblolly pine (86)
Bayou	Paleaquults	Longleaf pine-slash pine	Slash pine (85), longleaf (47), loblolly pine (65)
Beauregard	Paleudults	Loblolly pine-shortleaf pine	Loblolly pine (108)
Belden	Fluvaquents	Elm-ash-cottonwood	Sweetgum (100), yellow poplar (95), loblolly pine (100), E cottonwood (110), white oak (90), S red oak (100)
Benndale	Paleudults	Longleaf pine-slash pine	Slash pine (94), longleaf pine (79), loblolly pine (94)
Beulah	Dystrudepts	Oak-hickory	E cottonwood (100), water oak (90), cherrybark oak (90), willow oak (90), Nuttall oak (90)
Bibb	Fluvaquents	Oak-gum-cypress	Sweetgum (90), loblolly pine (90), water oak (90)
Bigbee	Quartzipsamments	Longleaf pine-slash pine	Loblolly pine (88)
Boswell	Paleudalfs	Longleaf pine-slash pine	Shortleaf pine (73), loblolly pine (83)
Bowdre	Hapludolls	Oak-gum-cypress	Sweetgum (95), E cottonwood (110), water oak (95), cherrybark oak (90)
Boykin	Paleudults	Loblolly pine-shortleaf pine	Shortleaf pine (80), slash pine (90), longleaf pine (80), loblolly pine (90)
Brandon	Hapludults	Oak-hickory	Yellow poplar (80), white oak (61), scarlet oak (69), S red oak (71), black oak (70)
Brantley	Hapludalfs	Oak-hickory	Shortleaf pine (75), loblolly pine (85)
Brewton	Paleudults	Longleaf pine-slash pine	Sweetgum (90), slash pine (88), longleaf pine (80), loblolly pine (90)
Bruin	Eutrudepts	Oak-hickory	Pecan (95), sugarberry (80), green ash (75), sweetgum (110), Am sycamore (118), E cottonwood (110), water oak (100), Am elm (80)
Bruno	Udifluvents	Oak-pine	Sweetgum (94), yellow poplar (94), loblolly pine (93), Am sycamore (100), E cottonwood (110), water oak (90), cherrybark oak (90), willow oak (90)
Buckatunna	Hapludults	Oak-pine	Sweetgum (85), yellow poplar (90), loblolly (90), white oak (80), water oak (90)
Bude	Fragiudalfs	Oak-pine	Sweetgum (90), loblolly pine (90), cherrybark oak (90)
Byram	Fragiudalfs	Oak-pine	Sweetgum (85), shortleaf pine (75), loblolly (85), white oak (80), S red oak (75), cherrybark oak (85)

(continued)

Table 16.4 (continued)

Soil series	Great group	Forest type	Tree indicator species (site index)
Cadeville	Hapludalfs	Oak-pine	Shortleaf pine (70), loblolly pine (80)
Cahaba	Hapludults	Loblolly pine-shortleaf pine	Sweetgum (90), shortleaf pine (69), loblolly pine (80), slash pine (91)
Caledonia	Paleudalfs	Oak-pine	Sweetgum (85), loblolly pine (90), cherrybark oak (90)
Calhoun	Glossaqualfs	Oak-hickory	shortleaf pine (84), loblolly pine (90)
Calloway	Fraglossudalfs	Oak-pine	Sweetgum (90), shortleaf pine (80), loblolly pine (80), water oak (90), cherrybark oak (80)
Cascilla	Dystrudepts	Oak-gum-cypress	Sweetgum (102), yellow poplar (115), loblolly pine (93), E cottonwood (110), water oak (104), cherrybark oak (112), Nuttall oak (114)
Catalpa	Hapludolls	Elm-ash-cottonwood	Green ash (90), sweetgum (100), yellow poplar (100), Am sycamore (100), E cottonwood (110)
Chastain	Endoaquepts	Oak-gum-cypress	Green ash (75), sweetgum (90), water tupelo (80), E cottonwood (85), water oak (90), cherrybark oak (90), willow oak (90), Nuttall oak (95)
Chenneby	Dystrudepts	Oak-gum-cypress	Sweetgum (100), yellow poplar (100), loblolly pine (100), Am sycmore (100), water oak (100)
Chickasawhay	Hapludults	Loblolly pine-shortleaf pine	Sweetgum (80), yellow poplar (90), shortleaf pine (70), slash pine (80), loblolly pine (80), Am sycamore (90)
Chicora	Hapludults	Oak-pine	Sweetgum (85), yellow poplar (90), loblolly pine (90), water oak (90)
Collins	Udifluvents	Elm-ash-cottonwood	Green ash (95), E cottonwood (115), cherrybark oak (110)
Columbus	Hapludults	Oak-pine	Sweetgum (85), yellow poplar (90), loblolly pine (90), water oak (90)
Commerce	Endoaquepts	Elm-ash-cottonwood	Green ash (120), E cottonwood (120), water oak (110), Nuttall oak (90)
Convent	Endoaquepts	Elm-ash-cottonwood	Green ash (80), sweetgum (110), E cottonwood (120), Nuttall oak (90)
Crevasse	Udipsamments	Elm-ash-cottonwood	White oak (100), sweetgum (90)
Croatan	Haplosaprists	Oak-gum-cypress	Sweetgum (70), pond pine (55), loblolly pine (70)
Cuthbert	Hapludults	Loblolly pine-shortleaf pine	Shortleaf pine (73), loblolly pine (83)
Daleville	Paleaquults	Oak-gum-cypress	Sweetgum (90), loblolly pine (95), water oak (85), willow oak (80)
Darco	Paleudults	Loblolly pine-shortleaf pine	Shortleaf pine (76), loblolly pine (81)
Dexter	Hapludalfs	Oak-hickory	Green ash (80), sweetgum (95), E cottonwood (100), water oak (90), cherrybark oak (100), willow oak (95), Shumard's oak (100), Nuttall oak (95)
Dogue	Hapludults	Oak-pine	Sweetgum (85), yellow poplar (90), loblolly pine (90), white oak (80), water oak (90)
Dorovan	Haplosaprists	Oak-gum-cypress	Black tupelo (70)
Dowling	Endoaquepts	Oak-gum-cypress	Green ash (94), overcup oak (96)
Dubbs	Hapludalfs	Oak-hickory	Green ash (60), sweetgum (75), E cottonwood (80), water oak (72), cherrybark oak (76), Nuttall oak (75)
Dulac	Fragiudalfs	Oak-pine	Sweetgum (80), shortleaf pine (75), loblolly pine (80), S red oak (70)
Dundee	Endoaqualfs	Oak-gum-cypress	Sweetgum (100), E cottonwood (100), water oak (95), cherrybark oak (105)
Escambia	Paleudults	Longleaf pine-slash pine	Sweetgum (90), slash pine (90), longleaf pine (80), loblolly pine (90)
Eustis	Paleudults	Longleaf pine-slash pine	Slash pine (80), longleaf pine (65), loblolly pine (80)
Eutaw	Dystraquerts	Oak-pine	E redcedar (45), sweetgum (80), loblolly pine (80)
Falaya	Fluvaquents	Elm-ash-cottonwood	Green ash (90), loblolly pine (90), E cottonwood (100), water oak (100), cherrybark oak (100), Nuttall oak (110)
Falkner	Paleudalfs	Loblolly pine-shortleaf pine	Sweetgum (90), shortleaf pine (75), loblolly pine (85)
Forestdale	Endoaqualfs	Oak-gum-cypress	Green ash (78), sweetgum (100), E cottonwood (100), water oak (90), cherrybark oak (94), willow oak (94), Nuttall oak (99)
Freeland	Fraglossudalfs	Oak-pine	Sweetgum (90), shortleaf pine (64), loblolly pine (84)
Freest	Paleudalfs	Loblolly pine-shortleaf pine	Shortleaf pine (80), slash pine (85), loblolly pine (90)
Freestone	Paleudalfs	Loblolly pine-shortleaf pine	Shortleaf pine (80), slash pine (85), loblolly pine (90)

(continued)

Table 16.4 (continued)

Soil series	Great group	Forest type	Tree indicator species (site index)
Frizzell	Hapludalfs	Oak-pine	Sweetgum (90), loblolly pine (90)
Frost	Glossaqualfs	Elm-ash-cottonwood	Slash pine (90), loblolly pine (90), cherrybark oak (85)
Fruitdale	Hapludults	Oak-pine	Loblolly pine (75), S red oak (70)
Gillsburg	Fluvaquents	Oak-gum-cypress	Green ash (90), sweetgum (90), yellow poplar (105), loblolly pine (90), Am sycamore (105), E cottonwood (100), water oak (100), cherrybark oak (100), Nuttall oak (110)
Greenville	Kandiudults	Oak-pine	Sweetgum (85), loblolly pine (90), cherrybark oak (90)
Grenada	Fraglossudalfs	Oak-pine	Sweetgum (80), shortleaf pine (75), loblolly pine (85), s red oak (80), cherrybark oak (85)
Griffith	Hapluderts	Elm-ash-cottonwood	Hackberry (100), green ash (95), sweetgum (95), yellow poplar (100), Am sycamore (100), E cottonwood (110), white oak (100)
Guyton	Glossaqualfs	Oak-gum-cypress	Slash pine (90), loblolly pine (85), willow oak (78)
Harleston	Paleudults	Loblolly pine-shortleaf pine	Sweetgum (75), shortleaf pine (80), loblolly pine (90)
Hatchie	Fraglossudalfs	Oak-pine	Sweetgum (90), loblolly pine (90), cherrybark oak (90)
Heidel	Paleudults	Loblolly pine-shortleaf pine	Shortleaf pine (72), slash pine (90), loblolly pine (90)
Henry	Fragiaqualfs	Elm-ash-cottonwood	Sweetgum (80), loblolly pine (80), water oak (80), willow oak (90)
Houlka	Epiaquerts	Elm-ash-cottonwood	Green ash (85), sweetgum (105), Am sycamore (100), E cottonwood (105), cherrybark oak (105), Shumard's oak (105), Nuttall oak (105)
Hyde	Umbraquults	Oak-gum-cypress	Green ash (90), sweetgum (90), loblolly pine (107), water oak (85), willow oak (80)
Ichusa	Dystruderts	Oak-hickory	Sweetgum (90), loblolly pine (90), white oak (80), water oak (80), cherrybark oak (90), Shumard's oak (85)
Irvington	Fragiudults	Longleaf pine-slash pine	Shortleaf pine (77), slash pine (90), longleaf pine (75), loblolly pine (85), water oak (85)
Iuka	Udifluvents	Oak-hickory	Sweetgum (100), loblolly pine (100), E cottonwood (105), water oak (100)
Izagora	Paleudults	Oak-pine	Sweetgum (90), yellow poplar (100), slash pine (90), loblolly pine (90), water oak (90)
Jena	Dystrudepts	Oak-pine	Sweetgum (90), loblolly pine (100), water oak (80)
Johns	Hapludults	Longleaf pine-slash pine	Sweetgum (90), slash pine (88), longleaf pine (61), loblolly pine (88)
Johnston	Humaquepts	Oak-gum-cypress	Sweetgum (94), yellow poplar (94), loblolly pine (106), water oak (103)
Kinston	Endoaquepts	Oak-hickory	Sweetgum (95), loblolly pine (100), E cottonwood (100), white oak (90), cherrybark oak (95)
Kipling	Paleudalfs	Oak-pine	Sweetgum (90), loblolly pine (90), white oak (80), water oak (80), cherrybark oak (90), Shumard's oak (85)
Kirkville	Dystrudepts	Oak-pine	Sweetgum (100), loblolly pine (95), water oak (100), cherrybark oak (100)
Kisatchie	Hapludalfs	Oak-pine	Slash pine (70), loblolly pine (70), shortleaf pine (55)
Kolin	Paleudalfs	Oak-pine	loblolly pine (85)
Lakeland	Quartzipsamments	Longleaf pine-slash pine	Slash pine (75), longleaf pine (60), loblolly pine (75)
Latonia	Hapludults	Longleaf pine-slash pine	Slash pine (90), longleaf pine (70), loblolly pine (90)
Lauderdale	Hapludults	Loblolly pine-shortleaf pine	Shortleaf pine (65), loblolly pine (70)
Lax	Fragiudults	Oak-hickory	Loblolly pine (80), S red oak (70)
Leaf	Albaquults	Oak-pine	Sweetgum (80), loblolly pine (80), water oak (80)
Leeper	Epiaquepts	Elm-ash-cottonwood	Green ash (90), sweetgum (95), Am sycamore (100), E cottonwood (110)
Lenoir	Paleaquults	Oak-pine	Sweetgum (90), loblolly pine (87), water oak (85)
Leon	Alaquods	Longleaf pine-slash pine	Slash pine (65), longleaf pine (60), loblolly pine (65), pond cypress (75)
Leverett	Glossudalfs	Oak-pine	Sweetgum (90), loblolly pine (90), cherrybark oak (90)
Lexington	Hapludalfs	Oak-hickory	Sweetgum (89), yellow poplar (90), shortleaf pine (70), loblolly pine (80), S red oak (70), cherrybark oak (80)

(continued)

Table 16.4 (continued)

Soil series	Great group	Forest type	Tree indicator species (site index)
Linker	Hapludults	Loblolly pine-shortleaf pine	Shortleaf pine (65), loblolly pine (65), VA pine (70)
Longview	Hapludalfs	Oak-hickory	Sweetgum (88), loblolly pine (86), water oak (82), cherrybark oak (88)
Loring	Fragiudalfs	Oak-hickory	Sweetgum (90), loblolly pine (95), S red oak (75), water oak (90), cherrybark oak (86)
Lorman	Hapludalfs	Oak-pine	Shortleaf pine (70), loblolly pine (80)
Louin	Dystruderts	Loblolly pine-shortleaf pine	Sweetgum (90), loblolly pine (90), white oak (80), water oak (80), cherrybark oak (90), Shumard's oak (85)
Lucedale	Paleudults	Longleaf pine-slash pine	Slash pine (90), longleaf pine (75), loblolly pine (90)
Lucy	Kandiudults	Longleaf pine-slash pine	Slash pine (84), longleaf pine (70), loblolly pine (80)
Luverne	Hapludults	Loblolly pine-shortleaf pine	Shortleaf pine (73), loblolly pine (81)
Maben	Hapludalfs	Loblolly pine-shortleaf pine	Shortleaf pine (73), loblolly pine (83)
Malbis	Paleudults	Longleaf pine-slash pine	Slash pine (90), longleaf pine (80), loblolly pine (90)
Mantachie	Endoaquepts	Elm-ash-cottonwood	Green ash (80), sweetgum (95), yellow poplar (95), loblolly pine (98), E cottonwood (90), cherrybark oak (100)
Marietta	Eutrudepts	Elm-ash-cottonwood	Green ash (90), sweetgum (100), yellow poplar (100), Am sycamore (105), E cottonwood (105)
Mashulaville	Fragiaquults	Oak-pine	Slash pine (90), loblolly pine (85), willow oak (78)
Mathiston	Fluvaquents	Oak-gum-cypress	Green ash (90), sweetgum (95), loblolly pine (95), cherrybark oak (100)
Maubila	Hapludults	Loblolly pine-shortleaf pine	Shortleaf pine (70), longleaf pine (65), loblolly pine (75)
Mayhew	Dystraquerts	Oak-pine	Sweetgum (90), loblolly pine (90), water oak (80)
McCrory	Natraqualfs	Oak-gum-cypress	Sweetgum (95), slash pine (80), loblolly pine (80), water oak (95)
McLaurin	Paleudults	Longleaf pine-slash pine	Slash pine (90), longleaf pine (72), loblolly pine (90)
McRaven	Endoaquepts	Elm-ash-cottonwood	Green ash (90), sweetgum (100), Am sycamore (110), E cottonwood (110), water oak (95), cherrybark oak (120), willow oak (95)
Memphis	Hapludalfs	Oak-pine	Sweetgum (90), loblolly pine (90), cherrybark oak (90)
Mhoon	Endoaquepts	Elm-ash-cottonwood	Green ash (90), Am sycamore (100), E cottonwood (110)
Mooreville	Dystrudepts	Elm-ash-cottonwood	Green ash (80), sweetgum (100), yellow poplar (100), loblolly pine (95), E cottonwood (105), cherrybark oak (100)
Morganfield	Udifluvents	Elm-ash-cottonwood	Green ash (90), sweetgum (110), yellow poplar (115), E cottonwood (120), water oak (105), Nuttall oak (100)
Myatt	Endoaquults	Oak-pine	Sweetgum (82), slash pine (92), loblolly pine (88), water oak (86)
Nahunta	Paleaquults	Oak-pine	Sweetgum (97), yellow poplar (95), loblolly pine (95), water oak (80)
Natchez	Eutrudepts	Oak-pine	Sweetgum (105), loblolly pine (90), E cottonwood (105)
Neshoba	Paleudults	Loblolly pine-shortleaf pine	Shortleaf pine (80), loblolly pine (90)
Newellton	Epiaquepts	Oak-hickory	Pecan (102), green ash (80), sweetgum (100), Am sycamore (108), E cottonwood (105), water oak (104), cherrybark oak (90), willow oak (104), Nuttall oak (104)
Nugent	Udifluvents	Oak-pine	Sweetgum (95), slash pine (90), loblolly pine (90), water oak (85), willow oak (85)
Oaklimeter	Dystrudepts	Oak-pine	Green ash (90), sweetgum (100), loblolly pine (90), E cottonwood (100), cherrybark oak (100), willow oak (100), Nuttall oak (100)
Ochlockonee	Udifluvents	Oak-hickory	Sweetgum (90), yellow poplar (110), slash pine (100), loblolly pine (100), E cottonwood (100), water oak (80)
Ocilla	Paleudults	Longleaf pine-slash pine	Slash pine (90), longleaf pine (77), loblolly pine (85)
Okeelala	Hapludalfs	Oak-pine	Longleaf pine (76), loblolly pine (89)
Oktibbeha	Dystruderts	Oak-pine	E redcedar (45), shortleaf pine (66), loblolly pine (76), S red oak (70)
Olivier	Fraglossudalfs	Oak-pine	Sweetgum (90), loblolly pine (85), S red oak (74), water oak (82), cherrybark oak (86)

(continued)

Table 16.4 (continued)

Soil series	Great group	Forest type	Tree indicator species (site index)
Olla	Hapludults	Longleaf pine-slash pine	Slash pine (86), longleaf pine (67), loblolly pine (86)
Openlake	Epiaquepts	Elm-ash-cottonwood	Green ash (90), sweetgum (95), Am sycamore (100), E cottonwood (110), cherrybark oak (100), Nuttall oak (100)
Ora	Fragiudults	Oak-pine	Sweetgum (80), shortleaf pine (69), loblolly pine (83)
Orangeburg	Kandiudults	Loblolly pine-shortleaf pine	Shortleaf pine (69), loblolly pine (80)
Osier	Psammaquents	Longleaf pine-slash pine	Slash pine (85), longleaf pine (69), loblolly pine (87)
Ouachita	Dystrudepts	Oak-pine	Sweetgum (100), loblolly pine (100), E cottonwood (100), cherrybark oak (100)
Ozan	Glossaqualfs	Oak-pine	Sweetgum (90), shortleaf pine (85), loblolly pine (95), water oak (90)
Pactolus	Quartzipsamments	Longleaf pine-slash pine	Slash pine (83), longleaf pine (70), loblolly pine (84)
Paden	Fragiudults	Oak-pine	Sweetgum (80), loblolly pine (80), S red oak (70)
Pamlico	Haplosaprists	Oak-gum-cypress	N/a
Pelahatchie	Hapludalfs	Oak-pine	Sweetgum (90), loblolly pine (90), white oak (80), water oak (80), cherrybark oak (90), Shumard's oak (85)
Peoria	Natraqualfs	Oak-gum-cypress	Sweetgum (80), shortleaf pine (80), loblolly pine (83), water oak (80), cherrybark oak (80), willow oak (80), Nuttall oak (80)
Petal	Paleudalfs	Loblolly pine-shortleaf pine	Shortleaf pine (80), slash pine (85), longleaf pine (75), loblolly pine (90)
Pheba	Fragiudults	Loblolly pine-shortleaf pine	Sweetgum (90), shortleaf pine (80), slash pine (90), loblolly pine (90)
Pikeville	Paleudults	Loblolly pine-shortleaf pine	Shortleaf pine (70), loblolly pine (80), VA pine (70)
Plummer	Paleaquults	Longleaf pine-slash pine	Slash pine (88), longleaf pine (70), loblolly pine (91)
Poarch	Paleudults	Longleaf pine-slash pine	Slash pine (90), longleaf pine (73), loblolly pine (90)
Pooleville	Glossudalfs	Oak-pine	Sweetgum (90), loblolly pine (85), water oak (80), cherrybark oak (90)
Prentiss	Fragiudults	Oak-pine	Sweetgum (90), shortleaf pine (79), loblolly pine (88), white oak (80), cherrybark oak (90)
Providence	Fragiudalfs	Oak-pine	Sweetgum (90), shortleaf pine (64), loblolly pine (84)
Quitman	Paleudults	Loblolly pine-shortleaf pine	Sweetgum (93), slash pine (90), loblolly pine (92)
Rattlesnake Falls	Quartzipsamments	Longleaf pine-slash pine	Shortleaf pine (75), slash pine (85), longleaf pine (75), loblolly pine (85)
Richland	Hapludalfs	Oak-hickory	Sweetgum (90), loblolly pine (85), S red oak (74), water oak (82), cherrybark oak (86)
Riedtown	Eutrudepts	Elm-ash-cottonwood	Green ash (90), sweetgum (105), yellow poplar (110), Am sycamore (110), E cottonwood (115), water oak (100), willow oak (100)
Robinsonville	Udifluvents	Elm-ash-cottonwood	Green ash (85), sweetgum (105), Am sycamore (115), E cottonwood (110)
Rosebloom	Endoaquepts	Elm-ash-cottonwood	Green ash (95), sweetgum (95), Am sycamore (80), E cottonwood (100), water oak (95), cherrybark oak (95), willow oak (90), Nuttall oak (95)
Rosella	Natraqualfs	Oak-gum-cypress	Sweetgum (75), loblolly pine (80), water oak (80), willow oak (80)
Ruston	Paleudults	Longleaf pine-slash pine	Hickory (80), sweetgum (85), slash pine (91), longleaf pine (76), shortleaf pine (69), loblolly pine (80, 91)
Rutan	Hapludults	Longleaf pine-slash pine	Slash pine (85), longleaf pine (70), loblolly pine (85)
Saffell	Hapludults	Loblolly pine-shortleaf pine	Shortleaf pine (60), loblolly pine (66)
Saucier	Paleudults	Longleaf pine-slash pine	Slash pine (80), longleaf pine (60), loblolly pine (80)
Savannah	Fragiudults	Loblolly pine-shortleaf pine	Sweetgum (85), shortleaf pine (76), slash pine (88), longleaf pine (78), loblolly pine (81, 85), S red oak (75)
Sessum	Dystraquerts	Oak-pine	E redcedar (45), sweetgum (80), loblolly pine (83), S red oak (70)
Sharkey	Epiaquerts	Oak-gum-cypress	Sweetgum (90), water oak (90), willow oak (100), Nuttall oak (90)
Shubuta	Paleudults	Longleaf pine-slash pine	Slash pine (90), longleaf pine (90), loblolly pine (90)

(continued)

Table 16.4 (continued)

Soil series	Great group	Forest type	Tree indicator species (site index)
Silverdale	Udifluvents	Oak-pine	Sweetgum (80), shortleaf pine (70), pin oak (80), black oak (70)
Siwell	Hapludalfs	Oak-pine	Sweetgum (85), yellow poplar (85), loblolly pine (85), white oak (80), cherrybark oak (85), Shumard's oak (90)
Smithdale	Hapludults	Longleaf pine-slash pine	Slash pine (85), longleaf pine (69), loblolly pine (86)
Smithton	Paleaquults	Oak-pine	Sweetgum (86), shortleaf pine (76), loblolly pine (86), water oak (85), cherrybark oak (85)
Steens	Endoaqualfs	Oak-pine	Sweetgum (85), loblolly pine (90), water oak (90)
Stough	Paleudults	Oak-pine	Sweetgum (85), slash pine (86), loblolly pine (90), water oak (80), cherrybark oak (85)
Sucarnoochee	Epiaquerts	Oak-gum-cypress	Green ash (90), sweetgum (100), Am sycamore (110), water oak (100)
Suffolk	Hapludults	Oak-pine	Longleaf pine (72), loblolly pine (82), S red oak (70)
Suggsville	Dystruderts	Oak-pine	E redcedar (55), shortleaf pine (80), loblolly pine (90), S red oak (80)
Susquehanna	Paleudalfs	Loblolly pine-shortleaf pine	Shortleaf pine (68), loblolly pine (78)
Sweatman	Hapludults	Loblolly pine-shortleaf pine	Shortleaf pine (69), loblolly pine (80)
Talla	Natrudalfs	Oak-pine	Sweetgum (90), shortleaf pine (80), loblolly pine (90), cherrybark oak (90)
Tensas	Epiaqualfs	Oak-gum-cypress	Green ash (80), sweetgum (100), water oak (95)
Tippah	Paleudalfs	Oak-pine	Sweetgum (90), yellow poplar (90), loblolly pine (78), white oak (80), cherrybark oak (95), shumard's oak (95)
Tippo	Glossudalfs	Oak-pine	Sweetgum (90), loblolly pine (90), cherrybark oak (80)
Toinette	Hapludults	Oak-pine	Yellow poplar (80), loblolly pine (80), VA pine (74), S red oak (74)
Trebloc	Paleaquults	Oak-gum-cypress	Sweetgum (90), loblolly pine (95), water oak (85), willow oak (80)
Troup	Kandiudults	Oak-pine	Longleaf pine (70), loblolly pine (80)
Tunica	Epiaquepts	Elm-ash-cottonwood	Green ash (100), sugarberry (100), E cottonwood (105), sweetgum (90), cherrybark oak (90), Nuttall oak (105)
Tuscumbia	Epiaquepts	Elm-ash-cottonwood	Green ash (95), sweetgum (85), E cottonwood (100)
Tutwiler	Hapludalfs	Oak-hickory	Sweetgum (100), E cottonwood (100), water oak (90), cherrybark oak (95)
Una	Epiaquepts	Oak-gum-cypress	Green ash (70), sweetgum (90), water tupelo (80), E cottonwood (85), water oak (90), cherrybark oak (90), willow oak (90), Nuttall oak (95)
Urbo	Epiaquepts	Oak-hickory	Green ash (95), sweetgum (95), water oak (95), cherrybark oak (95), willow oak (90), Nuttall oak (95)
Vaiden	Dystruderts	Oak-pine	Sweetgum (90), loblolly pine (90), white oak (80), water oak (80), cherrybark oak (90), Shumard's oak (85)
Vancleave	Fragiudults	Oak-pine	Sweetgum (90), slash pine (90), longleaf pine (70), loblolly pine (90), water oak (85)
Velda	Dystrudepts	Oak-gum-cypress	Sweetgum (90), loblolly pine (98), water oak (82)
Vicksburg	Udifluvents	Elm-ash-cottonwood	Green ash (90), sweetgum (100), loblolly pine (90), E cottonwood (110), cherrybark oak (110), Nuttall oak (100)
Vimville	Glossaqualfs	Oak-pine	Sweetgum (90), loblolly pine (95), water oak (85), willow oak (85)
Wadley	Paleudults	Longleaf pine-slash pine	Sand pine (75), slash pine (85), longleaf pine (79), loblolly pine (85)
Wanilla	Glossudalfs	Oak-pine	Sweetgum (90), loblolly pine (90)
Waverly	Endoaquepts	Elm-ash-cottonwood	Green ash (95), sweetgum (95), Am sycamore (80), E cottonwood (100), water oak (95), cherrybark oak (95), willow oak (90), Nuttall oak (95)
Wilcox	Dystruderts	Oak-pine	Shortleaf pine (68), slash pine (85), loblolly pine (81)
Williamsville	Hapludults	Loblolly pine-shortleaf pine	Shortleaf pine (80), loblolly pine (88)

Lucedale, Malbis, McLaurin, Poarch, Ruston, and Shubuta. The top soils for elm-ash-cottonwood production are the Adler, Belden, Catalpa, Collins, Commerce, Falaya, Griffith, Houlka, Leeper, Mantachie, Marietta, McRaven, Mooreville, Morganfield, Riedtown, Robinsonville, Rosebloom, Tunica, Vicksburg, and Waverly soil series. Kushla and Oldham (2020) provided general information on the management of forest soils in Mississippi.

16.5 Summary

For croplands and pastures in Mississippi, yield potential is based on yields reported in county soil survey reports and Land Capability Classes (LCC) reported in county soil survey reports and the Web Soil Survey. The LCC groups soils on the basis of their capability to produce common cultivated crops and pasture plants without deteriorating over time, with or without irrigation. Forest site quality is based on site index data of key indicator tree species that are provided in archived soil survey reports and in Web Soil Survey. Nearly one-half (45%) of Mississippi's soil series, accounting for 32% of the soil area, are "high-quality soils," i.e., are in Land Capability Classes 1 and 2. These soils are located primarily in the Southern Mississippi Valley Alluvium (MLRA 131A) and the Alabama and Mississippi Blackland Prairie (MLRA 135A), which have been designated as "prime farmland," because they contain the best combination of physical and chemical characteristics for producing food, feed, forage, and fiber, and oilseed crops. There is a strong correlation between forest type and soil taxon in Mississippi.

References

Broadfoot WM (1969) Problems in relating soil to site index for southern hardwoods. For Sci 15:354–364

Carmean WH, Hahn JT, Jacobs RD (1989) Site index curves for forest tree species in the eastern United States. US For Serv North Central For Exp Stn Gen Tech Rep NC-128, 153 pp

Clatterbuck WK (1987) Height growth and site index curves for cherrybark oak and sweetgum in mixed, even-aged stands on the minor bottoms of central Mississippi. South J Appl 11:219–222

Kushla JD, Oldham L (2020) Forest soils of Mississippi. Miss State Univ Ext, Publ 2822, 7 pp

Natural Resources Conservation Service (1997) Land capability class, by state

Oldham L (2012) Nutrient management guidelines for agronomic crops grown in Mississippi. Miss State Univ Exten, Publ 2647

Pettiet JV (1974) An interpretive evaluation of soils in the Yazoo—Mississippi Delta area for crop production. Miss Agr For Exp Sta Bull, 808

Pettry DE, Koos WM (1980) Soil productivity in Mississippi. Miss Agr For Exp Sta Infor Sheet No. 1301

Quandt A, Herrick J, Peacock G, Salley S, Buni A, Mkalawa CC, Neff J (2020) A standardized land capability classification system for land evaluation using mobile phone technology. http://jornada.nmsu.edu

17 Summary

- Mississippi is the 32nd largest state in the US with an area of 125,443 km^2. With a population of 3 million, Mississippi is the 35th most populous US state. In 2021, the gross domestic product for Mississippi was $104 billion, 37th in the US. Mississippi is famous for its writers and musicians, many of whom wrote or sung of the soil.
- Mississippi has a rich and long history of soils investigations that began in 1860 with a report on the geology and agriculture of the state of Mississippi by E. W. Hilgard, who is considered by some as "the father of modern soil science in the United States." Seventeen reconnaissance soil surveys were completed by the former Bureau of Soils in Mississippi between 1901 and 1910. Over 100 archived soil surveys have been conducted in the state. The last parcel of soil in the state was mapped in 2012.
- The Natchez soil series, a coarse-silty, mixed, superactive, thermic Typic Eutrudepts, is the official state soil of Mississippi.
- Mississippi has a favorable climate with abundant precipitation, mild temperatures, and a long growing season. Snowfall is rare. However, the coastal area is subject to tropical cyclones, and in 2005 sustained substantial damage from Hurricane Katrina, one of the strongest Atlantic storms on record.
- Forests cover 62% of Mississippi, including loblolly-shortleaf pine, oak-hickory, bottomland oak-gum-cypress, oak-pine, longleaf-slash pine, and elm-ash-cottonwood.
- Mississippi is divided into five Major Land Resource Areas, including from largest to smallest, the Southern Coastal Plain, the Southern Mississippi Loess, the Southern Mississippi Alluvium, the Alabama and Mississippi Blackland Prairie, the East Gulf Coast Flatwoods, and the Gulf Coast Marsh.
- Mississippi is contained almost entirely within the Coastal Plain so that the major geologic units are comprised of unconsolidated materials. However, Paleozoic shale and limestone is exposed in the northeastern part of the state, mudstones and sandstones are exposed on the Pontotoc Ridge, and marl, chalk, and sandstone layers are exposed in the Blackland Prairies.
- Alluvium is the dominant surficial deposit in Mississippi, followed by loess, marine deposits, and residuum.
- An ochric epipedon is present in 94% of the soil series. The argillic horizon is the dominant diagnostic subsurface horizon (61%), followed by the cambic, albic, fragipan, and natric. The histic epipedon is the thickest of the epipedons at 117 cm, followed by the mollic at 51 cm. The thickest subsurface horizon is the argillic (119 cm), followed by the cambic, natric, fragipan, glossic, and albic.
- Mississippi has soils representative of 8 orders, 17 suborders, 42 great groups, 98 subgroups, 187 families, and 232 soil series.
- On an area basis, Alfisols and Ultisols are the dominant orders in Mississippi, followed by Inceptisols, Vertisols, and Entisols, with Histosols, Mollisols, and Spodosols occupying about 1.1% of the soil area. The predominant suborders include Udults, Udalfs, Aquepts, Udepts, Aquerts, and Aqualfs. The dominant great groups are Fragiudalfs, Paleudults, Hapludults, Hapludalfs, Fragiudults, Endoaquepts, and Endoaqualfs.
- In addition to the early work of E. W. Hilgard, Mississippi pedologists have played an important role in our understanding of the origin of fragipans and bisequal soils (D. Lindbo and F. E. Rhoton), the distribution and behavior of expansive soils (D.E. Pettry), and the characteristics of Ultisols and related soils (T. O. Mabry).
- The dominant soil-forming processes in Mississippi are argilluviation, gleization, base cycling, cambisolization, and vertization.
- About 19% of the Mississippi's soil series are benchmark soils; 46% are rare (i.e., contain less than 10,000

D. Johnson et al., *The Soils of Mississippi*, World Soils Book Series,
https://doi.org/10.1007/978-3-031-36235-4_17

acres (40 km^2), 34% are endemic (i.e., are the single example at the family level), and 16% are endangered (rare and endemic).

- Key issues in Mississippi that affect, and/or are affected by, soils and land use are soil erosion and water quality, climate change, and loss of biodiversity.
- Agriculture is Mississippi's number one industry, employing approximately 17% of the state's work force. Soybeans is the state's most valuable crop ($1.8 billion in 2022), and Mississippi typically ranks as one of the top cotton-producing states. The top agricultural commodities include broilers (chicken), soybeans, forestry, corn, cotton, cattle and calves, and aquaculture (farm-raised catfish).
- About 77% of forest land in Mississippi is privately owned. Timber is the state's third largest commodity with a value exceeding $1.3 billion annually. The most important timber commodities are solid wood products (lumber, plywood, and poles), pulp and paper, and furniture.
- Mississippi has over 4000 km^2 of private grazing lands, primarily for cattle.
- A large portion of Mississippi's soil series have map units in Land Capability Classes 1 (7%) and 2 (38%), attesting to their high yield potential.

Appendix A: List of Soil Surveys

Soil survey name (Click links for online surveys)	Date	Archived PDF online	Web soil survey (generated from official soil data)
Adams County	1910	Yes	No
Adams County	1969	Yes	No
Adams County	Current	No	Yes
Alcorn County	1924	No	No
Alcorn County	1971	Yes	No
Alcorn County	Current	No	Yes
Amite County	1919	Yes	No
Amite County	1976	Yes	No
Amite County	Current	No	Yes
Attala County	2003	Yes	No
Attala County	Current	No	Yes
Benton County	1977	Yes	No
Benton County	Current	No	Yes
Biloxi Area	1904	Yes	No
Bolivar County	1958	Yes	No
Bolivar County	Current	No	Yes
Calhoun County	1965	Yes	No
Calhoun County	Current	No	Yes
Carroll County	1990	Yes	No
Carroll County	Current	No	Yes
Chickasaw County	1917	Yes	No
Chickasaw County	1974	Yes	No
Chickasaw County	Current	No	Yes
Choctaw County	1920	No	No
Choctaw County	1986	Yes	No
Choctaw County	Current	No	Yes
Claiborne County	1926	No	No
Claiborne County	1963	Yes	No
Claiborne County	Current	No	Yes
Clarke County	1915	Yes	No
Clarke County	1965	Yes	No
Clarke County	Current	No	Yes
Clay County	1909	Yes	No
Clay County	1976	Yes	No
Clay County	Current	No	Yes
Coahoma County	1916	Yes	No
Coahoma County	1959	Yes	No
Coahoma County	Current	No	Yes
Copiah County	1984	Yes	No
Copiah County	Current	No	Yes
Covington County	1919	Yes	No
Covington County	1965	Yes	No
Covington County	Current	No	Yes
Crystal Springs Area	1905	Yes	No
De Soto County	1959	Yes	No
De Soto County	Current	No	Yes
Forrest County	1911	Yes	No
Forrest County	1979	Yes	No
Forrest County	Current	No	Yes
Franklin County	1995	Yes	No
Franklin County	Current	No	Yes
George County	1925	No	No
George County	1971	Yes	No
George County	Current	No	Yes
Greene County	1932	No	No
Greene County	Current	No	Yes
Grenada County	1917	Yes	No
Grenada County	1967	Yes	No
Grenada County	Current	No	Yes
Hancock County	1933	Yes	No
Hancock County	1981	Yes	No
Hancock County	Current	No	Yes
Harrison County	1924	No	No
Harrison County	1975	Yes	No
Harrison County	Current	No	Yes
Hinds County	1918	Yes	No
Hinds County	1979	Yes	No
Hinds County	Current	No	Yes

(continued)

D. Johnson et al., *The Soils of Mississippi*, World Soils Book Series,
https://doi.org/10.1007/978-3-031-36235-4

(continued)

Soil survey name (Click links for online surveys)	Date	Archived PDF online	Web soil survey (generated from official soil data)
Holmes County	1909	Yes	No
Holmes County	Current	No	Yes
Humphreys County	1959	Yes	No
Humphreys County	Current	No	Yes
Issaquena County	1961	Yes	No
Issaquena County	Current	No	Yes
Itawamba County	1979	Yes	No
Itawamba County	Current	No	Yes
Jackson Area	1904	Yes	No
Jackson County	1927	No	No
Jackson County	1964	Yes	No
Jackson County	2006	Yes	No
Jackson County	Current	No	Yes
Jasper County	1907	Yes	No
Jasper County	1979	Yes	No
Jasper County	Current	No	Yes
Jefferson County	1980	Yes	No
Jefferson County	Current	No	Yes
Jefferson Davis County	1916	Yes	No
Jefferson Davis County	1976	Yes	No
Jefferson Davis County	Current	No	Yes
Jones County	1915	Yes	No
Jones County	1986	Yes	No
Jones County	Current	No	Yes
Kemper County	1999	Yes	No
Kemper County	Current	No	Yes
Lafayette County	1914	Yes	No
Lafayette County	1981	Yes	No
Lafayette County	Current	No	Yes
Lamar County	1919	No	No
Lamar County	1975	Yes	No
Lamar County	Current	No	Yes
Lauderdale County	1910	No	No
Lauderdale County	1983	Yes	No
Lauderdale County	Current	No	Yes
Lawrence County	1978	Yes	No
Lawrence County	Current	No	Yes
Leake County	2009	Yes	No
Leake County	Current	No	Yes
Lee County	1918	Yes	No
Lee County	1973	Yes	No
Lee County	Current	No	Yes
Leflore County	1959	Yes	No
Leflore County	2006	Yes	No
Leflore County	Current	No	Yes
Lincoln County	1913	Yes	No
Lincoln County	1963	Yes	No
Lincoln County	Current	No	Yes
Lowndes County	1911	Yes	No
Lowndes County	1979	Yes	No
Lowndes County	Current	No	Yes
Madison County	1920	Yes	No
Madison County	1984	Yes	No
Madison County	Current	No	Yes
Marion County	1938	Yes	No
Marion County	1985	Yes	No
Marion County	Current	No	Yes
Marshall County	1972	Yes	No
Marshall County	Current	No	Yes
McNeill Area	1903	Yes	No
Monroe County	1908	Yes	No
Monroe County	1966	Yes	No
Monroe County	Current	No	Yes
Montgomery County	1906	Yes	No
Montgomery County	1975	Yes	No
Montgomery County	Current	No	Yes
Neshoba County	1981	Yes	No
Neshoba County	Current	No	Yes
Newton County	1918	Yes	No
Newton County	1960	Yes	No
Newton County	Current	No	Yes
Noxubee County	1910	Yes	No
Noxubee County	1986	Yes	No
Noxubee County	Current	No	Yes
Oktibbeha County	1907	Yes	No
Oktibbeha County	1973	Yes	No
Oktibbeha County	Current	No	Yes
Panola County	1963	Yes	No
Panola County	Current	No	Yes
Pearl River County	1920	Yes	No
Pearl River County	1983	Yes	No
Pearl River County	Current	No	Yes
Perry County	1928	Yes	No
Perry County	2000	Yes	No
Perry County	Current	No	Yes
Pike County	1918	Yes	No
Pike County	1968	Yes	No
Pike County	Current	No	Yes
Pontotoc County	1906	Yes	No
Pontotoc County	1973	Yes	No

(continued)

(continued)

Soil survey name (Click links for online surveys)	Date	Archived PDF online	Web soil survey (generated from official soil data)
Pontotoc County	Current	No	Yes
Prentiss County	1908	Yes	No
Prentiss County	1957	No	No
Prentiss County	1997	Yes	No
Prentiss County	Current	No	Yes
Quitman County	1958	Yes	No
Quitman County	Current	No	Yes
Rankin County	1926	No	No
Rankin County	1987	Yes	No
Rankin County	Current	No	Yes
Scott County	2010	Yes	No
Scott County	Current	No	Yes
Scranton Area	1909	Yes	No
Sharkey County	1962	Yes	No
Sharkey County	Current	No	Yes
Simpson County	1919	No	No
Simpson County	1996	Yes	No
Simpson County	Current	No	Yes
Smedes Area	1902	Yes	No
Smith County	1920	No	No
Smith County	2001	Yes	No
Smith County	Current	No	Yes
Stone County	2008	Yes	No
Stone County	Current	No	Yes
Sunflower County	1959	Yes	No
Sunflower County	Current	No	Yes
Tallahatchie County	1970	Yes	No
Tallahatchie County	Current	No	Yes
Tate County	1967	Yes	No
Tate County	Current	No	Yes
Tippah County	1966	Yes	No

Soil survey name (Click links for online surveys)	Date	Archived PDF online	Web soil survey (generated from official soil data)
Tippah County	Current	No	Yes
Tishomingo County	1944	Yes	No
Tishomingo County	1983	Yes	No
Tishomingo County	Current	No	Yes
Tunica County	1956	Yes	No
Tunica County	Current	No	Yes
Union County	1979	Yes	No
Union County	Current	No	Yes
Walthall County	1968	Yes	No
Walthall County	Current	No	Yes
Warren County	1914	Yes	No
Warren County	1964	Yes	No
Warren County	Current	No	Yes
Washington County	1961	Yes	No
Washington County	Current	No	Yes
Wayne County	1913	Yes	No
Wayne County	2009	Yes	No
Wayne County	Current	No	Yes
Webster County	1978	Yes	No
Webster County	Current	No	Yes
Wilkinson County	1913	Yes	No
Wilkinson County	Current	No	Yes
Winston County	1913	Yes	No
Winston County	2007	Yes	No
Winston County	Current	No	Yes
Yalobusha County	1978	Yes	No
Yalobusha County	Current	No	Yes
Yazoo Area	1901	Yes	No
Yazoo County	1975	Yes	No
Yazoo County	Current	No	Yes

Appendix B: Soil-Forming Factors for Soil Series in Mississippi

Series name	Depth class[a]	Drainage class[b]	MAAT (°C)	MAP (mm)	Max slope (%)	Vegetation	Parent material	Composition	Landform
ADATON	VD	PD	20	1295	2	Oak-pine	Alluvium	Silty	Alluvial terraces, uplands
ADLER	VD	MWD	18.9	1397	2	Elm-ash-cottonwood	Alluvium	Silty	Flood plains
ALAGA	VD	ED	18.3	1346	8	Longleaf pine-slash pine	Marine, alluvium	Sandy	Alluvial terraces, marine terraces
ALLIGATOR	VD	PD	17	1371	1	Elm-ash-cottonwood	Alluvium	Clayey	Flood plains, backswamps, meander belts
ALMO	VD	PD	15.6	1270	2	Oak-hickory	Alluvium	Loamy	Alluvial terraces, floodplains
ANGIE	VD	MWD	19	1473	12	Loblolly pine-shortleaf pine	Marine, alluvium	Loamy, clayey	Uplands, marine terraces
ANNEMAINE	VD	MWD	18.3	1346	12	Loblolly pine-shortleaf pine	Alluvium (stratified)	Clayey, loamy	Alluvial terraces
ARAT	VD	VPD	20	1345	0.5	Oak-gum-cypress	Alluvium	Silty clay loam	Backswamps
ARIEL	D	WD	18.9	1448	2	Oak-gum-cypress	Alluvium	Silty	Flood plains, alluvial terraces (low)
ARKABUTLA	VD	SPD	16.7	1220	2	Oak-gum-cypress	Alluvium	Silty	Flood plains
ARUNDEL	MD	WD	17.8	1320	60	Loblolly pine-shortleaf pine	Marine	Clayey	Uplands (dissected)
ASKEW	VD	MWD	15.6	1194	8	Oak-hickory	Alluvium	Loamy	Alluvial terraces (low)
ATMORE	VD	PD	16.7	1473	8	Longleaf pine-slash pine	Marine	Sandy, loamy	Interstream divides
ATWOOD	VD	WD	17	1450	17	Oak-hickory	Loess/marine	Silty/loamy	Hillslopes
AXIS	D	VPD	19.4	1600	2	Salt-tolerant marsh	Marine	Loamy	Coastal marshes
BAMA	VD	WD	19.4	1600	15	Loblolly pine-shortleaf pine	Marine, alluvium	Loamy	Marine terraces
BASIN	D	SPD	19.4	1575	5	Longleaf pine-slash pine	Alluvium	Loamy	Alluvial terraces, uplands
BASSFIELD	D	WD	18.9	1524	5	Oak-pine	Alluvium, marine	Loamy/sandy	Alluvial terraces, marine terraces
BAXTERVILLE	D	MWD	18.9	1498	8	Longleaf pine-slash pine	Marine	Loamy	Uplands, marine terraces
BAYOU	VD	PD	19.4	1600	2	Longleaf pine-slash pine	Marine	Loamy	Flatwoods
BEAUREGARD	VD	MWD	18.3	1334	5	Loblolly pine-shortleaf pine	Alluvium	Loamy	Alluvial terraces
BELDEN	D	SPD	17.2	1295	2	Elm-ash-cottonwood	Alluvium	Silty	Flood plains
BENNDALE	VD	WD	18.9	1575	12	Longleaf pine-slash pine	Alluvium	Loamy	Alluvial terraces (high), marine terraces
BEULAH	VD	SED	17.8	1219	8	Oak-hickory	Alluvium	Loamy, sandy	Flood plains, alluvial terraces (low)
BIBB	VD	PD	18.3	1372	2	Oak-gum-cypress	Alluvium	Loamy, sandy	Flood plains
BIGBEE	VD	ED	17.8	1320	5	Longleaf pine-slash pine	Alluvium	Sandy	Flood plains, natural levees
BINNSVILLE	D	WD	17.8	1320	17	Redcedar, osage orange, grasses	Residuum	Chalk	Uplands

(continued)

D. Johnson et al., *The Soils of Mississippi,* World Soils Book Series,
https://doi.org/10.1007/978-3-031-36235-4

(continued)

Series name	Depth class[a]	Drainage class[b]	MAAT (°C)	MAP (mm)	Max slope (%)	Vegetation	Parent material	Composition	Landform
BOHICKET	VD	VPD	20	1320	2	Cordgrass	Marine	Clayey	Tidal marshes
BONN	D	PD	19.7	1384	1	Poor-quality forest	Loess/alluvium	Silty	Alluvial terraces, floodplains
BOSWELL	VD	MWD	18.9	1346	17	Longleaf pine-slash pine	Marine	Clayey	Uplands
BOWDRE	D	SPD	15	1295	8	Oak-gum-cypress	Alluvium	Clayey/loamy	Flood plains
BOYKIN	D	WD	20	1575	20	Loblolly pine-shortleaf pine	Marine	Sandy, loamy	Uplands, marine terraces
BRANDON	VD	WD	14.4	1372	50	Oak-hickory	Loess/marine, alluvium	Silty/gravelly	Upland
BRANTLEY	VD	WD	19.4	1295	60	Oak-hickory	Residuum	Loamy, clayey	Ridgetops, side slopes
BREWTON	D	SPD	18.9	1473	4	Longleaf pine-slash pine	Marine	Loamy	Uplands, marine terraces
BROOKSVILLE	D	SPD	17.2	1220	5	Redcedar, osage orange, grasses	Residuum	Clay/calcareous	Uplands
BRUIN	D	MWD	18.9	1320	3	Oak-hickory	Alluvium	Silty	Flood plains (natural levees)
BRUNO	VD	ED	15.6	1220	5	Oak-pine	Alluvium	Sandy	Flood plains
BUCKATUNNA	VD	MWD	19.4	1525	8	Oak-pine	Alluvium	Clayey	Alluvial terraces
BUDE	VD	SPD	16.7	1448	5	Oak-pine	Loess/alluvium	Silty/loamy	Uplands, alluvial terraces
BYRAM	VD	MWD	18.3	1295	15	Oak-pine	Loess/marine	Silty/clay	Uplands, alluvial terraces
CADEVILLE	D	MWD	18.6	1283	40	Oak-pine	Alluvium, marine	Clayey	Uplands
CAHABA	VD	WD	18.3	1345	12	Loblolly pine-shortleaf pine	Alluvium	Loamy, sandy	Alluvial terraces
CALEDONIA	VD	WD	16	1375	8	Oak-pine	Marine	Loamy	Marine terraces
CALHOUN	VD	PD	19.4	1473	1	Oak-hickory	Loess	Silty	Alluvial terraces, floodplains
CALLOWAY	VD	SPD	17.8	1270	5	Oak-pine	Loess	Silty	Uplands, alluvial terraces
CASCILLA	D	WD	17.2	1219	2	Oak-gum-cypress	Alluvium	Silty	Flood plains (natural levees)
CATALPA	VD	SPD	17.2	1295	3	Elm-ash-cottonwood	Alluvium	Silty clay	Flood plains, alluvial terraces (low)
CHASTAIN	VD	PD	16.7	1143	2	Oak-gum-cypress	Alluvium	Clayey/sandy	Flood plains
CHENNEBY	VD	SPD	17.2	1320	3	Oak-gum-cypress	Alluvium	Loamy, silty	Flood plains
CHICKASAWHAY	VD	MWD	18.3	1397	5	Loblolly pine-shortleaf pine	Alluvium, marine	Clayey, loamy	Alluvial terraces, marine terraces
CHICORA	VD	MWD	17.2	1397	5	Oak-pine	Marine	Loamy	Alluvial terraces
COLLINS	VD	MWD	17.8	1270	2	Elm-ash-cottonwood	Alluvium	Silty	Flood plains
COLUMBUS	VD	MWD	17.2	1295	2	Oak-pine	Marine	Loamy	Alluvial terraces (low)
COMMERCE	D	SPD	18.9	1320	5	Elm-ash-cottonwood	Alluvium	Loamy	Alluvial plains
CONVENT	VD	SPD	19.4	1524	3	Elm-ash-cottonwood	Alluvium	Loamy	Flood plains
CREVASSE	VD	ED	19.4	1422	5	Elm-ash-cottonwood	Alluvium	Sandy	Flood plains, point bar deposits
CROATAN	VD	VPD	18	1295	2	Oak-gum-cypress	Organic/marine	Organic/loamy	Depressions
CUTHBERT	MD	WD	18.3	1220	40	Loblolly pine-shortleaf pine	Marine	Loamy, sandy (sandstone, shale)	Uplands
DALEVILLE	VD	PD	17.8	1320	2	Oak-gum-cypress	Marine, alluvium	Loamy	Uplands, alluvial terraces
DARCO	VD	SED	18.5	1138	25	Loblolly pine-shortleaf pine	Residuum	Sandy, loamy	Uplands, marine terraces
DEERFORD	VD	SPD	21	1397	2	N/a	Marine	Silty	Coastal plain
DEMOPOLIS	S	WD	17.8	1473	35	Redcedar, osage orange, grasses	Residuum	Chalk	Uplands
DEXTER	VD	WD	18.3	1320	8	Oak-hickory	Loess/alluvium	Silty/loamy, sandy	Alluvial terraces
DOGUE	VD	MWD	16.1	1270	15	Oak-pine	Alluvium/marine	Clayey	Alluvial terraces, marine terraces

(continued)

(continued)

Series name	Depth class[a]	Drainage class[b]	MAAT (°C)	MAP (mm)	Max slope (%)	Vegetation	Parent material	Composition	Landform
DOROVAN	VD	VPD	19.4	1448	1	Oak-gum-cypress	Organic/alluvium	Organic/sand	Depressions
DOWLING	VD	VPD	21.1	1625	1	Oak-gum-cypress	Alluvium	Clayey	Oxbows, backswamps
DUBBS	VD	WD	17.8	1219	8	Oak-hickory	Alluvium	Loamy	Alluvial terraces (low)
DUCKSTON	VD	PD	16.7	1345	2	Waxmyrtle, black willow, blueberry, cordgrass	Eolian	Sandy	Interdunes
DULAC	VD	MWD	15.6	1345	12	Oak-pine	Loess/marine	Silty/acid clay	Uplands, alluvial terraces
DUNDEE	VD	PD	20	1295	2	Oak-gum-cypress	Alluvium	Silty	Alluvial terraces
ESCAMBIA	VD	SPD	18.9	1473	8	Longleaf pine-slash pine	Marine	Sandy, loamy	Uplands
EUSTIS	D	SED	20.6	1320	12	Longleaf pine-slash pine	Alluvium, marine	Sandy	Uplands (dissected)
EUTAW	VD	PD	15.6	1270	2	Oak-pine	Marine	Clayey	Uplands
FALAYA	VD	SPD	16.1	1345	2	Elm-ash-cottonwood	Alluvium	Silty	Flood plains
FALKNER	VD	SPD	16.7	1219	8	Loblolly pine-shortleaf pine	Loess/marine	Silty/clayey	Marine terraces, uplands
FORESTDALE	VD	PD	16.7	1345	8	Oak-gum-cypress	Alluvium	Clayey, silty	Alluvial terraces (low)
FREELAND	VD	MWD	16.1	1370	12	Oak-pine	Loess/alluvium	Silty/loamy	Alluvial terraces
FREEST	VD	MWD	16.7	1320	8	Loblolly pine-shortleaf pine	Loess/marine	Loamy/clayey	Marine terraces, uplands
FREESTONE	VD	MWD	19.4	1092	5	Loblolly pine-shortleaf pine	Alluvium	Loamy, clayey	Alluvial terraces
FRIZZELL	D	SPD	18.3	1345	3	Oak-pine	Alluvium	Silty	Alluvial terraces (mid-Pleistocene)
FROST	VD	PD	19.4	1525	5	Elm-ash-cottonwood	Alluvium	Silty	Drainageways
FRUITDALE	VD	WD	18.9	1575	15	Oak-pine	Marine	Loamy	Alluvial terraces, marine terraces
GILLSBURG	D	SPD	18.9	1448	2	Oak-gum-cypress	Alluvium	Silty	Flood plains
GREENVILLE	VD	WD	18.3	1500	18	Oak-pine	Marine	Clayey	Marine terrace
GRENADA	VD	MWD	17.8	1320	12	Oak-pine	Loess	Silty	Upland
GRIFFITH	VD	MWD	17.8	1320	2	Elm-ash-cottonwood	Alluvium	Clayey	Flood plains
GUYTON	VD	PD	17	1397	1	Oak-gum-cypress	Alluvium	Loamy	Flood plains
HANDSBORO	D	VPD	20	1473	1	Cordgrass, marsh aster, sea lavender	Organic		Depressions
HARLESTON	VD	MWD	18.3	1575	12	Loblolly pine-shortleaf pine	Alluvium, marine	Sandy, loamy	Marine terraces, alluvial terraces, uplands
HATCHIE	VD	SPD	15.6	1270	2	Oak-pine	Loess/alluvium	Silty/loamy	Alluvial terraces
HEIDEL	VD	WD	16.7	1320	35	Loblolly pine-shortleaf pine	Marine	Loamy	Uplands
HENRY	VD	PD	15.6	1245	2	Elm-ash-cottonwood	Loess	Silty	Uplands
HOULKA	VD	SPD	17	1430	2	Elm-ash-cottonwood	Alluvium	Clayey	Flood plains
HOUSTON	VD	MWD	19	1295	8	Redcedar, osage orange, grasses	Residuum	Clays, chark	Uplands
HYDE	VD	VPD	16.7	1320	2	Oak-gum-cypress	Marine	Loamy	Marine terrace
ICHUSA	VD	SPD	17.2	1320	8	Oak-hickory	Marine	Clayey	Uplands
IRVINGTON	VD	MWD	19.4	1702	8	Longleaf pine-slash pine	Marine	Loamy	Uplands, interstream divides
IUKA	VD	MWD	18.9	1372	2	Oak-hickory	Alluvium	Loamy, sandy	Flood plains
IZAGORA	D	MWD	18.3	1345	8	Oak-pine	Alluvium, marine	Loamy, clayey	Alluvial terraces, marine terraces
JENA	D	WD	18.3	1422	1	Oak-pine	Alluvium	Loamy	Flood plains (natural levees)
JOHNS	VD	MWD, SPD	17.2	1168	2	Longleaf pine-slash pine	Marine, alluvium	Sandy	Alluvial terraces
JOHNSTON	VD	VPD	18	1270	2	Oak-gum-cypress	Alluvium		Flood plains
KINSTON	VD	PD	16.7	1219	2	Oak-hickory	Marine	Clayey/sandy	Flood plains
KIPLING	VD	SPD	17.2	1320	40	Oak-pine	Marine	Clayey (acidic)	Marine terraces, uplands
KIRKVILLE	VD	MWD	17.2	1345	2	Oak-pine	Alluvium	Loamy	Flood plains
KISATCHIE	MD	WD	18.3	1422	40	Oak-pine	Marine	Clayey/siltstone	Uplands
KOLIN	VD	SPD	19	1448	5	Oak-pine	Alluvium	Loamy/clayey	Alluvial terraces

(continued)

(continued)

Series name	Depth class[a]	Drainage class[b]	MAAT (°C)	MAP (mm)	Max slope (%)	Vegetation	Parent material	Composition	Landform
LAKELAND	VD	ED	19	1320	12	Longleaf pine-slash pine	Eolian, marine	Sandy	Uplands
LATONIA	D	WD	16.7	1220	5	Longleaf pine-slash pine	Alluvium, marine	Loamy/sandy	Alluvial terraces, marine terraces
LAUDERDALE	S	WD	17.8	1320	30	Loblolly pine-shortleaf pine	Residuum	Sandstone, shale	Uplands (dissected)
LAX	VD	MWD	14.4	1370	20	Oak-hickory	Loess/alluvium	Silty/gravelly residuum	Uplands
LEAF	VD	PD	20	1450	2	Oak-pine	Alluvium	Clayey	Flood plains; alluvial terraces (low)
LEEPER	VD	SPD	18	1400	3	Elm-ash-cottonwood	Alluvium	Clayey	Flood plains
LENOIR	VD	SPD	17.2	1220	2	Oak-pine	Marine, alluvium	Clayey	Uplands, alluvial terraces
LEON	VD	VPD, PD	20	1650	5	Longleaf pine-slash pine	Marine	Sandy	Alluvial terraces, upland flats
LEVERETT	D	WD	17.8	1345	5	Oak-pine	Alluvium	Silty	Flood plains, alluvial terraces (low)
LEXINGTON	VD	WD	15.6	1295	30	Oak-hickory	Loess/marine	Silty/loamy, sandy	Uplands
LINKER	MD	WD	17	1540	15	Loblolly pine-shortleaf pine	Residuum	Sandstone	Hillslopes
LONGVIEW	D	SPD	20	1295	5	Oak-hickory	Alluvium	Loamy	Uplands
LORING	VD	MWD	16.7	1257	20	Oak-hickory	Loess	Silty	Uplands, alluvial terraces
LORMAN	VD	MWD	18.9	1372	40	Oak-pine	Residuum	Clayey, silty	Uplands (dissected)
LOUIN	D	SPD	17.2	1320	2	Loblolly pine-shortleaf pine	Alluvium	Clayey	Uplands, alluvial terraces (gilgai)
LUCEDALE	D	WD	19.4	1550	15	Longleaf pine-slash pine	Marine	Loamy	Uplands
LUCY	VD	WD	18	1345	45	Longleaf pine-slash pine	Marine, alluvium	Sandy, loamy	Uplands
LUVERNE	VD	WD	15.6	1400	45	Loblolly pine-shortleaf pine	Marine	Stratif sands, silts, clays	Uplands
MABEN	VD	WD	20	1295	60	Loblolly pine-shortleaf pine	Residuum/shale	Sandy to clayey (shale)	Hillslopes
MALBIS	VD	MWD, WD	19.4	1650	12	Longleaf pine-slash pine	Alluvium, marine	Loamy, sandy clays	Interfluves, uplands
MANTACHIE	VD	SPD	17.2	1345	3	Elm-ash-cottonwood	Alluvium	Loamy	Flood plains
MARIETTA	D	MWD	17.8	1295	2	Elm-ash-cottonwood	Alluvium	Loamy	Flood plains
MASHULAVILLE	D	PD	17.8	1500	3	Oak-pine	Alluvium	Loamy	Alluvial terraces, uplands
MATHISTON	D	SPD	20	1295	2	Oak-gum-cypress	Alluvium	Silty	Flood plains
MAUBILA	VD	MWD	17.2	1345	45	Loblolly pine-shortleaf pine	Marine	Clayey	Uplands
MAUREPAS	VD	VPD	20	1625	1	Baldcypress, marsh plants	Organic		Depressions, backswamps
MAYHEW	D	PD	16.7	1220	12	Oak-pine	Marine	Clayey (shale)	Uplands
MAYTAG	VD	WD	18.3	1270	12	Redcedar, hackberry, grasses	Residuum	Marl, chalk	Uplands
MCCRORY	VD	PD	16	1270	2	Oak-gum-cypress	Alluvium	Loamy	Alluvial terraces (low)
MCLAURIN	VD	WD	17.8	1550	8	Longleaf pine-slash pine	Alluvium	Loamy	Alluvial terraces (dissected), marine terraces
MCRAVEN	D	SPD	18.3	1450	2	Elm-ash-cottonwood	Alluvium	Silty	Flood plains, alluvial terraces (low)
MEMPHIS	VD	WD	18	1220	50	Oak-pine	Loess	Silty	Alluvial terraces, uplands
MHOON	VD	PD	18.9	1320	5	Elm-ash-cottonwood	Alluvium	Silty	Alluvial plains
MOOREVILLE	VD	MWD	17.2	1345	2	Elm-ash-cottonwood	Alluvium	Loamy	Flood plains
MORGANFIELD	D	WD	19.4	1397	2	Elm-ash-cottonwood	Alluvium	Silty	Flood plains, drainageways
MYATT	VD	PD	16.7	1345	2	Oak-pine	Alluvium, marine	Fine	Alluvial terraces, uplands

(continued)

(continued)

Series name	Depth class[a]	Drainage class[b]	MAAT (°C)	MAP (mm)	Max slope (%)	Vegetation	Parent material	Composition	Landform
NAHUNTA	VD	SPD	17.2	1345	2	Oak-pine	Marine	Loamy, silty	Uplands
NATCHEZ	D	WD	17.2	1320	60	Oak-pine	Loess	Silty	Uplands (dissected)
NESHOBA	D	WD	17.8	1372	8	Loblolly pine-shortleaf pine	Marine	Clayey	Uplands
NEWELLTON	VD	SPD	18.9	1320	5	Oak-hickory	Alluvium	Clayey/loamy	Flood plains (natural levees)
NEWHAN	VD	ED	17.2	1370	30	Wax myrtle, yaupon, holly, live oak, bayberry, sea-oats, seacoast bluestem, Am beachgrass	Eolian	Sandy	Dunes
NUGENT	D	ED	20	1473	2	Oak-pine	Alluvium	Sandy	Flood plains (natural levees)
OAKLIMETER	VD	MWD	16.7	1400	2	Oak-pine	Alluvium	Silty	Flood plains, alluvial terraces (low)
OCHLOCKONEE	VD	WD	15	1397	3	Oak-hickory	Alluvium	Loamy	Flood plains
OCILLA	VD	SPD	19	1270	10	Longleaf pine-slash pine	Marine	Sandy, loamy	Uplands, alluvial terraces
OKEELALA	VD	WD	15.6	1345	60	Oak-pine	Marine	Loamy, sandy	Uplands (dissected)
OKOLONA	D	WD	17.2	1220	5	Redcedar, osage orange	Residuum	Marl, chalk	Uplands
OKTIBBEHA	VD	MWD	17.2	1346	30	Oak-pine	Residuum	Clayey/chalk	Uplands
OLIVIER	VD	SPD	20	1400	5	Oak-pine	Loess	Silty	Alluvial terraces
OLLA	D	WD	17.8	1320	60	Longleaf pine-slash pine	Alluvium, marine	Loamy	Uplands (dissected)
OPENLAKE	VD	SPD	15	1295	2	Elm-ash-cottonwood	Alluvium	Clayey	Flood plains
ORA	VD	MWD	18.9	1524	12	Oak-pine	Marine, alluvial	Loamy	Alluvial terraces, marine terraces, uplands
ORANGEBURG	VD	WD	18.3	1320	25	Loblolly pine-shortleaf pine	Marine	Loamy, clayey	Uplands
OSIER	VD	PD	19.4	1170	2	longleaf pine-slash pine	Alluvium	Sandy	Alluvial terraces (low)
OUACHITA	D	WD	17	1270	3	Oak-pine	Alluvium	Loamy	Flood plains, natural levees
OZAN	D	PD	17.8	1245	1	Oak-pine	Alluvium	Loamy	Flood plains, alluvial terraces (low)
PACTOLUS	VD	MWD, SPD	17.2	1220	6	Longleaf pine-slash pine	Alluvium, marine	Sandy	Marine terraces
PADEN	VD	MWD	16.1	1500	12	Oak-pine	Loess/alluvium, residuum	Silty/clayey	Alluvial terraces
PAMLICO	VD	VPD	17.2	1220	1	Oak-gum-cypress	Organic/sand	Organic/sandy	Flood plains, bays, depressions
PELAHATCHIE	D	MWD	18.3	1295	5	Oak-pine	Loess/residuum	Silty/clayey	Uplands
PEORIA	D	PD	19	1450	2	Oak-gum-cypress	Alluvium	Silty	Flood plains, alluvial terraces (low)
PETAL	D	MWD	16.7	1320	20	Loblolly pine-shortleaf pine	Marine	Loamy, clayey	Uplands
PHEBA	D	SPD	18.9	1473	3	Loblolly pine-shortleaf pine	Marine	Loamy	Uplands, terraces
PIKEVILLE	VD	WD	18.3	1320	35	Loblolly pine-shortleaf pine	Marine	Gravelly, loamy	Uplands
PLUMMER	VD	PD, VPD	19	1300	5	Longleaf pine-slash pine	Alluvium, marine	Sandy	Alluvial terraces, marine terraces
POARCH	VD	MWD	19	1420	8	Longleaf pine-slash pine	Marine	Sandy, loamy	Uplands
POOLEVILLE	D	SPD	16.7	1220	2	Oak-pine	Alluvium	Silty	Uplands, alluvial terraces (high)
PRENTISS	D	MWD	17.8	1345	8	Oak-pine	Alluvium, marine	Loamy	Alluvial terraces, uplands
PRIM	S	WD	17.8	1420	40	E redcedar, hardwoods	Residuum	Limestone, chalk	Uplands (dissected)
PROVIDENCE	VD	MWD	19.4	1448	15	Oak-pine	Loess/alluvium	Silty/sandy, loamy	Uplands, alluvial terraces
QUITMAN	VD	SPD	19	1370	5	Loblolly pine-shortleaf pine	Marine, alluvium	Loamy	Marine terraces, alluvial terraces, uplands
RATTLESNAKE FORKS	VD	SED	19	1525	35	Longleaf pine-slash pine	Alluvium, marine	Sandy	Uplands

(continued)

(continued)

Series name	Depth class[a]	Drainage class[b]	MAAT (°C)	MAP (mm)	Max slope (%)	Vegetation	Parent material	Composition	Landform
RICHLAND	VD	WD	12.8	1092	40	Oak-hickory	Colluvium	Loamy	Uplands
RIEDTOWN	D	MWD	18.3	1448	2	Elm-ash-cottonwood	Alluvium	Silty	Flood plains, alluvial terraces (low)
ROBINSONVILLE	VD	WD	19.4	1400	5	Elm-ash-cottonwood	Alluvium	Loamy	Flood plains
ROSEBLOOM	D	PD	18.3	1345	2	Elm-ash-cottonwood	Alluvium	Silty	Flood plains
ROSELLA	D	PD	19	1500	2	Oak-gum-cypress	Alluvium, marine	Loamy	Alluvial terraces
RUSTON	VD	WD	18.3	1500	8	Longleaf pine-slash pine	Marine, alluvium	Loamy	Uplands
RUTAN	VD	WD, SED	19	1525	35	Longleaf pine-slash pine	Marine, alluvium	Loamy, sandy	Marine terraces (dissected)
SAFFELL	VD	WD	17.2	1270	60	Loblolly pine-shortleaf pine	Marine	Loamy, gravelly	Uplands
SAUCIER	VD	MWD	20	1500	12	Longleaf pine-slash pine	Marine	Loamy/clayey	Uplands
SAVANNAH	VD	MWD	19	1370	15	Loblolly pine-shortleaf pine	Marine	Loamy	Uplands, marine terraces
SESSUM	D	PD	20	1295	5	Oak-pine	Residuum	Clayey	Uplands
SHARKEY	VD	PD, VPD	18.3	1397	5	Oak-gum-cypress	Alluvium	Clayey	Flood plains, alluvial terraces (low)
SHUBUTA	VD	WD	16.7	1345	12	Longleaf pine-slash pine	Marine, alluvium	Clayey	Uplands
SILVERDALE	D	MWD	15.6	1270		Oak-pine	Alluvium	Sandy/loamy	Natural levees
SIWELL	D	MWD	18.3	1295	15	Oak-pine	Loess/alluvium	Silty/clayey (calcareous)	Uplands
SMITHDALE	VD	WD	17.2	1450	60	Longleaf pine-slash pine	Marine	Loamy	Uplands
SMITHTON	VD	PD	19	1295	3	Oak-pine	Alluvium	Loamy	Alluvial terraces
STEENS	VD	SPD	17.2	1295	2	Oak-pine	Alluvium, marine	Loamy	Alluvial terraces, marine terraces, uplands
STOUGH	VD	SPD	19	1422	5	Oak-pine	Alluvium, marine	Loamy	Alluvial terraces, inerfluves
SUCARNOOCHEE	VD	SPD	18	1400	2	Oak-gum-cypress	Alluvium	Clayey (alkaline)	Flood plains
SUFFOLK	VD	WD	20	1320	50	Oak-pine	Marine, alluvial		Alluvial terraces, uplands
SUGGSVILLE	D	WD	17.8	1500	35	Oak-pine	Residuum	Clayey/chalk	Uplands
SUMTER	MD	MWD, WD	19.4	1295	40	E redcedar, oaks	Residuum	Loamy, clayey (chalk)	Uplands
SUSQUEHANNA	D	SPD	20	1550	17	Loblolly pine-shortleaf pine	Marine, alluvium	Silty clay, clayey	Uplands (dissected)
SWEATMAN	VD	WD	17.2	1270	35	Loblolly pine-shortleaf pine	Marine	Clayey, sandy, loamy	Uplands
TALLA	D	SPD	16.7	1220	2	Oak-pine	Alluvium	Loamy	Alluvial terraces (older)
TEKSOB	VD	WD	16.7	1295	8	Hardwoods	Alluvium	Loamy	Alluvial terraces
TENSAS	VD	SPD	19	1320	8	Oak-gum-cypress	Alluvium	Clayey/loamy	Natural levees
TIPPAH	D	MWD	16.7	1220	12	Oak-pine	Alluvium	Silty/clayey	Alluvial terraces, uplands
TIPPO	VD	SPD	17.8	1345	5	Oak-pine	Alluvium	Silty	Flood plains, alluvial terraces (low)
TOINETTE	VD	WD, SED	19	1625	60	Oak-pine	Alluvium, marine	Sandy, loamy	Hillslopes
TREBLOC	VD	PD	19	1500	2	Oak-gum-cypress	Marine, alluvium	Silty clay	Alluvial terraces
TRINITY	VD	MWD	17	1200	3	Elm, hackberry, oak, ash	Alluvium	Clayey (calcareous)	Flood plains, alluvial plains (dissected)
TROUP	VD	SED	17	1320	15	Oak-pine	Marine	Sandy, loamy	Ridges, hillslopes
TUNICA	D	PD	16.5	1370	3	Elm-ash-cottonwood	Alluvium	Clayey/loamy	Flood plains (natural levees)
TUSCUMBIA	D	PD	17.8	1295	2	Elm-ash-cottonwood	Alluvium	Clayey	Flood plains
TUTWILER	VD	WD	19	1345	5	Oak-hickory	Alluvium	Loamy	Natural levees, alluvial terraces (low)
UNA	D	PD	17.2	1320	4	Oak-gum-cypress	Alluvium	Clayey	Flood plains
URBO	D	SPD	16.7	1220	3	Oak-hickory	Alluvium	Clayey	Flood plains

(continued)

(continued)

Series name	Depth class[a]	Drainage class[b]	MAAT (°C)	MAP (mm)	Max slope (%)	Vegetation	Parent material	Composition	Landform
VAIDEN	VD	SPD	17.2	1346	17	Oak-pine	Alluvium	Clayey/chalk	Uplands, alluvial terraces (old)
VANCLEAVE	VD	MWD	19.4	1420	8	Oak-pine	Marine	Loamy	Marine terraces
VELDA	D	WD	19	1500	2	Oak-gum-cypress	Alluvium	Silty	Flood plains, alluvial terraces (low)
VERDUN	D	SPD	19	1370	1	Poor-quality hardwoods	Alluvium	Silty	Alluvial terraces, uplands
VICKSBURG	D	WD	17.8	1270	3	Elm-ash-cottonwood	Alluvium	Silty	Flood plains
VIMVILLE	VD	PD	17.8	1320	2	Oak-pine	Marine, alluvium	Loamy	Uplands
WADLEY	VD	WD, SED	20	1400	40	Longleaf pine-slash pine	Marine	Sandy, loamy	Uplands
WANILLA	D	SPD	19	1500	4	Oak-pine	Alluvium, marine	Loamy	Alluvial terraces, marine terraces
WATSONIA	S	WD	17.2	1345	25	Loblolly pine, E redcedar, hardwoods	Residuum	Clayey/chalk	Uplands
WAVERLY	VD	PD	18.3	1370	2	Elm-ash-cottonwood	Alluvium	Silty	Flood plains
WEYANOKE	VD	WD	19	1575	3	N/a	Alluvium	Silty	Flood plains, alluvial terraces
WILCOX	D	SPD	17.2	1345	35	Oak-pine	Residuum/shale	Clayey	Uplands
WILLIAMSVILLE	D	WD	17.8	1370	40	Loblolly pine-shortleaf pine	Marine	Clayey	Uplands

[a]Soil depth class: *VD* very deep (> 150 cm); *D* deep (100–150 cm); *MD* moderately deep (50–100 cm); *S* shallow (25–50 cm); and *VS* very shallow (< 25 cm)

[b]Drainage class: *ED* excessively drained; *SED* somewhat excessively drained; *WD* well drained; *MWD* moderately well drained; *SPD* somewhat poorly drained

PD poorly drained; *VPD* very poorly drained

Appendix C: Thicknesses of Diagnostic Horizons in Mississippi Soil Series

Series name	Ochric	Mollic	Histic	Umbric	Albic	Argillic	Cambic	Fragipan	Glossic	Kandic	Natric	Spodic	Plinthite	Sulfidic materials	None	To para-lithic	To Lithic
ADATON	15					152											
ADLER	18						23										
ALAGA	15														x		
ALLIGATOR	18						147										
ANGIE	25					140											
ANNEMAINE	23					71											
ARAT	15														x		
ARIEL	12						**140**										
ARKABUTLA	15						**124**										
ARUNDEL	15					81										97	
ASKEW	38					61											
ATMORE	76					**127**							**127**				
ATWOOD	15					**186**											
AXIS	30													114	x		
BAMA	23					**152**											
BASIN	23					**130**							**74**				
BASSFIELD	30					76											
BAYOU	46					**122**											
BAXTERVILLE	23					**150**							56				
BEAUREGARD	23					**206**							41				
BELDEN	20														x		
BENNDALE	12					160											
BEULAH	20						51										
BIBB	30														x		
BIGBEE	20														x		
BINNSVILLE		20													x	30	
BOHICKET	25													x	x		
BONN	2.5				18				20		76						
BOSWELL	12					**165**											
BOWDRE		41													x		
BOYKIN	61					**142**											
BRANDON	23					43											
BRANTLEY	15					119											
BREWTON	15					84											
BROOKSVILLE		41					91										
BRUIN	28						25										
BRUNO	20														x		

(continued)

D. Johnson et al., *The Soils of Mississippi,* World Soils Book Series,
https://doi.org/10.1007/978-3-031-36235-4

(continued)

Series name	Ochric	Mollic	Histic	Umbric	Albic	Argillic	Cambic	Fragipan	Glossic	Kandic	Natric	Spodic	Plinthite	Sulfidic materials	None	To para-lithic	To Lithic
BUCKATUNNA	18					109											
BUDE	36					**94**		**94**									
BYRAM	10					160		71									
CADEVILLE	18					97											
CAHABA	23					112											
CALEDONIA	18					**200**											
CALHOUN	43				36	33			33								
CALLOWAY	18				41	**109**		58	33								
CASCILLA	18						99										
CATALPA		51					102										
CHASTAIN	12						119										
CHENNEBY	41						99										
CHICKASAWHAY	18					86											
CHICORA	25					155											
COLLINS	18														x		
COLUMBUS	15					56											
COMMERCE	18						**185**										
CONVENT	10						122										
CREVASSE	10														x		
CROATAN			71												x		
CUTHBERT	20					66											
DALEVILLE	41					**137**											
DARCO	117					**101**											
DEERFORD	25				10				61		61						
DEMOPOLIS	15														x	33	
DOGUE	25					94											
DOROVAN			180												x		
DOWLING	10						**190**										
DUBBS	12					89											
DUCKSTON	7.5														x		
DULAC	15					137		36									
DUNDEE	12					61											
ESCAMBIA	15					**168**							**122**				
EUSTIS	61					132											
EUTAW	12														x		
FALAYA	25														x		
FALKNER	15					**150**											
FORESTDALE	15					**137**											
FREEST	15					**188**											
FREESTONE	28					**201**			61								
FRIZZELL	5					104			84								
FROST	56				30	216			61								
FRUITDALE	48					132											
GILLSBURG	10						74										
GRENADA	12				7.5	**91**	41	**91**	66								
GRIFFITH		84					**84**										
GUYTON	58				43	47			41								
HANDSBORO			94												x		
HARLESTON	23					**160**											
HEIDEL	10					**175**											
HENRY	79				56	48		48									

(continued)

(continued)

Series name	Ochric	Mollic	Histic	Umbric	Albic	Argillic	Cambic	Fragipan	Glossic	Kandic	Natric	Spodic	Plinthite	Sulfidic materials	None	To para-lithic	To Lithic
HOULKA	20						**145**										
HOUSTON		107													x		
HYDE				38		91											
ICHUSA	10					74											
IRVINGTON	15					**167**		71					**155**				
IUKA	18														x		
IZAGORA	28					132											
JENA	12						107										
JOHNS	38					43											
JOHNSTON				76											x		
KINSTON	30						91										
KIPLING	7.5					130											
KIRKVILLE	23						132										
KISATCHIE	15				7.5	46										61	
KOLIN	20				7	**183**			13								
LAKELAND	7.5														x		
LATONIA	10					61											
LAUDERDALE	7.5					33										41	
LAX	25					**127**		20									
LEAF	18				10	170											
LEEPER	20						66										
LENOIR	20					170											
LEON	38											107					
LEVERETT	23					**188**			**122**								
LEXINGTON	18					**165**											
LINKER	10					79											89
LONGVIEW	10					**160**											
LORING	18					109		30									
LORMAN	12					**71**											
LOUIN	5						147										
LUCEDALE	20					**183**											
LUCY	61					**117**				**117**							
LUVERNE	18					58											
MABEN	12					46											
MALBIS	18					**163**							**114**				
MANTACHIE	12						20										
MARIETTA	25						91										
MASHULAVILLE	12				53	**117**		**117**									
MATHISTON	15						20										
MAUBILA	7.5					79											
MAUREPAS			163												x		
MAYHEW	18					147											
MAYTAG	7.5														x		
MCCRORY	38				20	140					127						
MCLAURIN	20					**132**											
MCRAVEN	12						**188**										
MEMPHIS	23					107											
MHOON	15						81										
MOOREVILLE	12						114										
MORGANFIELD	20														x		
MYATT	25					102											

(continued)

(continued)

Series name	Ochric	Mollic	Histic	Umbric	Albic	Argillic	Cambic	Fragipan	Glossic	Kandic	Natric	Spodic	Plinthite	Sulfidic materials	None	To para-lithic	To Lithic
NAHUNTA	20					127											
NATCHEZ	23						46										
NESHOBA	12					165											
NEWELLTON	18						48										
NEWHAN	5														x		
NUGENT	7.5														x		
OAKLIMETER	28						58										
OCHLOCKONEE	15														x		
OCILLA	71					99											
OKEELALA	28					94											
OKOLONA		46					76										165
OKTIBBEHA	10					23	**170**										
OLIVIER	23				7.5	109		56	18								
OLLA	15					81											
OPENLAKE	18						155										
ORA	18					124		76									
OSIER	7.5														x		
OUACHITA	48						127										
OZAN	38					104			36								
PACTOLUS	20														x		
PADEN	20					**208**		30	15								
PAMLICO			76												x		
PELAHATCHIE	15					94											
PEORIA	23				69				91		**91**						
PETAL	20					**145**											
PHEBA	7.5					76		**111**									
PIKEVILLE	30					135											
PLUMMER	127				104	73											
POARCH	30					**137**							**81**				
POOLEVILLE	10					**178**			152								
PRENTISS	18					**168**		**119**	48								
PRIM		18													x	38	
PROVIDENCE	18					**185**		76									
QUITMAN	30					**183**		130									
RATTLESNAKE FORKS	86														x		
RIEDTOWN	18						**185**										
ROBINSONVILLE	18														x		
ROSEBLOOM	20						102										
ROSELLA	7.5				23				66		**135**						
RUSTON	18				7.5	**198**											
RUTAN	18					91											
SAFFELL	20					84											
SAUCIER	12					**152**							64				
SAVANNAH	25					**198**		112									
SESSUM	15					94											
SHARKEY	25						**188**										
SHUBUTA	20					**157**											
SILVERDALE	20														x		
SIWELL	10					89											
SMITHDALE	28					175											

(continued)

(continued)

Series name	Ochric	Mollic	Histic	Umbric	Albic	Argillic	Cambic	Fragipan	Glossic	Kandic	Natric	Spodic	Plinthite	Sulfidic materials	None	To para-lithic	To Lithic
SMITHTON	25				23	**157**											
STEENS	38					**127**											
STOUGH	36					144		144									
SUCARNOOCHEE	25						**140**										
SUFFOLK	41					78											
SUGGSVILLE	2.5					**104**										107	
SUMTER	25														x	71	
SUSQUEHANNA	12					**183**											
SWEATMAN	15					79											
TALLA	15				46	**140**					**104**						
TEKSOB	51					51											
TENSAS	10					**36**											
TIPPAH	15					**137**											
TIPPO	15					**163**			**127**								
TOINETTE	79					86											
TREBLOC	12				7.5	**152**											
TRINITY	41						**150**										
TROUP	112					**91**				**91**							
TUNICA	8						57										
TUSCUMBIA	10														x		
TUTWILER	15					36											
UNA	12						**140**										
URBO	23						**157**										
VAIDEN	10					56	**137**										
VANCLEAVE	28					185		135					135				
VELDA	18						91										
VERDUN	10								15		112						
VICKSBURG	18														x		
VIMVILLE	10					109			**135**								
WADLEY	183					**20**											
WANILLA	7.5					**180**			**180**								
WATSONIA	7.5						33									41	
WAVERLY	12						**190**										
WEYANOKE	7.5						61										
WILCOX	10					33										163	
WILLIAMSVILLE	30					**178**											
ALMO	18					**81**		89									
DEXTER	25					56											
FREELAND	33					**167**		66	12								
GREENVILLE	22									**180**							
HATCHIE	18					**117**		**86**									
ORANGEBURG	18					**165**				**165**							
RICHLAND	20					119											
Average	24	51	117	57	30	120	107	81	65	138	101	107	97	114		68	127
Standard deviation	21	30	51	27	25	48	51	36	49	41	27		38			43	54
Number	203	7	4	2	21	132	48	20	22	2	6	1	10	2	40	8	3
% of all soil series	94	3.2	1.9	0.9	9.7	61.1	22.2	9.3	10.2	0.9	2.8	0.5	4.6	0.9	17.3	3.7	1.4

Bold face = minimum depth
Yellow highlight = estimated by JGB

Appendix D: Classification and reas of Soil Series Recognized in Mississippi

Series name	State TL	Other states using	MLRAs using MS	Order	Suborder	Great group	Subgroup	Particle-size class	Mineralogy class	CEC activity class	Reaction class	STR	SMR	Other family class
ADATON	MS	AL, AR, OK, TN, TX	133C	Alfisols	Aqualfs	Endoaqualfs	Typic Endoaqualfs	Fine-silty	Mixed	Active		Thermic	Aquic	
ADLER	MS	KY, MO, TN	134	Inceptisols	Udepts	Eutrudepts	Fluvaquentic Eutrudepts	Coarse-silty	Mixed	Superactive		Thermic	Udic	
ALAGA	AL	AR, FL, LA, MS, NC, SC, TX, VA	133C, 152A	Entisols	Psamments	Quartzipsamments	Typic Quartzipsamments					Thermic	Udic	Coated
ALLIGATOR	MS	AR, KY, LA, MO, TN	131A	Vertisols	Aquerts	Dystraquerts	Chromic Dystraquerts	Very-fine	Smectitic			Thermic	Aquic	
ALMO	TN		133C, 134	Alfisols	Aqualfs	Fragiaqualfs	Typic Fragiaqualfs	Fine-silty	Mixed	Active		Thermic	Aquic	
ANGIE	LA	AL, AR, FL, MS, SC, TX, VA	133C	Ultisols	Udults	Paleudults	Aquic Paleudults	Fine	Mixed	Semiactive		Thermic	Udic	
ANNEMAINE	AL	MS	133C	Ultisols	Udults	Hapludults	Aquic Hapludults	Fine	Mixed	Semiactive		Thermic	Udic	
ARAT	LA		151	Entisols	Aquents	Hydraquents	Typic Hydraquents	Fine-silty	Siliceous	Superactive	Nonacid	Thermic	Aquic	
ARIEL	MS	AR	134	Inceptisols	Udepts	Dystrudepts	Fluventic Dystrudepts	Coarse-silty	Mixed	Active		Thermic	Udic	
ARKABUTLA	MS	AR, KY, LA, OK, TN	134	Inceptisols	Aquepts	Endoaquepts	Fluventic Endoaquepts	Fine-silty	Mixed	Active	Acid	Thermic	Aquic	
ARUNDEL	MS	AL, GA, TN	133C	Ultisols	Udults	Hapludults	Typic Hapludults	Fine	Smectitic			Thermic	Udic	
ASKEW	AR	MO, MS, TN	131A	Alfisols	Udalfs	Hapludalfs	Aquic Hapludalfs	Fine-silty	Mixed	Active		Thermic	Udic	
ATMORE	AL	MS	133C, 152A	Ultisols	Aquults	Paleaquults	Plinthic Paleaquults	Coarse-loamy	Siliceous	Semiactive		Thermic	Aquic	
ATWOOD	MS		133C, 134	Alfisols	Udalfs	Paleudalfs	Typic Paleudalfs	Fine-silty	Mixed	Semiactive		Thermic	Udic	
AXIS	AL	FL, MD, MS, NC, VA	152A	Entisols	Aquents	Sulfaquents	Typic Sulfaquents	Coarse-loamy	Mixed	Superactive	Nonacid	Thermic	Aquic	
BAMA	AL	FL, MS, VA	133C	Ultisols	Udults	Paleudults	Typic Paleudults	Fine-loamy	Siliceous	Subactive		Thermic	Udic	
BASIN	MS	AL	133C, 152A	Ultisols	Udults	Paleudults	Fragiaquic Paleudults	Coarse-loamy	Siliceous	Semiactive		Thermic	Udic	
BASSFIELD	MS	AL, LA	133C	Ultisols	Udults	Hapludults	Typic Hapludults	Coarse-loamy	Siliceous	Semiactive		Thermic	Udic	
BAXTERVILLE	MS	AL	133C	Ultisols	Udults	Paleudults	Plinthic Paleudults	Fine-loamy	Siliceous	Subactive		Thermic	Udic	
BAYOU	AL		133C, 152A	Ultisols	Aquults	Paleaquults	Typic Paleaquults	Coarse-loamy	Siliceous	Semiactive		Thermic	Aquic	
BEAUREGARD	LA	MS	133C,152A	Ultisols	Udults	Paleudults	Plinthaquic Paleudults	Fine-silty	Siliceous	Superactive		Thermic	Udic	
BELDEN	MS		133C, 135A	Entisols	Aquents	Fluvaquents	Aeric Fluvaquents	Fine-silty	Mixed	Active	Nonacid	Thermic	Aquic	
BENNDALE	MS	AL	133C	Ultisols	Udults	Paleudults	Typic Paleudults	Coarse-loamy	Siliceous	Semiactive		Thermic	Udic	
BEULAH	AR	KY, MO, MS, TN	131A	Inceptisols	Udepts	Dystrudepts	Typic Dystrudepts	Coarse-loamy	Mixed	Active		Thermic	Udic	
BIBB	AL	AR, FL, GA, KY, LA, MD, MS, NC, OK, SC, TN, TX, VA	133C,152A	Entisols	Aquents	Fluvaquents	Typic Fluvaquents	Coarse-loamy	Siliceous	Active	Acid	Thermic	Aquic	
BIGBEE	MS	AL, FL, GA, LA, TN	133C	Entisols	Psamments	Quartzipsamments	Typic Quartzipsamments					Thermic	Udic	Coated
BINNSVILLE	MS	AL, GA	135A	Mollisols	Rendolls	Haprendolls	Typic Haprendolls	Clayey	Carbonatic			Thermic	Udic	Shallow
BOHICKET	SC	FL, GA, MS, NC, VA	152A	Entisols	Aquents	Sulfaquents	Typic Sulfaquents	Fine	Mixed	Superactive	Nonacid	Thermic	Aquic	

(continued)

D. Johnson et al., *The Soils of Mississippi*, World Soils Book Series,
https://doi.org/10.1007/978-3-031-36235-4

(continued)

Series name	State TL	Other states using	MLRAs using MS	Order	Suborder	Great group	Subgroup	Particle-size class	Mineralogy class	CEC activity class	Reaction class	STR	SMR	Other family class
BONN	LA	AR, MS, TN, TX	133C, 134	Alfisols	Aqualfs	Natraqualfs	Glossic Natraqualfs	Fine-silty	Mixed	Superactive		Thermic	Aquic	
BOSWELL	MS	AL, AR, FL, GA, LA, TN	133C	Alfisols	Udalfs	Paleudalfs	Vertic Paleudalfs	Fine	Mixed	Active		Thermic	Udic	
BOWDRE	MS	AR, IL, KY, MO, TN	131A	Mollisols	Udolls	Hapludolls	Fluvaquentic Hapludolls	Clayey over loamy	Smectitic			Thermic	Udic	
BOYKIN	TX	LA	133C	Ultisols	Udults	Paleudults	Arenic Paleudults	Loamy	Siliceous	Active		Thermic	Udic	
BRANDON	KY	AR, IL, MO, MS, TN	134	Ultisols	Udults	Hapludults	Typic Hapludults	Fine-silty	Mixed	Semiactive		Thermic	Udic	
BRANTLEY	AL	FL, MS, TN	133C, 135A	Alfisols	Udalfs	Hapludalfs	Ultic Hapludalfs	Fine	Mixed	Active		Thermic	Udic	
BREWTON	AL	MS	133C	Ultisols	Udults	Paleudults	Fragiaquic Paleudults	Coarse-loamy	Siliceous	Subactive		Thermic	Udic	
BROOKSVILLE	MS	AL	135A	Vertisols	Uderts	Hapluderts	Aquic Hapluderts	Fine	Smectitic			Thermic	Udic	
BRUIN	LA	MS	131A	Inceptisols	Udepts	Eutrudepts	Oxyaquic Eutrudepts	Coarse-silty	Mixed	Superactive		Thermic	Udic	
BRUNO	TN	AR, KY, LA, MS	131A	Entisols	Fluvents	Udifluvents	Typic Udifluvents	Sandy	Mixed			Thermic	Udic	
BUCKATUNNA	MS		133C	Ultisols	Udults	Hapludults	Aquic Hapludults	Fine	Mixed	Semiactive		Thermic	Udic	
BUDE	MS	AR, LA, TN	133C, 134	Alfisols	Udalfs	Fragiudalfs	Aquic Fragiudalfs	Fine-silty	Mixed	Active		Thermic	Udic	
BYRAM	MS		134	Alfisols	Udalfs	Fragiudalfs	Typic Fragiudalfs	Fine-silty	Mixed	Active		Thermic	Udic	
CADEVILLE	LA	AL, MS, OK	133C	Alfisols	Udalfs	Hapludalfs	Albaquic Hapludalfs	Fine	Mixed	Active		Thermic	Udic	
CAHABA	AL	AR, FL, GA, LA, MS, OK, SC, VA	133C	Ultisols	Udults	Hapludults	Typic Hapludults	Fine-loamy	Siliceous	Semiactive		Thermic	Udic	
CALEDONIA	MS		133C	Alfisols	Udalfs	Paleudalfs	Typic Paleudalfs	Fine-loamy	Siliceous	Semiactive		Thermic	Udic	
CALHOUN	LA	AR, KY, MO, MS, TN, TX	131A, 134	Alfisols	Aqualfs	Glossaqualfs	Typic Glossaqualfs	Fine-silty	Mixed	Active		Thermic	Aquic	
CALLOWAY	MS	AR, KY, LA, TN	134	Alfisols	Udalfs	Fraglossudalfs	Aquic Fraglossudalfs	Fine-silty	Mixed	Active		Thermic	Udic	
CASCILLA	MS	KY, LA, TN	134	Inceptisols	Udepts	Dystrudepts	Fluventic Dystrudepts	Fine-silty	Mixed	Active		Thermic	Udic	
CATALPA	MS	AL, AR	135A	Mollisols	Udolls	Hapludolls	Fluvaquentic Hapludolls	Fine	Smectitic			Thermic	Udic	
CHASTAIN	SC	AL, GA, MS, NC, VA	133C	Inceptisols	Aquepts	Endoaquepts	Fluvaquentic Endoaquepts	Fine	Mixed	Semiactive	Acid	Thermic	Aquic	
CHENNEBY	AL	GA, MS, NC, SC, TN, VA	133C, 134	Inceptisols	Udepts	Dystrudepts	Fluvaquentic Dystrudepts	Fine-silty	Mixed	Active		Thermic	Udic	
CHICKASAWHAY	MS		133C	Ultisols	Udults	Hapludults	Aquic Hapludults	Fine	Mixed	Semiactive		Thermic	Udic	
CHICORA	MS		133C	Ultisols	Udults	Hapludults	Aquic Hapludults	Fine-loamy	Siliceous	Semiactive		Thermic	Udic	
COLLINS	MS	AR, KY, MO, TN	131A, 133C, 134	Entisols	Fluvents	Udifluvents	Aquic Udifluvents	Coarse-silty	Mixed	Active	Acid	Thermic	Udic	
COLUMBUS	MS	AL	133C	Ultisols	Udults	Hapludults	Aquic Hapludults	Fine-loamy	Siliceous	Semiactive		Thermic	Udic	
COMMERCE	LA	AR, KY, MO, MS, TN	131A	Inceptisols	Aquepts	Endoaquepts	Fluvaquentic Endoaquepts	Fine-silty	Mixed	Superactive	Nonacid	Thermic	Aquic	
CONVENT	LA	AR, KY, MO, MS, TN	131A	Inceptisols	Aquepts	Endoaquepts	Fluvaquentic Endoaquepts	Coarse-silty	Mixed	Superactive	Nonacid	Thermic	Aquic	
CREVASSE	MS	AR, KY, LA, MO, TN	131A	Entisols	Psamments	Udipsamments	Typic Udipsamments		Mixed			Thermic	Udic	
CROATAN	NC	FL, GA, MS	133C	Histosols	Saprists	Haplosaprists	Terric Haplosaprists	Loamy	Siliceous		Dysic	Thermic	Aquic	
CUTHBERT	TX	AR, MS, TN	133C	Ultisols	Udults	Hapludults	Typic Hapludults	Fine	Mixed	Semiactive		Thermic	Udic	
DALEVILLE	MS	SC, VA	133C	Ultisols	Aquults	Paleaquults	Typic Paleaquults	Fine-loamy	Siliceous	Active		Thermic	Aquic	
DARCO	TX	AR, MS	133C	Ultisols	Udults	Paleudults	Grossarenic Paleudults	Loamy	Siliceous	Semiactive		Thermic	Udic	
DEERFORD	LA	AL, MS	134	Alfisols	Aqualfs	Natraqualfs	Albic Glossic Natraqualfs	Fine-silty	Mixed	Superactive		Thermic	Aquic	
DEMOPOLIS	AL	AR, MS	135A	Entisols	Orthents	Udorthents	Typic Udorthents	Loamy	Carbonatic			Thermic	Udic	Shallow
DEXTER	LA	AR, TN	133C, 134	Alfisols	Udalfs	Hapludalfs	Ultic Hapludalfs	Fine-silty	Mixed	Active		Thermic	Udic	
DOGUE	VA	GA, NC	133C	Ultisols	Udults	Hapludults	Aquic Hapludults	Fine	Mixed	Semiactive		Thermic	Udic	
DOROVAN	MS	AL, FL, NC, SC, VA	133C, 152A	Histosols	Saprists	Haplosaprists	Typic Haplosaprists				Dysic	Thermic	Aquic	
DOWLING	LA	AR, MS, TN	131A	Inceptisols	Aquepts	Endoaquepts	Vertic Endoaquepts	Very-fine	Smectitic		Nonacid	Thermic	Aquic	
DUBBS	MS	AR, KY, LA, MO, TN	131A	Alfisols	Udalfs	Hapludalfs	Typic Hapludalfs	Fine-silty	Mixed	Active		Thermic	Udic	

(continued)

(continued)

Series name	State TL	Other states using	MLRAs using MS	Order	Suborder	Great group	Subgroup	Particle-size class	Mineralogy class	CEC activity class	Reaction class	STR	SMR	Other family class
DUCKSTON	NC	AL, FL, GA, VA	152A	Entisols	Aquents	Psammaquents	Typic Psammaquents		Siliceous			Thermic	Aquic	
DULAC	TN	AL, MS	133C	Alfisols	Udalfs	Fragiudalfs	Oxyaquic Fragiudalfs	Fine-silty	Mixed	Semiactive		Thermic	Udic	
DUNDEE	MS	AR, KY, LA, MO, TN	131A	Alfisols	Aqualfs	Endoaqualfs	Typic Endoaqualfs	Fine-silty	Mixed	Active		Thermic	Aquic	
ESCAMBIA	AL	FL, MS	133C, 152A	Ultisols	Udults	Paleudults	Plinthaquic Paleudults	Coarse-loamy	Siliceous	Semiactive		Thermic	Udic	
EUSTIS	FL	AL, GA, LA, MO, MS, NC, SC, TN, TX	133C	Ultisols	Udults	Paleudults	Psammentic Paleudults	Sandy	Siliceous			Thermic	Udic	
EUTAW	MS	AL, AR, GA	133C, 135A	Vertisols	Aquerts	Dystraquerts	Chromic Dystraquerts	Very-fine	Smectitic			Thermic	Aquic	
FALAYA	TN	AL, AR, KY, LA, MO, MS	131A, 133C, 134	Entisols	Aquents	Fluvaquents	Aeric Fluvaquents	Coarse-silty	Mixed	Active	Acid	Thermic	Aquic	
FALKNER	MS	AL, AR, LA, TN	133C, 134	Alfisols	Udalfs	Paleudalfs	Aquic Paleudalfs	Fine-silty	Siliceous	Active		Thermic	Udic	
FORESTDALE	MS	AR, KY, LA, MO, TN	131A	Alfisols	Aqualfs	Endoaqualfs	Typic Endoaqualfs	Fine	Smectitic			Thermic	Aquic	
FREELAND	TN	AR, KY	133C	Alfisols	Udalfs	Fraglossudalfs	Oxyaquic Fraglossudalfs	Fine-silty	Siliceous	Active		Thermic	Udic	
FREEST	MS	AL	133C, 135A	Alfisols	Udalfs	Paleudalfs	Aquic Paleudalfs	Fine-loamy	Siliceous	Active		Thermic	Udic	
FREESTONE	TX	MS, OK	133C	Alfisols	Udalfs	Paleudalfs	Glossaquic Paleudalfs	Fine-loamy	Siliceous	Semiactive		Thermic	Udic	
FRIZZELL	LA	MS, OK	133C	Alfisols	Udalfs	Hapludalfs	Glossaquic Hapludalfs	Coarse-silty	Siliceous	Active		Thermic	Udic	
FROST	LA	MS	134	Alfisols	Aqualfs	Glossaqualfs	Typic Glossaqualfs	Fine-silty	Mixed	Active		Thermic	Aquic	
FRUITDALE	AL	MS	133C	Ultisols	Udults	Hapludults	Typic Hapludults	Fine-loamy	Siliceous	Subactive		Thermic	Udic	
GILLSBURG	MS		133C, 134	Entisols	Aquents	Fluvaquents	Aeric Fluvaquents	Coarse-silty	Mixed	Active	Acid	Thermic	Aquic	
GREENVILLE	AL	AR, FL, GA, SC	133C	Ultisols	Udults	Kandiudults	Rhodic Kandiudults	Fine	Kaolinitic			Thermic	Udic	
GRENADA	MS	AR, KY, LA, TN	134	Alfisols	Udalfs	Fraglossudalfs	Oxyaquic Fraglossudalfs	Fine-silty	Mixed	Active		Thermic	Udic	
GRIFFITH	MS		135A	Vertisols	Uderts	Hapluderts	Aquic Hapluderts	Fine	Smectitic			Thermic	Udic	
GUYTON	LA	AR, MS, OK, TN, TX	133C, 151, 152A	Alfisols	Aqualfs	Glossaqualfs	Typic Glossaqualfs	Fine-silty	Siliceous	Active		Thermic	Aquic	
HANDSBORO	MS	FL, SC	152A, 151	Histosols	Hemists	Sulfihemists	Typic Sulfihemists				Euic	Thermic	Aquic	
HARLESTON	MS	AL, AR, LA	133C	Ultisols	Udults	Paleudults	Aquic Paleudults	Coarse-loamy	Siliceous	Semiactive		Thermic	Udic	
HATCHIE	TN	AR	133C	Alfisols	Udalfs	Fraglossudalfs	Aquic Fraglossudalfs	Fine-silty	Siliceous	Active		Thermic	Udic	
HEIDEL	MS	AL	133C	Ultisols	Udults	Paleudults	Typic Paleudults	Coarse-loamy	Siliceous	Subactive		Thermic	Udic	
HENRY	TN	AR, KY, MS	134	Alfisols	Aqualfs	Fragiaqualfs	Typic Fragiaqualfs	Coarse-silty	Mixed	Active		Thermic	Aquic	
HOULKA	MS	AL	133C, 135A	Vertisols	Aquerts	Epiaquerts	Aeric Epiaquerts	Fine	Smectitic			Thermic	Aquic	
HOUSTON	AL	AR, MS	135A	Vertisols	Uderts	Hapluderts	Oxyaquic Hapluderts	Very-fine	Smectitic			Thermic	Udic	
HYDE	NC	AL, MS, SC, VA	152A	Ultisols	Aquults	Umbraquults	Typic Umbraquults	Fine-silty	Mixed	Active		Thermic	Aquic	
ICHUSA	MS	AL	133C, 135A	Vertisols	Uderts	Dystruderts	Aquic Dystruderts	Fine	Smectitic			Thermic	Udic	
IRVINGTON	AL	GA, MS	133C, 152A	Ultisols	Udults	Fragiudults	Plinthic Fragiudults	Fine-loamy	Siliceous	Subactive		Thermic	Udic	
IUKA	AL	AR, FL, GA, KY, LA, MS, NC, OK, TN, TX, VA	133C, 152A	Entisols	Fluvents	Udifluvents	Aquic Udifluvents	Coarse-loamy	Siliceous	Active	Acid	Thermic	Udic	
IZAGORA	AL	FL, GA, MS, NC, SC, VA	133C	Ultisols	Udults	Paleudults	Aquic Paleudults	Fine-loamy	Siliceous	Semiactive		Thermic	Udic	
JENA	LA	MS	133C	Inceptisols	Udepts	Dystrudepts	Fluventic Dystrudepts	Coarse-loamy	Siliceous	Active		Thermic	Udic	
JOHNS	NC	FL, GA, MS, SC, VA	133C	Ultisols	Udults	Hapludults	Aquic Hapludults	Fine-loamy over sandy or sandy-skeletal	Siliceous	Semiactive		Thermic	Udic	
JOHNSTON	NC	AL, FL, GA, MS, SC, VA	133C	Inceptisols	Aquepts	Humaquepts	Cumulic Humaquepts	Coarse-loamy	Siliceous	Active	Acid	Thermic	Aquic	
KINSTON	NC	AL, FL, GA, MS, SC, TN, VA	133C	Inceptisols	Aquepts	Endoaquepts	Fluvaquentic Endoaquepts	Fine-loamy	Siliceous	Semiactive	Acid	Thermic	Aquic	
KIPLING	MS	AL, AR	135A	Alfisols	Udalfs	Paleudalfs	Vertic Paleudalfs	Fine	Smectitic			Thermic	Udic	
KIRKVILLE	MS	AL, AR	133C	Inceptisols	Udepts	Dystrudepts	Fluvaquentic Dystrudepts	Coarse-loamy	Siliceous	Active		Thermic	Udic	

(continued)

(continued)

Series name	State TL	Other states using	MLRAs using MS	Order	Suborder	Great group	Subgroup	Particle-size class	Mineralogy class	CEC activity class	Reaction class	STR	SMR	Other family class
KISATCHIE	LA	MS, TX	133C	Alfisols	Udalfs	Hapludalfs	Typic Hapludalfs	Fine	Smectitic			Thermic	Udic	
KOLIN	LA	MS	133C	Alfisols	Udalfs	Paleudalfs	Glossaquic Paleudalfs	Fine-silty	Siliceous	Active		Thermic	Udic	
LAKELAND	FL	AL, GA, LA, MD, MS, NC, NJ, SC, VA	133C, 152A	Entisols	Psamments	Quartzipsamments	Typic Quartzipsamments					Thermic	Udic	Coated
LATONIA	MS	AL, AR, LA	133C, 152A	Ultisols	Udults	Hapludults	Typic Hapludults	Coarse-loamy	Siliceous	Semiactive		Thermic	Udic	
LAUDERDALE	MS	AL	133C	Ultisols	Udults	Hapludults	Typic Hapludults	Loamy	Mixed	Active		Thermic	Udic	Shallow
LAX	TN	KY, MS	133C, 134	Ultisols	Udults	Fragiudults	Typic Fragiudults	Fine-silty	Mixed	Semiactive		Thermic	Udic	
LEAF	MS	AL, AR, FL, GA, LA, NC, SC, VA	133C, 152A	Ultisols	Aquults	Albaquults	Typic Albaquults	Fine	Mixed	Active		Thermic	Aquic	
LEEPER	MS	AL, AR	133C, 135A	Inceptisols	Aquepts	Epiaquepts	Vertic Epiaquepts	Fine	Smectitic		Nonacid	Thermic	Aquic	
LENOIR	NC	AL, MS, SC, VA	133C	Ultisols	Aquults	Paleaquults	Aeric Paleaquults	Fine	Mixed	Semiactive		Thermic	Aquic	
LEON	FL	AL, GA, MD, MS, NC, SC, VA	133C, 152A	Spodosols	Aquods	Alaquods	Aeric Alaquods	Sandy	Siliceous			Thermic	Aquic	
LEVERETT	MS		134	Alfisols	Udalfs	Glossudalfs	Haplic Glossudalfs	Coarse-silty	Mixed	Active		Thermic	Udic	
LEXINGTON	TN	AR, KY, LA, MS	133C, 134	Alfisols	Udalfs	Hapludalfs	Ultic Hapludalfs	Fine-silty	Mixed	Active		Thermic	Udic	
LINKER	AR	AL, GA, KY, MS, OK, TN	133C	Ultisols	Udults	Hapludults	Typic Hapludults	Fine-loamy	Siliceous	Semiactive		Thermic	Udic	
LONGVIEW	MS	TN	133C	Alfisols	Udalfs	Hapludalfs	Glossaquic Hapludalfs	Fine-silty	Siliceous	Active		Thermic	Udic	
LORING	TN	AR, KY, LA, MO, MS	134	Alfisols	Udalfs	Fragiudalfs	Oxyaquic Fragiudalfs	Fine-silty	Mixed	Active		Thermic	Udic	
LORMAN	MS	AL	133C, 134	Alfisols	Udalfs	Hapludalfs	Chromic Vertic Hapludalfs	Fine	Smectitic			Thermic	Udic	
LOUIN	MS	AR	135A	Vertisols	Uderts	Dystruderts	Aquic Dystruderts	Fine	Smectitic			Thermic	Udic	
LUCEDALE	MS	AL, LA	133C	Ultisols	Udults	Paleudults	Rhodic Paleudults	Fine-loamy	Siliceous	Subactive		Thermic	Udic	
LUCY	AL	AR, FL, GA, LA, MS, NC, SC, TN, VA	133C	Ultisols	Udults	Kandiudults	Arenic Kandiudults	Loamy	Kaolinitic			Thermic	Udic	
LUVERNE	AL	AR, LA, MS, TN	133C	Ultisols	Udults	Hapludults	Typic Hapludults	Fine	Mixed	Semiactive		Thermic	Udic	
MABEN	MS	TX	133C	Alfisols	Udalfs	Hapludalfs	Ultic Hapludalfs	Fine	Mixed	Active		Thermic	Udic	
MALBIS	AL	FL, LA, MS, TX, VA	133C	Ultisols	Udults	Paleudults	Plinthic Paleudults	Fine-loamy	Siliceous	Subactive		Thermic	Udic	
MANTACHIE	MS	AL, AR, FL, KY, LA, NC, TN, TX, VA	133C	Inceptisols	Aquepts	Endoaquepts	Fluventic Endoaquepts	Fine-loamy	Siliceous	Active	Acid	Thermic	Aquic	
MARIETTA	MS	AL, AR, TX	133C, 135A	Inceptisols	Udepts	Eutrudepts	Fluvaquentic Eutrudepts	Fine-loamy	Siliceous	Active		Thermic	Udic	
MASHULAVILLE	MS	AL, AR	133C	Ultisols	Aquults	Fragiaquults	Typic Fragiaquults	Coarse-loamy	Siliceous	Active		Thermic	Aquic	
MATHISTON	MS	TX	133C	Entisols	Aquents	Fluvaquents	Aeric Fluvaquents	Fine-silty	Siliceous	Active	Acid	Thermic	Aquic	
MAUBILA	AL	FL, GA	133C	Ultisols	Udults	Hapludults	Aquic Hapludults	Fine	Mixed	Subactive		Thermic	Udic	
MAUREPAS	LA		131A	Histosols	Saprists	Haplosaprists	Typic Haplosaprists				Euic	Hyperthermic	Aquic	
MAYHEW	MS	AL, AR, LA	133C, 135A	Vertisols	Aquerts	Dystraquerts	Chromic Dystraquerts	Fine	Smectitic			Thermic	Aquic	
MAYTAG	AL	MS	133C, 135A	Vertisols	Uderts	Hapluderts	Oxyaquic Hapluderts	Fine	Smectitic			Thermic	Udic	
MCCRORY	AR		131A	Alfisols	Aqualfs	Natraqualfs	Albic Glossic Natraqualfs	Fine-loamy	Mixed	Active		Thermic	Aquic	
MCLAURIN	MS	AL, AR, LA, VA	133C	Ultisols	Udults	Paleudults	Typic Paleudults	Coarse-loamy	Siliceous	Subactive		Thermic	Udic	
MCRAVEN	MS		134	Inceptisols	Aquepts	Endoaquepts	Aeric Endoaquepts	Coarse-silty	Mixed	Active	Nonacid	Thermic	Aquic	
MEMPHIS	MS	AR, KY, LA, MO, TN	134	Alfisols	Udalfs	Hapludalfs	Typic Hapludalfs	Fine-silty	Mixed	Active		Thermic	Udic	
MHOON	LA	AR, KY, MO, MS, TN	131A	Inceptisols	Aquepts	Endoaquepts	Fluvaquentic Endoaquepts	Fine-silty	Mixed	Superactive	Nonacid	Thermic	Aquic	
MOOREVILLE	MS	AL, TX	133C, 135A	Inceptisols	Udepts	Dystrudepts	Fluvaquentic Dystrudepts	Fine-loamy	Siliceous	Active		Thermic	Udic	
MORGANFIELD	MS	AR, KY, LA, TN	134	Entisols	Fluvents	Udifluvents	Typic Udifluvents	Coarse-silty	Mixed	Active	Nonacid	Thermic	Udic	

(continued)

(continued)

Series name	State TL	Other states using	MLRAs using MS	Order	Suborder	Great group	Subgroup	Particle-size class	Mineralogy class	CEC activity class	Reaction class	STR	SMR	Other family class
MYATT	AL	AR, FL, GA, LA, MS, NC, SC, VA	133C, 152A	Ultisols	Aquults	Endoaquults	Typic Endoaquults	Fine-loamy	Siliceous	Active		Thermic	Aquic	
NAHUNTA	NC	AL, MS, VA	133C	Ultisols	Aquults	Paleaquults	Aeric Paleaquults	Fine-silty	Siliceous	Subactive		Thermic	Aquic	
NATCHEZ	MS	AR, LA, TN	134	Inceptisols	Udepts	Eutrudepts	Typic Eutrudepts	Coarse-silty	Mixed	Superactive		Thermic	Udic	
NESHOBA	MS		133C	Ultisols	Udults	Paleudults	Rhodic Paleudults	Fine	Mixed	Semiactive		Thermic	Udic	
NEWELLTON	LA	AR, MS, OK	131A	Inceptisols	Aquepts	Epiaquepts	Fluvaquentic Epiaquepts	Clayey over loamy	Smectitic over mixed	Superactive	Nonacid	Thermic	Aquic	
NEWHAN	NC	FL, SC, VA	152A	Entisols	Psamments	Quartzipsamments	Typic Quartzipsamments					Thermic	Udic	Uncoated
NUGENT	MS	AL, AR, LA, TN, TX	133C, 152A	Entisols	Fluvents	Udifluvents	Typic Udifluvents	Sandy	Siliceous			Thermic	Udic	
OAKLIMETER	MS	AR, MO, TN	131A, 133C, 134	Inceptisols	Udepts	Dystrudepts	Fluvaquentic Dystrudepts	Coarse-silty	Mixed	Active		Thermic	Udic	
OCHLOCKONEE	AL	AR, GA, KY, LA, MS, OK, SC, TN, TX, VA	133C	Entisols	Fluvents	Udifluvents	Typic Udifluvents	Coarse-loamy	Siliceous	Active	Acid	Thermic	Udic	
OCILLA	GA	AL, FL, MS, NC, SC, VA	133C	Ultisols	Udults	Paleudults	Aquic Arenic Paleudults	Loamy	Siliceous	Semiactive		Thermic	Udic	
OKEELALA	MS	AL, FL	133C	Alfisols	Udalfs	Hapludalfs	Ultic Hapludalfs	Fine-loamy	Siliceous	Semiactive		Thermic	Udic	
OKOLONA	MS	AL	135A	Vertisols	Uderts	Hapluderts	Oxyaquic Hapluderts	Fine	Smectitic			Thermic	Udic	
OKTIBBEHA	AL	AR, FL, GA, LA, MS, TN	133C, 135A	Vertisols	Uderts	Dystruderts	Chromic Dystruderts	Very-fine	Smectitic			Thermic	Udic	
OLIVIER	LA	MS	134	Alfisols	Udalfs	Fraglossudalfs	Aquic Fraglossudalfs	Fine-silty	Mixed	Active		Thermic	Udic	
OLLA	LA	TN	133C	Ultisols	Udults	Hapludults	Typic Hapludults	Fine-loamy	Siliceous	Active		Thermic	Udic	
OPENLAKE	TN	KY, MS	131A	Inceptisols	Aquepts	Epiaquepts	Vertic Epiaquepts	Fine	Smectitic		Nonacid	Thermic	Aquic	
ORA	MS	AL, AR, LA	133C	Ultisols	Udults	Fragiudults	Typic Fragiudults	Fine-loamy	Siliceous	Semiactive		Thermic	Udic	
ORANGEBURG	GA	AL, AR, FL, LA, NC, SC, VA	133C	Ultisols	Udults	Kandiudults	Typic Kandiudults	Fine-loamy	Kaolinitic			Thermic	Udic	
OSIER	GA	AL, DE, FL, LA, MD, MS, NC, SC, TX, VA	133C	Entisols	Aquents	Psammaquents	Typic Psammaquents		Siliceous			Thermic	Aquic	
OUACHITA	AR	LA, MS	152A	Inceptisols	Udepts	Dystrudepts	Fluventic Dystrudepts	Fine-silty	Siliceous	Active		Thermic	Udic	
OZAN	AR	MS, TX	133C	Alfisols	Aqualfs	Glossaqualfs	Typic Glossaqualfs	Coarse-loamy	Siliceous	Subactive		Thermic	Aquic	
PACTOLUS	NC	AL, FL, SC, VA	133C	Entisols	Psamments	Quartzipsamments	Aquic Quartzipsamments					Thermic	Aquic	Coated
PADEN	TN	MS	133C	Ultisols	Udults	Fragiudults	Glossic Fragiudults	Fine-silty	Mixed	Semiactive		Thermic	Udic	
PAMLICO	NC	AL, FL, MS, SC, VA	133C, 152A	Histosols	Saprists	Haplosaprists	Terric Haplosaprists	Sandy or sandy-skeletal	Siliceous		Dysic	Thermic	Aquic	
PELAHATCHIE	MS		135A	Alfisols	Udalfs	Hapludalfs	Aquic Hapludalfs	Fine-silty	Mixed	Active		Thermic	Udic	
PEORIA	MS		133C, 134	Alfisols	Aqualfs	Natraqualfs	Albic Glossic Natraqualfs	Fine-silty	Mixed	Superactive		Thermic	Aquic	
PETAL	MS		133C, 152A	Alfisols	Udalfs	Paleudalfs	Aquic Paleudalfs	Fine-loamy	Siliceous	Active		Thermic	Udic	
PHEBA	MS	AL, AR, LA	133C	Ultisols	Udults	Fragiudults	Glossaquic Fragiudults	Coarse-silty	Siliceous	Active		Thermic	Udic	
PIKEVILLE	AL	AR, MS, TN	133C	Ultisols	Udults	Paleudults	Typic Paleudults	Fine-loamy	Siliceous	Subactive		Thermic	Udic	
PLUMMER	GA	AL, FL, MS, NC, SC, TX, VA	133C	Ultisols	Aquults	Paleaquults	Grossarenic Paleaquults	Loamy	Siliceous	Subactive		Thermic	Aquic	
POARCH	AL	FL, MS	133C, 152A	Ultisols	Udults	Paleudults	Plinthic Paleudults	Coarse-loamy	Siliceous	Semiactive		Thermic	Udic	
POOLEVILLE	MS		133C, 134	Alfisols	Udalfs	Glossudalfs	Aquic Glossudalfs	Fine-silty	Siliceous	Active		Thermic	Udic	
PRENTISS	MS	AL, AR, LA	133C	Ultisols	Udults	Fragiudults	Glossic Fragiudults	Coarse-loamy	Siliceous	Semiactive		Thermic	Udic	
PRIM	AL		133C, 135A	Mollisols	Rendolls	Haprendolls	Typic Haprendolls	Loamy-skeletal	Carbonatic			Thermic	Udic	Shallow
PROVIDENCE	MS	AR, KY, LA, TN	133C, 134	Alfisols	Udalfs	Fragiudalfs	Oxyaquic Fragiudalfs	Fine-silty	Mixed	Active		Thermic	Udic	
QUITMAN	MS	AL	133C	Ultisols	Udults	Paleudults	Aquic Paleudults	Fine-loamy	Siliceous	Semiactive		Thermic	Udic	
RATTLESNAKE FORKS	AL	MS	133C	Entisols	Psamments	Quartzipsamments	Lamellic Quartzipsamments					Thermic	Udic	Coated

(continued)

(continued)

Series name	State TL	Other states using	MLRAs using MS	Order	Suborder	Great group	Subgroup	Particle-size class	Mineralogy class	CEC activity class	Reaction class	STR	SMR	Other family class
RICHLAND	OH		134	Alfisols	Udalfs	Hapludalfs	Oxyaquic Hapludalfs	Fine-loamy	Mixed	Active		Mesic	Udic	
RIEDTOWN	MS		134	Inceptisols	Udepts	Eutrudepts	Fluvaquentic Eutrudepts	Coarse-silty	Mixed	Active		Thermic	Udic	
ROBINSONVILLE	MS	AR, KY, LA, OK, TN, TX	131A, 135A	Entisols	Fluvents	Udifluvents	Typic Udifluvents	Coarse-loamy	Mixed	Superactive	Nonacid	Thermic	Udic	
ROSEBLOOM	MS	KY, LA, TN	134	Inceptisols	Aquepts	Endoaquepts	Fluvaquentic Endoaquepts	Fine-silty	Mixed	Active	Acid	Thermic	Aquic	
ROSELLA	MS		133C	Alfisols	Aqualfs	Natraqualfs	Albic Glossic Natraqualfs	Fine-loamy	Siliceous	Active		Thermic	Aquic	
RUSTON	LA	AL, AR, FL, KY, MS, NC, OK, TN, TX, VA	133C	Ultisols	Udults	Paleudults	Typic Paleudults	Fine-loamy	Siliceous	Semiactive		Thermic	Udic	
RUTAN	AL	MS	133C	Ultisols	Udults	Hapludults	Typic Hapludults	Coarse-loamy	Siliceous	Semiactive		Thermic	Udic	
SAFFELL	AR	AL, IL, KY, MO, MS, OK, TN, TX	133C, 134	Ultisols	Udults	Hapludults	Typic Hapludults	Loamy-skeletal	Siliceous	Semiactive		Thermic	Udic	
SAUCIER	MS	AL, LA	133C, 152A	Ultisols	Udults	Paleudults	Plinthaquic Paleudults	Fine-loamy	Siliceous	Subactive		Thermic	Udic	
SAVANNAH	MS	AL, AR, LA, TN, VA	133C	Ultisols	Udults	Fragiudults	Typic Fragiudults	Fine-loamy	Siliceous	Semiactive		Thermic	Udic	
SESSUM	MS	AR	135A	Vertisols	Aquerts	Dystraquerts	Chromic Dystraquerts	Fine	Smectitic			Thermic	Aquic	
SHARKEY	LA	AR, KY, MO, MS, TN	131A	Vertisols	Aquerts	Epiaquerts	Chromic Epiaquerts	Very-fine	Smectitic			Thermic	Aquic	
SHUBUTA	MS	AL, AR, FL, LA, TN	133C	Ultisols	Udults	Paleudults	Typic Paleudults	Fine	Mixed	Semiactive		Thermic	Udic	
SILVERDALE	MO	MS	131A	Entisols	Fluvents	Udifluvents	Aquic Udifluvents	Sandy over loamy	Mixed	Superactive	Nonacid	Thermic	Udic	
SIWELL	MS		134	Alfisols	Udalfs	Hapludalfs	Typic Hapludalfs	Fine-silty	Mixed	Active		Thermic	Udic	
SMITHDALE	MS	AL, AR, KY, LA, MO, OK, TN, TX	133C	Ultisols	Udults	Hapludults	Typic Hapludults	Fine-loamy	Siliceous	Subactive		Thermic	Udic	
SMITHTON	AR	AL, LA, MS	133C	Ultisols	Aquults	Paleaquults	Typic Paleaquults	Coarse-loamy	Siliceous	Semiactive		Thermic	Aquic	
STEENS	MS	AL, TN	133C	Alfisols	Aqualfs	Endoaqualfs	Aeric Endoaqualfs	Fine-loamy	Siliceous	Semiactive		Thermic	Aquic	
STOUGH	MS	AL, AR, LA	133C, 152A	Ultisols	Udults	Paleudults	Fragiaquic Paleudults	Coarse-loamy	Siliceous	Semiactive		Thermic	Udic	
SUCARNOOCHEE	AL		135A	Vertisols	Aquerts	Epiaquerts	Chromic Epiaquerts	Fine	Smectitic			Thermic	Aquic	
SUFFOLK	VA	AL, GA, NC, SC	133C	Ultisols	Udults	Hapludults	Typic Hapludults	Fine-loamy	Siliceous	Semiactive		Thermic	Udic	
SUGGSVILLE	AL	MS	133C, 135A	Vertisols	Uderts	Dystruderts	Chromic Dystruderts	Very-fine	Smectitic			Thermic	Udic	
SUMTER	AL	AR, GA, MS, OK, TN, TX	133C, 135A	Inceptisols	Udepts	Eutrudepts	Rendollic Eutrudepts	Fine-silty	Carbonatic			Thermic	Udic	
SUSQUEHANNA	MS	AL, AR, FL, GA, LA, TN, VA	133C	Alfisols	Udalfs	Paleudalfs	Vertic Paleudalfs	Fine	Smectitic			Thermic	Udic	
SWEATMAN	MS	LA, TN	133C	Ultisols	Udults	Hapludults	Typic Hapludults	Fine	Mixed	Semiactive		Thermic	Udic	
TALLA	MS		133C, 134	Alfisols	Udalfs	Natrudalfs	Glossaquic Natrudalfs	Fine-loamy	Siliceous	Active		Thermic	Udic	
TEKSOB	AR	KY, MO, MS, TN	131A	Alfisols	Udalfs	Hapludalfs	Typic Hapludalfs	Fine-loamy	Mixed	Active		Thermic	Udic	
TENSAS	LA	MS	131A	Alfisols	Aqualfs	Epiaqualfs	Chromic Vertic Epiaqualfs	Fine	Smectitic			Thermic	Aquic	
TIPPAH	MS	AR, LA, TN	133C, 134	Alfisols	Udalfs	Paleudalfs	Aquic Paleudalfs	Fine-silty	Mixed	Active		Thermic	Udic	
TIPPO	MS		134	Alfisols	Udalfs	Glossudalfs	Aquic Glossudalfs	Coarse-silty	Mixed	Active		Thermic	Udic	
TOINETTE	AL	KY, MS, TN	133C	Ultisols	Udults	Hapludults	Arenic Hapludults	Loamy	Siliceous	Subactive		Thermic	Udic	
TREBLOC	MS	AR, NC	133C	Ultisols	Aquults	Paleaquults	Typic Paleaquults	Fine-silty	Siliceous	Active		Thermic	Aquic	
TRINITY	TX	AL, AR, OK	133C	Vertisols	Uderts	Hapluderts	Typic Hapluderts	Very-fine	Smectitic			Thermic	Udic	
TROUP	AL	FL, GA, MS, NC, SC	133C	Ultisols	Udults	Kandiudults	Grossarenic Kandiudults	Loamy	Kaolinitic			Thermic	Udic	
TUNICA	MS	AR, KY, LA, MO, TN	131A	Inceptisols	Aquepts	Epiaquepts	Vertic Epiaquepts	Clayey over loamy	Smectitic over mixed	Superactive	Nonacid	Thermic	Aquic	
TUSCUMBIA	MS	AL, AR, OK, TN	135A	Inceptisols	Aquepts	Epiaquepts	Vertic Epiaquepts	Fine	Mixed	Active	Nonacid	Thermic	Aquic	

(continued)

(continued)

Series name	State TL	Other states using	MLRAs using MS	Order	Suborder	Great group	Subgroup	Particle-size class	Mineralogy class	CEC activity class	Reaction class	STR	SMR	Other family class
TUTWILER	MS	AR	131A	Alfisols	Udalfs	Hapludalfs	Typic Hapludalfs	Coarse-silty	Mixed	Active		Thermic	Udic	
UNA	MS	AL, AR, LA, NC, TN	133C, 135A	Inceptisols	Aquepts	Epiaquepts	Typic Epiaquepts	Fine	Mixed	Active	Acid	Thermic	Aquic	
URBO	MS	AL, AR, LA, TN, TX	133C, 135A	Inceptisols	Aquepts	Epiaquepts	Vertic Epiaquepts	Fine	Mixed	Active	Acid	Thermic	Aquic	
VAIDEN	AL	LA, MS	133C, 135A	Vertisols	Uderts	Dystruderts	Aquic Dystruderts	Very-fine	Smectitic			Thermic	Udic	
VANCLEAVE	MS		133C, 152A	Ultisols	Udults	Fragiudults	Plinthic Fragiudults	Coarse-loamy	Siliceous	Semiactive		Thermic	Udic	
VELDA	MS		133C, 134	Inceptisols	Udepts	Dystrudepts	Fluventic Dystrudepts	Coarse-silty	Siliceous	Active		Thermic	Udic	
VERDUN	LA	MS, TX	134	Alfisols	Aqualfs	Natraqualfs	Glossic Natraqualfs	Fine-silty	Mixed	Superactive		Thermic	Aquic	
VICKSBURG	MS	KY, LA, TN	131A, 134	Entisols	Fluvents	Udifluvents	Typic Udifluvents	Coarse-silty	Mixed	Active	Acid	Thermic	Udic	
VIMVILLE	MS	TX	133C	Alfisols	Aqualfs	Glossaqualfs	Typic Glossaqualfs	Fine-loamy	Siliceous	Active		Thermic	Aquic	
WADLEY	FL	AL, MS	133C,152A	Ultisols	Udults	Paleudults	Grossarenic Paleudults	Loamy	Siliceous	Subactive		Thermic	Udic	
WANILLA	MS		133C	Alfisols	Udalfs	Glossudalfs	Aquic Glossudalfs	Coarse-loamy	Siliceous	Active		Thermic	Udic	
WATSONIA	AL	LA	135A	Vertisols	Uderts	Hapluderts	Leptic Hapluderts	Clayey	Smectitic			Thermic	Udic	Shallow
WAVERLY	MS	AR, KY, LA, MO, TN	134	Inceptisols	Aquepts	Endoaquepts	Fluvaquentic Endoaquepts	Coarse-silty	Mixed	Active	Acid	Thermic	Aquic	
WEYANOKE	LA	MS	134	Inceptisols	Udepts	Eutrudepts	Dystric Eutrudepts	Coarse-silty	Mixed	Active		Thermic	Udic	
WILCOX	AL	AR, FL, GA, LA, MS, TN	133C, 135A	Vertisols	Uderts	Dystruderts	Chromic Dystruderts	Very-fine	Smectitic			Thermic	Udic	
WILLIAMSVILLE	MS	AL	133C	Ultisols	Udults	Hapludults	Typic Hapludults	Fine	Mixed	Active		Thermic	Udic	

Appendix E: Benchmark, Endemic, Rare, and Endangered Soils in Mississippi

Series name	Benchmark	Endemic[a]	Rare[b]	Endangered[c]
ADATON	No	No	No	No
ADLER	No	No	No	No
ALAGA	No	No	No	No
ALLIGATOR	Yes	No	No	No
ANGIE	No	No	Yes	No
ANNEMAINE	No	No	Yes	No
ARAT	No	Yes	Yes	Yes
ARIEL	No	Yes	No	No
ARKABUTLA	No	Yes	No	No
ARUNDEL	No	Yes	No	No
ASKEW	No	Yes	No	No
ATMORE	No	Yes	No	No
ATWOOD	No	Yes	No	No
AXIS	No	Yes	Yes	Yes
BAMA	No	No	Yes	No
BASIN	No	No	Yes	No
BASSFIELD	No	No	No	No
BAXTERVILLE	No	No	Yes	No
BAYOU	No	No	Yes	No
BEAUREGARD	No	Yes	Yes	Yes
BELDEN	No	Yes	No	No
BENNDALE	No	No	No	No
BEULAH	No	Yes	Yes	Yes
BIBB	No	Yes	No	No
BIGBEE	No	No	No	No
BINNSVILLE	No	Yes	Yes	Yes
BOHICKET	Yes	No	Yes	No
BONN	No	No	Yes	No
BOSWELL	No	No	No	No
BOWDRE	No	Yes	No	No
BOYKIN	No	No	Yes	No
BRANDON	No	No	Yes	No
BRANTLEY	No	No	Yes	No
BREWTON	No	No	Yes	No
BROOKSVILLE	No	No	No	No
BRUIN	No	Yes	No	No
BRUNO	No	No	No	No
BUCKATUNNA	No	No	Yes	No
BUDE	No	Yes	No	No
BYRAM	No	No	No	No
CADEVILLE	No	No	No	No
CAHABA	No	No	No	No
CALEDONIA	No	No	Yes	No
CALHOUN	Yes	No	Yes	No
CALLOWAY	Yes	No	Yes	No
CASCILLA	No	No	No	No
CATALPA	No	Yes	No	No
CHASTAIN	Yes	Yes	Yes	Yes
CHENNEBY	No	Yes	No	No
CHICKASAWHAY	No	No	Yes	No
CHICORA	No	No	Yes	No
COLLINS	Yes	Yes	No	No
COLUMBUS	No	No	Yes	No
COMMERCE	Yes	Yes	No	No
CONVENT	No	Yes	Yes	Yes
CREVASSE	No	No	No	No
CROATAN	No	Yes	No	No
CUTHBERT	No	No	No	No
DALEVILLE	No	No	Yes	No
DARCO	Yes	No	Yes	No
DEERFORD	No	No	Yes	No
DEMOPOLIS	No	Yes	Yes	Yes
DOGUE	No	No	Yes	No
DOROVAN	No	No	No	No
DOWLING	Yes	Yes	No	No
DUBBS	Yes	No	No	No
DUCKSTON	No	No	Yes	No
DULAC	No	No	No	No
DUNDEE	Yes	No	No	No
ESCAMBIA	No	Yes	No	No
EUSTIS	No	No	Yes	No
EUTAW	No	No	Yes	No

(continued)

D. Johnson et al., *The Soils of Mississippi*, World Soils Book Series,
https://doi.org/10.1007/978-3-031-36235-4

(continued)

Series name	Benchmark	Endemic[a]	Rare[b]	Endangered[c]
FALAYA	Yes	No	No	No
FALKNER	No	No	No	No
FORESTDALE	Yes	Yes	No	No
FREEST	No	No	No	No
FREESTONE	Yes	Yes	Yes	Yes
FRIZZELL	No	Yes	Yes	Yes
FROST	No	No	Yes	No
FRUITDALE	No	No	Yes	No
GILLSBURG	No	No	No	No
GRENADA	Yes	Yes	No	No
GRIFFITH	No	No	No	No
GUYTON	Yes	No	No	No
HANDSBORO	No	No	No	No
HARLESTON	No	No	No	No
HEIDEL	No	Yes	No	No
HENRY	Yes	Yes	Yes	Yes
HOULKA	No	No	No	No
HOUSTON	Yes	Yes	No	No
HYDE	No	Yes	Yes	Yes
ICHUSA	No	No	No	No
IRVINGTON	No	Yes	Yes	Yes
IUKA	No	Yes	No	No
IZAGORA	No	No	Yes	No
JENA	No	No	No	No
JOHNS	Yes	Yes	Yes	Yes
JOHNSTON	Yes	Yes	No	No
KINSTON	No	Yes	No	No
KIPLING	No	No	No	No
KIRKVILLE	No	No	No	No
KISATCHIE	No	No	Yes	No
KOLIN	Yes	No	Yes	No
LAKELAND	Yes	No	Yes	No
LATONIA	No	No	No	No
LAUDERDALE	No	Yes	Yes	Yes
LAX	No	Yes	Yes	Yes
LEAF	No	No	Yes	No
LEEPER	No	No	No	No
LENOIR	No	No	Yes	No
LEON	Yes	No	Yes	No
LEVERETT	No	Yes	Yes	Yes
LEXINGTON	No	No	No	No
LINKER	Yes	No	Yes	No
LONGVIEW	No	No	No	No
LORING	Yes	No	No	No
LORMAN	No	No	No	No
LOUIN	No	No	No	No
LUCEDALE	No	No	Yes	No
LUCY	No	No	No	No
LUVERNE	No	No	No	No
MABEN	No	No	No	No
MALBIS	No	No	No	No
MANTACHIE	Yes	Yes	No	No
MARIETTA	No	No	No	No
MASHULAVILLE	No	No	Yes	No
MATHISTON	No	Yes	No	No
MAUBILA	No	No	Yes	No
MAUREPAS	No	No	Yes	No
MAYHEW	No	No	No	No
MAYTAG	No	No	Yes	No
MCCRORY	No	Yes	Yes	Yes
MCLAURIN	No	No	No	No
MCRAVEN	No	No	Yes	No
MEMPHIS	Yes	No	No	No
MHOON	No	Yes	Yes	Yes
MOOREVILLE	No	No	No	No
MORGANFIELD	No	No	No	No
MYATT	No	Yes	No	No
NAHUNTA	No	No	Yes	No
NATCHEZ	No	Yes	No	No
NESHOBA	No	Yes	Yes	Yes
NEWELLTON	No	Yes	No	No
NEWHAN	No	No	Yes	No
NUGENT	No	Yes	No	No
OAKLIMETER	No	No	No	No
OCHLOCKONEE	No	Yes	No	No
OCILLA	No	Yes	Yes	Yes
OKEELALA	No	Yes	No	No
OKOLONA	No	No	No	No
OKTIBBEHA	No	No	No	No
OLIVIER	No	No	Yes	No
OLLA	No	No	Yes	No
OPENLAKE	No	No	Yes	No
ORA	No	No	No	No
OSIER	No	No	Yes	No
OUACHITA	Yes	Yes	Yes	Yes
OZAN	No	No	Yes	No
PADEN	No	Yes	No	No
PAMLICO	No	Yes	Yes	Yes
PELAHATCHIE	No	No	Yes	No
PEORIA	No	No	Yes	No
PETAL	No	No	No	No
PHEBA	Yes	Yes	Yes	Yes

(continued)

(continued)

Series name	Benchmark	Endemic[a]	Rare[b]	Endangered[c]
PIKEVILLE	No	No	Yes	No
PLUMMER	No	No	Yes	No
POARCH	Yes	No	Yes	No
POOLEVILLE	No	No	Yes	No
PRENTISS	No	No	No	No
PRIM	No	Yes	Yes	Yes
PROVIDENCE	Yes	No	No	No
QUITMAN	No	No	No	No
RATTLESNAKE FORKS	No	No	Yes	No
RIEDTOWN	No	No	No	No
ROBINSONVILLE	No	Yes	Yes	Yes
ROSEBLOOM	No	Yes	No	No
ROSELLA	No	Yes	Yes	Yes
RUSTON	Yes	No	No	No
RUTAN	No	No	Yes	No
SAFFELL	Yes	No	No	No
SAUCIER	No	No	No	No
SAVANNAH	Yes	No	No	No
SESSUM	No	No	Yes	No
SHARKEY	Yes	Yes	No	No
SHUBUTA	No	No	No	No
SILVERDALE	No	Yes	Yes	Yes
SIWELL	No	No	Yes	No
SMITHDALE	No	No	No	No
SMITHTON	Yes	No	No	No
STEENS	No	Yes	Yes	Yes
STOUGH	No	No	No	No
SUCARNOOCHEE	No	No	Yes	No
SUFFOLK	No	No	Yes	No
SUGGSVILLE	No	No	Yes	No
SUMTER	Yes	No	No	No
SUSQUEHANNA	Yes	No	No	No
SWEATMAN	No	No	No	No
TALLA	No	Yes	Yes	Yes
TEKSOB	No	No	Yes	No
TENSAS	No	No	No	No
TIPPAH	No	No	No	No
TIPPO	No	Yes	Yes	Yes
TOINETTE	No	No	No	No
TREBLOC	No	No	No	No
TRINITY	Yes	No	Yes	No
TROUP	Yes	No	Yes	No
TUNICA	Yes	No	No	No
TUSCUMBIA	No	Yes	Yes	Yes
TUTWILER	No	Yes	Yes	Yes
UNA	No	No	No	No
URBO	No	Yes	No	No
VAIDEN	Yes	Yes	No	No
VANCLEAVE	No	Yes	No	No
VELDA	No	Yes	Yes	Yes
VERDUN	No	No	Yes	No
VICKSBURG	No	Yes	No	No
VIMVILLE	No	No	Yes	No
WADLEY	No	No	No	No
WANILLA	No	Yes	Yes	Yes
WATSONIA	No	Yes	Yes	Yes
WAVERLY	No	No	No	No
WEYANOKE	No	Yes	Yes	Yes
WILCOX	No	No	No	No
WILLIAMSVILLE	No	No	No	No

[a]Only soil series in the family
[b]Occupies less than 100 km^2
cEndemic and rare

Appendix F: Land Use of Mississippi Soil Series

Series name	Land capability class (no irrigation)	Land capability (with irrigation)	Land use[a]	Land use details
ADATON	3w	3w	C, F, P	Crops: soybeans, cotton, corn, pasture, and hay; forestry: hardwood and pines
ADLER	2w		C, F, P	Crops: cotton, soybeans, corn, small grains, and hay; forestry: bottomland hardwoods
ALAGA	3s, 7s		C, F, P	Limited crops: corn, soybeans, pasture; forestry: longleaf and loblolly pines, oaks
ALLIGATOR	3e, 3w		C, F	Crops: soybeans, rice, cotton, and wheat; forestry: bottomland hardwoods (baldcypress, ash, tupelo gum, swamp maple, oaks, hickories, sweetgum)
ALMO	3w		C, P	Crops: soybeans, cotton; pasture
ANGIE	2e, 4e		F, P	Forestry: hardwoods and pines; pasture
ANNEMAINE	2w, 3e	2w	C, F	Commonly produced crops; forestry: hardwoods and pines
ARAT	7w		W	Wildlife and recreation
ARIEL	2w	2w	C, F	Crops: cotton, corn, soybeans; forestry: bottomland hardwoods with pines
ARKABUTLA	2w, 4w	4w	C, F, P	Crops: cotton, soybeans, corn, small grains, pasture, hay; forestry: bottomland hardwoods
ARUNDEL	7e		P, F	Pasture; forestry: pine forest
ASKEW	2e		C, F	Crops: cotton, soybeans, grain sorgum, wheat; forestry: hardwoods (cottonwood, oaks, pecan, ash, sweetgum)
ATMORE	4w		P, F, C	Pasture; forestry: pines; pasture; crops: soybeans, cotton, corn, wheat
ATWOOD	2e, 3e, 6e		C, P, F	Crops: cotton, soybeans, corn, peach orchards; pasture; forestry: hardwoods and shortleaf pine
AXIS	7w		W	Wetland wildlife
BAMA	1		C, F	Crops: corn, cotton, soybeans, wheat, pecans; forestry: pines with scattered oak
BASIN	2w		F, C, P	Small areas for crops: soybeans, cotton, corn, oats; pastures; forestry: pines
BASSFIELD	2s, 2e		C, F, P	Crops: cotton, corn, soybeans, wheat, pecans; forestry: pines with scattered oak and hardwoods; some pastures
BAXTERVILLE	2e		C, F, P	Crops: corn, cotton, small grains; pasture; forestry: pines and hardwoods
BAYOU	4w		F, P	Forestry: longleaf pine; some pasture
BEAUREGARD	2w		F, P	Pasture; forestry: pines
BELDEN	2w		C, F, P	Crops: cotton, corn, soybeans; pasture, hay; forestry: hardwoods and loblolly pine
BENNDALE	2e, 4e		C, F, P	Crops: corn, soybeans; pasture: hay; forestry: pines
BEULAH	2s, 3e		C, F	Crops: soybeans, cotton, wheat; forestry: hardwoods
BIBB	5w	5w	F, P, C	Pastures; forestry: hardwoods; crops: soybeans, cotton, corn, wheat
BIGBEE	5w		C, F, P	Crops: truck crops; pasture; pine plantations
BINNSVILLE	2e, 6e	2e	P	Native grasses or pastures
BOHICKET	8w		W	Wildlife
BONN	4s		P, F	Pasture; poor-quality woodlands
BOSWELL	3e, 6e, 7e	3e	F, P, C	Pasture; forestry: pines; crops: soybeans, cotton, corn, oats

(continued)

D. Johnson et al., *The Soils of Mississippi,* World Soils Book Series,
https://doi.org/10.1007/978-3-031-36235-4

(continued)

Series name	Land capability class (no irrigation)	Land capability (with irrigation)	Land use[a]	Land use details
BOWDRE	2w		C, F, P	Crops: soybeans, small grain, cotton, corn; pasture; bottomland hardwoods
BOYKIN	4s, 6s	4s	F, P	Pasture; forestry: pines with hardwoods
BRANDON	4e		F	Forestry: hardwoods
BRANTLEY	6e, 7e	7e	F, P	Pasture; forestry: pines and hardwoods
BREWTON	3w		F, C, P	Small areas for crops, pastures; forestry: pines with some hardwoods
BROOKSVILLE	2e, 2w, 3e	2e, 2w	C, P	Crops: cotton, soybeans, sorghum, small grains, hay; pasture
BRUIN	1, 2e, 5w		C	Crops: cotton, corn, wheat, soybeans
BRUNO	3s, 5w		F, P, C	Grain for pastures; forestry: hardwoods and pines; crops: soybeans, cotton, corn, wheat, oats
BUCKATUNNA	4w	4w	F	Forestry: hardwoods and pines
BUDE	2w, 3e		C, F, P	Crops: corn, cotton, soybeans, hay; pasture; forestry: hardwoods and pines
BYRAM	2e, 3e		F, D, C	Crops: soybeans, cotton, corn; forestry: oaks, hickory, sweetgum, shortleaf and loblolly pines
CADEVILLE	4e, 6e, 7e		F	Forestry: pine and hardwoods
CAHABA	2e, 2w, 7e		C, F, P	Crops: corn, cotton; pasture; forestry: pines and hardwoods
CALEDONIA	1, 2e		C, F, P	Crops: cotton, corn, soybeans; pasture; forestry: oaks and shortleaf pine
CALHOUN	3w, 5w		F, P, C	Pasture; forestry: pines and hardwoods; crops: soybeans, corn, cotton, wheat, oats
CALLOWAY	2e		C, P	Crops: cotton, corn, soybeans; pasture; minor forestry
CASCILLA	1, 2w, 4w		C, F	Crops: corn, cotton, soybeans; forestry: bottomland hardwoods
CATALPA	2w, 4w		C, P, F	Crops: cotton, corn, soybeans; pasture; forestry: hardwoods
CHASTAIN	3w		F, P, C	Pastures; forestry: bottomland hardwoods; crops: soybeans, cotton, corn, oats
CHENNEBY	2w, 4w	2w	C, F, P	Crops: cotton, corn, soybeans; pasture; forestry: bottomland hardwoods and pine
CHICKASAWHAY	3e	3e	F, P	Common crops; pasture; forestry: hardwoods and pines
CHICORA	2w, 3e	2w, 3e	F, P	Forestry: pines and hardwoods
COLLINS	2w		C, P	Crops: cotton, corn, soybeans, small grains; pasture; bottomland hardwoods
COLUMBUS	2w, 5w		C, F, P	Crops: soybeans; pasture; forestry: pine and hardwoods
COMMERCE	2w, 2e		C, P, F	Crops: cotton, soybeans, corn, wheat; pasture; forestry: hardwoods
CONVENT	3w, 5w		C, P, F	Crops: cotton, sugarcane, small grain, soybeans, corn, truck crops; pasture; forestry: hardwoods
CREVASSE	5w		P, C	Flood protection; pasture; forests: cottonwood, elm, hackberry, pecan, sycamore, willow; crops: soybeans, cotton, cotton, oats
CROATAN	7w		F, P	Pastures: small grain; forestry: hardwoods (water & swamp tupelo, baldcypress)
CUTHBERT	7e		F, P	Pasture; forestry: pines and hardwoods
DALEVILLE	5w	5w	C, F, P	Crops: soybeans; pasture: hay; forestry: bottomland hardwoods and pine
DARCO	3s		F, P, C	Pasture; forestry: loblolly pine, shortleaf pine, southern red oak, hickory; crops: soybeans, corn, oats
DEERFORD	3w		P, C	Pasture; a few areas cropped: soybeans, cotton
DEMOPOLIS	6e		P	Small areas cropped; pastures; scattered redcedar, osage orange and ash
DEXTER	1, 3e		C, P, F	Crops: soybeans, cotton, cotton oats
DOGUE	2e		C, F	Crops: corn, small grain, soybeans, tobacco, peanuts, truck crops; forestry: hardwoods and pines
DOROVAN	7w		F, W	Forestry: bottomland hardwoods, scattered pine
DOWLING	4w, 5w, 7w		F, C	Forestry: baldcypress, water tupelo, red maple; crops: soybeans, cotton, corn, sorghum, oats, wheat
DUBBS	1, 2e		C, F, P	Crops: cotton, corn, soybeans; pasture; forestry: bottomland hardwoods
DUCKSTON	7w	7w	W	
DULAC	3e, 6e, 7e		C, F, P	Crops: corn, cotton, soybeans; pasture; forestry: oak, gum, pine, hickory, beech, elm

(continued)

(continued)

Series name	Land capability class (no irrigation)	Land capability (with irrigation)	Land use[a]	Land use details
DUNDEE	2e, 2w		C, F, P	Crops: cotton, corn, soybeans, small grain; pasture; forestry: bottomland hardwoods
ESCAMBIA	2w		C, F, P	Crops: vegetables; pasture; forestry: pines and oak
EUSTIS	7e, 7s		C, F	Crops: corn, peanuts, melons, tung, peaches, tobacco; forestry: pines and oaks
EUTAW	3w		F, P	Some cropping and pasture; forestry: oaks, sweetgum and some pines
FALAYA	2w, 4w		C, P	Crops: corn, cotton, soybeans, small grains; pasture; forestry: hardwoods
FALKNER	2w, 4e	2w	C, P, F	Crops: soybeans, cotton, corn; pasture; hardwoods and pines
FORESTDALE	3w		C, P, F	Crops: cotton, soybeans, small grains; pasture; forestry: bottomland hardwoods
FREELAND	4e		C, P	Crops: soybeans, corn, small grains; pasture: hay; forestry: pine, hardwoods
FREEST	2e, 3e	2e	C, F, P	Crops: cotton, corn, soybeans; pasture grasses; forestry: pine and hardwoods
FREESTONE	2w, 3e		C, F, P	Pasture; forestry: oaks, hickory, sweetgum, pine plantations; crops: soybeans, cotton, corn, oats
FRIZZELL	2e		C, F, P	Forestry: oak and pine; crops: soybeans, cotton, corn
FROST	4w		F, P	Crops: corn, oats; pasture; forestry: hardwoods
FRUITDALE	7e	7e	F	Forestry: pine plantations
GILLSBURG	2w		F, P, C	Pasture; forestry: bottomland hardwoods; crops: soybeans, cotton, corn, wheat
GREENVILLE	1, 3e		C, F, P	Crops: cotton, corn, small grains, soybeans, truck crops, orchards; pasture; forestry: oak, pine
GRENADA	2e, 6e, 7e		C, P	Crops: cotton, corn, soybeans; pasture; forestry: some hardwoods (north) and pines (south)
GRIFFITH	2w	2w	C, P	Crops: cotton, corn, soybeans, hay; pasture
GUYTON	4w, 5w	4w	C, F, P	Small areas in crops and pastures; forestry: sweetgum dominant
HANDSBORO	8w		W	Waterfowl habitat
HARLESTON	2w, 3e		F, C	Forestry: flatwoods (loblolly pine, shortleaf pine, slash pine); crops: soybeans, cotton, corn, oats, wheat
HATCHIE	2w, 3e		C, P	Crops: soybeans, corn, cotton; pasture
HEIDEL	6e, 7e		C, F, P	Mainly forestry: loblolly pine, shortleaf pine, slash pine; some pasture; some crops: corn, soybeans
HENRY	3w		C, F, P	Crops: corn, cotton; pasture; forestry: oaks, sweetgum, sycamore, hackberry, elm
HOULKA	2w, 4w		C, P, F	Crops: cotton, corn, soybeans, small grains; pasture; forestry: hardwoods or loblolly pine plantations
HOUSTON	2e, 2s		C, P	Crops: soybeans, cotton, corn, oats
HYDE	6w		F, P	Pasture grasses; forestry: sweetgum, swamp tupelo, water tupelo, red maple baldcypress, pond pine, water oak, willow oak
ICHUSA	3e, 4e	3e	C, F, P	Crops: cotton, soybeans, grain sorghum, corn; pasture; forestry: loblolly pine, shortleaf pine, post oak, sweetgum red oak, post oak, hickory
IRVINGTON	2e		C, F, P	Crops: potatoes, corn, small grains, soybeans; forestry: pines, oaks, sweetgum, other hardwoods
IUKA	2w, 5w		C, F, P	Crops: corn, soybeans, small grains, truck crops; pasture (hay); forestry: water oak, willow, beech, sweetgum, hickory, maple
IZAGORA	2w, 3e	2w, 3e	F, P	Pasture (hay); forestry: hardwoods (oaks gums, poplar, hickory)
JENA	2w, 5w	5w	F, P, C	Crops: soybeans, cotton, corn, oats, wheat; pastures; forestry: hardwoods and pine
JOHNS	2w		C, F, P	Crops: cotton, soybeans; pasture (hay); forestry: loblolly and longleaf pines and hardwoods
JOHNSTON	4w, 7w	7w	F, C, P	Crops: corn, soybeans; pasture; forests: water tupelo, swamp tupelo, sweetgum, yellow poplar, green ash, water oak, baldcypress
KINSTON	6w	6w	F, P	Pasture; forests: sweetgum, black tupelo, water oak, poplar, hickories
KIPLING	3w, 7e	3w	C, P, F	Crops: cotton, soybeans, small grains; pasture (hay); forestry: hardwoods, shortleaf pine, loblolly pine

(continued)

Series name	Land capability class (no irrigation)	Land capability (with irrigation)	Land use[a]	Land use details
KIRKVILLE	2w, 5w	2w	C, P, F	Crops: cotton, corn, soybeans, oats; pasture; forestry: bottomland hardwoods (cherrybark oak, sweetgum, water oak)
KISATCHIE	6e		F	Forestry: southern pine and hardwoods
KOLIN	4e		F, C, P	Crops: soybeans, cotton, corn, wheat; forestry: hardwoods and pines
LAKELAND	4s, 6s		P, F	Improved pasture; forestry: oaks, scattered longleaf pine
LATONIA	2w, 2s		F, C, P	Some crops: soybeans, cotton, corn, oats, wheat; pasture; forestry: southern pines
LAUDERDALE	4s, 7e		F	Forestry: loblolly and shortleaf pines
LAX	2e		C, P, F	Crops: corn, tobacco, small grains, soybeans; pasture (hay); forestry: hardwoods (oaks, hickories, beech)
LEAF	3w, 4w	4w	F, C, P	Small areas for corn and small grains; forestry: bottomland hardwoods and pines
LEEPER	2w	2w	C, P, F	Crops: soybeans, cotton, corn, oats, wheat; pastures; forestry: hardwoods
LENOIR	3w		C, F, P	Crops: corn, soybeans, tobacco, cotton, small grains, peanuts, truck crops; pasture; forests: hardwoods and pines
LEON	4w		F, P	Forestry: longleaf pine, slash pine, water oak, saw palmetto
LEVERETT	1, 4w		C, P, F	Crops: cotton, corn, soybeans; pasture; forestry: hardwoods and pines
LEXINGTON	3e, 6e		C, P, F	Crops: cotton, corn, soybeans, small grain; pasture (hay); forestry: oaks, yellow poplar, hickory, gum, beech, elm
LINKER	6e		P, F	Pasture; forestry: red oak, post oak, blackjack oak, sweetgum, blackgum, hickory, shortleaf pine
LONGVIEW	2e	2e	C, P, F	Crops: corn, soybeans, grain sorghum; pasture (hay); forestry: sweetgum, hickory, gum, oaks, pines
LORING	2e, 3e, 4e, 6e		C, P	Crops: cotton, small grains, soybeans; pasture (hay)
LORMAN	6e, 7e	6e	F, P	Forestry: pines and hardwoods; some crops: soybeans
LOUIN	3w	3w	F, P	Pasture (hay); forestry: loblolly and shortleaf pines
LUCEDALE	1		C, F, P	Crops: cotton, soybeans, corn, small grains; forestry: hardwoods and pines
LUCY	2e, 4s, 6e		C, F, P	Crops: peanuts, corn, cotton, soybeans; pasture (hay); forestry: pines
LUVERNE	4e, 6e, 7e		F, C, P	Forestry: hardwoods, pines; crops: corn, cotton, hay, truck crops
MABEN	7e		F, C, P	Crops: corn, cotton; pasture: hay; forestry: hardwoods, pines
MALBIS	1, 3e	3e	C, F	Crops: corn, cotton, potatoes, peanuts, soybeans; forestry: loblolly pine, slash pine, shortleaf pine, longleaf pine, sweetgum, red oak
MANTACHIE	2w, 5w	5w	C, P, F	Crops: cotton, soybeans, corn, small grains; pasture (hay); forestry: bottomland hardwoods or loblolly, slash pine plantations
MARIETTA	2w		C, P, F	Mostly cropped for cotton soybeans, corn, small grains; some pasture and forestry (hardwoods)
MASHULAVILLE	3w, 4w		F, P, C	Mainly forestry: water oak, sweetgum, loblolly pine; some crops: soybeans; some pasture: hay
MATHISTON	2w, 4w	2w	C, P	Crops: cotton, corn, soybeans, small grains; pasture: hay; forestry: bottomland hardwoods
MAUBILA	7e	7e	F	Forestry: longleaf, loblolly, shortleaf pines; southern red oak, white oak, hickory, sweetgum
MAUREPAS	8w		W	Wetland wildlife
MAYHEW	3e, 3w	3w	C, P, F	Crops: soybeans, corn, cotton, sweet potatoes; pasture: hay; forestry: oaks, sweetgum, shortleaf and loblolly pines
MAYTAG	4e		C, P	Crops: soybeans, grain sorghum; pasture; forestry: some hackberry and E redcedar
MCCRORY	4w		F	Forestry: oaks, gum, hackberry, elm
MCLAURIN	2e, 3e	2e	C, P, F	Crops: cotton, corn, soybeans, small grains; pasture; forestry: longleaf pine, slash pine, loblolly pine, shortleaf pine
MCRAVEN	2w		C, P, F	Mainly crops: soybeans, cotton, corn; pasture: hay; forestry: small areas in bottomland hardwoods

Series name	Land capability class (no irrigation)	Land capability (with irrigation)	Land use[a]	Land use details
MEMPHIS	2e, 3e, 6e, 7e		C, F, P	Crops: cotton, soybeans, small grains; pasture; forestry: hardwoods and pines
MHOON	2w		C, P	Crops: cotton, soybeans, sugarcane, small grain, corn; pasture
MOOREVILLE	2w, 5w		F, C, P	Crops: soybeans, corn, cotton; pasture; forestry: bottomland hardwoods, loblolly pine plantations
MORGANFIELD	2w		C, P, F	Crops: cotton, corn, soybeans, small grains; pastures; forestry: bottomland hardwoods
MYATT	3w, 4w		F, P	Forestry: water oak, maple, sweetgum, ash, cypress, bay, some pines
NAHUNTA	2w, 5w		P, F	Pasture; forestry: loblolly pine, hardwoods
NATCHEZ	6e, 7e		F	Forestry: hardwoods and loblolly pine
NESHOBA	2e, 3e, 4e		C, P, F	Mainly crops: cotton, corn, soybeans; pastures: hay; forestry: some pines, hardwoods
NEWELLTON	3e, 3w, 5w		C, P	Crops: soybeans, cotton, small grains; pasture; some forestry
NEWHAN	8s		D	Developed for urban beach cottages
NUGENT	2e, 3s, 5w		F	Forestry: hardwoods, pines
OAKLIMETER	2w	2w	C, P, F	Crops: cotton, corn, soybeans; pasture (hay); forestry: hardwoods and loblolly pine
OCHLOCKONEE	2w		C, P, F	Crops: corn, sorghum, soybeans, cotton, small grains, vegetable; pastures: hay; forestry: bottomland hardwoods
OCILLA	4w	4w	F, P, C	Mainly forestry; some cleared for grazing and cropping (soybeans and corn)
OKEELALA	6e, 7e	6e	F, P	Forestry: loblolly pine, hardwoods; pasture: hay
OKOLONA	2e, 2s, 3e	2e, 2s	C, P	Crops: cotton, soybeans, sorghum, small grains; pasture: hay; forestry: E redcedar and osage orange
OKTIBBEHA	3e, 4e, 6e	3e	F, P, C	Crops: cotton, corn, soybeans; pasture: hay; forestry: loblolly pine, shortleaf pine, southern red oak, post oak, sweetgum, hickory
OLIVIER	2e, 2w		C, F	Crops: soybeans, cotton, corn, oats; forestry: hardwoods, pines
OLLA	7e, 7s	7s	F, W, C	Loblolly pine, shortleaf pine, southern red oak, white oak, beech, hickory; crops: soybeans, cotton, corn
OPENLAKE	2w		P, F	Pasture; forestry: bottomland forest
ORA	2e, 3e, 4e	3e	C, P, F	Crops: cotton, corn, small grains, truck crops; pasture (hay); forestry: loblolly, shortleaf, longleaf pines, oaks, hickory, sweetgum
ORANGEBURG	7e		C, P, F	Crops: cotton, corn, tobacco, peanuts
OSIER	5w		F	Forestry: sweetgum, blackgum, water oak, red maple, slash pine, longleaf pine
OUACHITA	4w		P, F	Forests: bottomland hardwoods (gum, oak, maple, ash, pine)
OZAN	3w		F, P	Forestry: shortleaf pine, loblolly pine, sweetgum, oaks; pastures
PACTOLUS	3s	3s	D	Mostly developed in MS; cropped elsewhere: corn, soybeans, peanuts
PADEN	2e, 2w		C, P	Crops: cotton, corn, soybeans; pasture: hay
PAMLICO	7w		F, W	Forestry: pond pine, black tupelo, sweetbay, baldcypress, sweetgum
PELAHATCHIE	2e, 2w	2e, 2w	C	Crops: soybeans, cotton, corn
PEORIA	3w		F, P	Forestry: hardwoods; pasture
PETAL	4e	4e	F, P	Mainly forestry: pines, hardwoods; some pasture
PHEBA	3w		F, C, P	Mainly forestry: loblolly, shortleaf, slash pines, sweetgum; some crops: corn, cotton, small grains, grain sorghum; pasture
PIKEVILLE	4e		F, P	Mainly forestry: loblolly pine, longleaf pine, water oak, shortleaf pine, sweetgum; some pasture
PLUMMER	4w, 5w	5w	D, F, P	Mainly developed; slash, loblolly, and longleaf pine, swamp tupelo, baldcypress; some pasture
POARCH	1, 3e		C, P	Crops: corn, cotton, soybeans, peanuts, small grains, truck crops
POOLEVILLE	2w		C, P, F	Mainly crops: soybeans, cotton; pasture; forestry: bottomland hardwoods

(continued)

Series name	Land capability class (no irrigation)	Land capability (with irrigation)	Land use[a]	Land use details
PRENTISS	2e, 2w	2e, 2w	C, P, F	Crops: cotton, corn, soybeans, small grain; pasture (hay); forestry: pines and hardwoods
PRIM	6s, 7s		P	Pasture; forestry: redcedar, hardwoods
PROVIDENCE	2e, 3e, 4e, 6e, 7e		C, P, F	Crops: corn, soybeans, cotton, small grains, annual legumes, truck crops, orchards; pasture (hay); forestry: loblolly pine, shortleaf pine, Shumard's oak, sweetgum, yellow poplar
QUITMAN	2e, 2w		C, F, P	Mostly pastures and forests; some areas cleared for cotton, corn, soybean; forestry: hardwoods (sweetgum, Am sycamore) and pines (loblolly, slash)
RATTLESNAKE FORKS	7e		F, P	Small areas cleared for crops, pasture; forestry: oaks and pines
RICHLAND	3e, 4e		C, P, F	Crops: corn, small grain; pasture: hay; forestry: yellow poplar, beech, maple, hickory, walnut
RIEDTOWN	1, 2w		C, P, F	Crops: soybeans, sorghum, cotton, corn; pasture: hay; forestry: E cottonwood, green ash, Am sycamore, water oak, willow oak, yellow poplar, cherrybark oak, loblolly pine, sweetgum
ROBINSONVILLE	1, 2w, 4w, 5w		C, P, F	Crops: cotton, corn, soybeans; pasture: hay; forestry: hardwoods
ROSEBLOOM	3w, 5w, 6w	5w	F, C	Forestry: bottomland hardwoods; crops: soybeans, cotton, corn, rice, oats, wheat
ROSELLA	3w		F, P, C	Mainly forestry: water oak, sweetgum, maple, ash, pines; some pasture: hay; crops: soybeans, cotton, corn
RUSTON	2e, 3e, 4e, 7e	2e, 3e, 4e	F, P, C	Small acreage for crops; pastures common; forestry: southern pines
RUTAN	4e, 7e	4e	F, P	Mainly forestry: pines, oaks; small areas for cropland, pasture
SAFFELL	7e		F, P, C	Forestry: shortleaf pine, loblolly pine, upland oaks; pastures; crops: soybeans, cotton, worn, oats
SAUCIER	3e, 4e		F, C, P	Crops: soybeans, cotton, cotton; pasture; mainly forestry: pines and hardwoods
SAVANNAH	2e, 3e	2e, 3e	C, F, P	Crops: cotton, corn, soybeans, small grains, truck crops; pasture; forestry: pines and hardwoods
SESSUM	4w		C, P, F	Crops: soybeans, grain sorghum; pasture: hay; forestry: oaks, sweetgum, loblolly pine
SHARKEY	3e, 3w, 4w, 5w		C, P, F	Mainly crops: corn, wheat, grain sorghum, small grains; pastures; forestry: bottomland hardwoods
SHUBUTA	2e, 4e	2e	F, C, P	Forestry: hardwoods and pine; crops: corn, cotton, soybeans, truck crops; pasture: hay
SILVERDALE	3s, 4w		C, F	Crops: soybeans, wheat, cotton; forestry: bottomland hardwoods
SIWELL	3e, 6e		D, P	Mainly developed; some pasture; forestry: cherrybark oak, white oak, post oak, so. Red oak, hickory, sweetgum, shortleaf pine, loblolly pine
SMITHDALE	4e, 6e, 7e	7e	F, C, P	Some cropped and pastures; mainly forestry: pines
SMITHTON	4w, 5w		F, P	Some cropped for soybeans, small grains, pasture; forestry: pines and hardwoods
STEENS	2w		C, F, P	Crops: soybeans; pasture: hay; forestry: pines, hardwoods
STOUGH	2w	2w	F, C, P	Crops: cotton, corn, small grains, wheat, truck crops; pasture; forestry: hardwoods and some pines
SUCARNOOCHEE	4w		P	Mainly pasture; crops: soybeans, corn, grain sorghum, cotton; forestry: hardwoods
SUFFOLK	1		C, F, P	Crops: corn, cotton, peanuts, soybeans, tobacco, small grains, truck crops; forestry: pines, hardwoods
SUGGSVILLE	4e, 7e		F, P	Pasture; forestry: longleaf, loblolly, sortleaf pines, oaks, yellow poplar
SUMTER	3e, 4e, 7e	3e	P, F, C	Mainly pasture; forestry: some redcedar; crops: cotton, corn
SUSQUEHANNA	6e		F, P, C	Pasture; forestry: hardwoods and pines; crops: soybeans, cotton, corn, wheat
SWEATMAN	6e, 7e		F, C, P	Forestry: loblolly pine, shortleaf pine, southern red oak, post oak, sweetgum, hickory; crops: corn, cotton, soybeans; pasture: hay

(continued)

(continued)

Series name	Land capability class (no irrigation)	Land capability (with irrigation)	Land use[a]	Land use details
TALLA	2w		C, P, F	Crops: cotton, corn, soybeans; pasture: hay; forestry: green ash, E cottonwood, cherrybark oak, loblolly pine, sweetgum, yellow poplar
TEKSOB	1, 4e		P	Mainly pasture
TENSAS	3w		C, F, P	Crops: cotton, soybeans, rice, wheat, corn; pasture; forestry: bottomland hardwoods
TIPPAH	2e, 3e, 4e		C, P, F	Crops: cotton, soybeans, corn; pasture: hay; forestry: oaks, hickory, elm, poplar, shortleaf pine, loblolly pine
TIPPO	2w		C, P, F	Mainly crops: cotton, corn, soybeans; pastures: hay; forestry: some hardwoods, pine
TOINETTE	7e		F, P	Forestry: longleaf pine, slash pine, live oak; some pasture
TREBLOC	5w, 6w	6w	F, P, C	Forestry: water oak, maple, sweetgum, ash, cypress, pine; cleared: pasture (hay); crops: soybeans, cotton, corn, wheat
TRINITY	2w		P, C	Crops: cotton, corn, sorghum, small grains; pasture
TROUP	6s, 7e		F, C, P	Mainly forestry: pine, hardwoods; some areas for pasture and crops: peanuts, watermelons, vegetables
TUNICA	3w, 5w		C, P, F	Crops: cotton, soybeans, small grains, corn; pasture (hay); forestry: bottomland hardwoods
TUSCUMBIA	3w, 4w		F, P, C	Forestry: bottomland hardwoods; pasture: hay; crops: corn, soybeans
TUTWILER	1, 2e		C, P, F	Crops: cotton, corn, soybeans; pasture; forestry: hardwoods
UNA	3w, 4w, 5w, 6w		F, P	Forestry: bottomland hardwoods; pasture: hay
URBO	2w, 4w, 5w	2w	F, C, P	Crops: cotton, corn, soybeans, small grains; forestry: floodplain hardwoods
VAIDEN	3e		C, P, F	Crops: cotton, soybeans, corn, grain sorghum; pasture (hay); forestry: loblolly pine, shortleaf pine, post oak, sweetgum, southern red oak, hickory
VANCLEAVE	2w		F, C	Forestry: loblolly pine, slash pine, oaks, blackgum, sweetgum; crops: cotton, corn
VELDA	2w, 4w		C, P, F	Crops: cotton, corn, soybeans; pasture: grasses; forestry: bottomland hardwoods (cherrybark oak, sweetgum, water oak, loblolly pine, yellow poplar)
VERDUN	4s		P	Predominantly pasture
VICKSBURG	1, 2w		C, P, F	Crops: cotton, corn, soybeans, small grains; pasture: hay; forestry: bottomland hardwoods
VIMVILLE	3w		F, P	Forestry: water oak, willow oak, sweetgum, loblolly pine
WADLEY	3s, 6s, 7e	3s	F, P, C	Mainly forestry: live oak, turkey oak, bluejack oak, chapman oak, longleaf pine; some pasture; some crops: cotton, corn
WANILLA	3w		P, F	Forestry: hardwoods (cherrybark oak, sweetgum), loblolly pine); pastures
WATSONIA	6e		P	Pasture
WAVERLY	3w		F, C, P	Crops: cotton, corn, soybean; pasture (hay); forestry: bottomland hardwoods
WEYANOKE	2e		P	Mainly pasture
WILCOX	3e, 3w, 7e	3e	F, C	Forestry: hardwoods, pines; crops: cotton, corn, soybeans; crops: soybeans, cotton, corn, oats, wheat
WILLIAMSVILLE	6e, 7e		P, F	Pasture: grasses; forestry: lobolly pine, redcedar, hardwoods

[a]Land use: *C* agricultural crops; *P* pasture; *F* forestry; *W* wildlife; *D* developed land

Index

A

Albic horizon, 12, 13, 59, 67, 69, 71, 74, 81, 96, 97, 105, 110, 115, 134, 139
Alcorn University, 4
Alfisols, 12–16, 47, 49, 51, 53, 54, 56, 59, 65, 113–115, 121, 125, 137–139, 143, 167, 187–193
Alluvial fan, 51
Alluvium, 29, 31–33, 35, 39, 40, 47–54, 56, 67, 70, 72, 73, 75, 77–79, 82–91, 93–95, 97, 101, 109, 110, 113–115, 121, 123, 125–127, 129, 131–133, 137, 138, 143, 152, 165, 167, 173–179
Aquic soil moisture regime, 39, 50, 52, 56, 76, 78, 84, 86, 88, 91, 93, 110, 115, 132
Aquults, 109, 110, 187–192
Argillic horizon, 12, 13, 59, 67, 69–74, 78–81, 88, 90, 91, 93, 94, 96, 97, 103, 105, 109, 110, 113, 115, 125, 126, 167
Argilluviation, 21, 110, 115, 137, 139, 140, 167
Ash, green, 49, 137, 158, 201, 204
Association, soil, 7, 8, 12, 15, 16, 20, 39, 44, 47–55

B

Baldcypress, 24, 49, 137, 176, 199–201
Base cycling, 137, 140, 167
Benchmark soil, 141, 167
Biodiversity, loss, 143, 146, 148, 168
Black Prairie, 16, 29, 33, 51
Bottomland hardwoods, 27, 49, 75, 82, 83, 85, 87, 88, 90, 91, 121, 123, 125, 126, 129, 132, 133, 148, 158, 199–205
Brown Loam Region, 47
Buckshot soils, 49
Bureau of Soils, 8, 17, 167

C

Cambic horizon, 12, 13, 59, 75, 76, 82, 84–88, 90, 91, 97, 105, 107, 121, 123, 125, 129, 138
Cambisolization, 121, 129, 132, 133, 137, 138, 140, 167
Catena, soil, 29
Catfish, farming, 168
Cation-Exchange Capacity (CEC), 12, 13
CEC activity class, 187
Cherrybark oak, 47, 49, 158, 202, 204, 205
Coefficient of Linear Extensibility (COLE), 125
Colluvium, 178
Complex, soil, 12
Consociation, soil, 12
Cottonwood, eastern, 49
Crawdad land, 49
Cropland, 113, 115, 143, 147, 148, 151, 165, 204

D

Delta, 1, 4, 5, 7, 10, 15, 16, 18, 22, 24, 29, 33, 47–49, 73, 76, 79, 81, 83, 143–145, 147, 148
De Soto, Hernando, 1
Developed land, 23, 146, 205
Diagnostic horizons, 5, 13, 59, 67, 107, 137, 181
Droughts, 21, 33, 49, 148
Dystraquerts, 49, 50, 87, 88, 92, 93, 125, 126, 187, 189, 190, 192
Dystrudepts, 24, 39, 47, 50, 59, 60, 82, 83, 86, 107, 121, 123, 138, 152, 153, 158, 187–191, 193
Dystruderts, 52, 53, 90, 91, 98, 99, 107, 125, 126, 138, 189–193

E

Ecoregions, 7, 16, 18, 20, 26, 30, 33, 39
Endangered soil, 5, 141, 195
Endemic soil, 141
Endoaqualfs, 49, 51, 60, 78, 79, 82, 83, 97, 113, 115, 167, 187, 189, 192
Endoaquepts, 24, 40, 49, 50, 52, 59, 60, 75, 76, 80, 81, 97, 121, 123, 138, 152, 153, 158, 167, 188–190, 192, 193
Entisols, 12–16, 39, 47, 49, 54, 59, 65, 121, 129, 131, 132, 137, 139, 140, 167, 187–193
Eolian deposits, 129, 131, 175–177
Epiaquepts, 24, 50, 52, 59, 91, 100, 101, 107, 121, 123, 138, 158, 190–193
Epiaquerts, 50, 52, 53, 60, 85, 86, 89, 90, 107, 125, 126, 189, 192
Epipedons, 12, 13, 59, 65, 67, 69–76, 78–80, 82–91, 93, 94, 96, 97, 103, 105, 107, 109, 110, 113, 115, 121, 123, 125, 126, 129, 131, 133, 134, 167
Eutrudepts, 12, 17, 19, 20, 25, 47, 51, 52, 59, 90, 96, 97, 107, 121, 123, 138, 152, 158, 167, 187, 188, 190–193
Extragrade subgroups, 61

F

Family, soil, 11, 141, 197
Ferralitization, 137, 139, 140
Flatwoods, 14, 15, 24, 29, 33, 39, 53–56, 67, 69, 92, 109, 129, 131, 133, 134, 139, 167, 201
Flooding, 1, 10, 21, 32, 33, 48, 148
Fluvaquents, 32, 40, 47, 50, 52, 53, 59, 60, 83, 84, 87, 88, 107, 129, 131, 153, 187, 189, 190
Fluvents, 61, 129, 131, 188–193
Forestland, 151
Fragification, 110, 115, 137, 139, 140
Fragipan, 12, 13, 29, 59, 65, 67, 73, 74, 94, 96, 97, 105, 109, 110, 113–115, 139, 167, 181
Fragiudalfs, 24, 29, 39, 47, 52, 59, 60, 67, 70, 71, 97, 109, 113–115, 167, 188–191
Fragiudults, 9, 24, 39, 53, 59, 73, 74, 78, 79, 97, 109, 110, 113, 158, 167, 189–193

D. Johnson et al., *The Soils of Mississippi*, World Soils Book Series,
https://doi.org/10.1007/978-3-031-36235-4

Fraglossudalfs, 24, 47, 59, 94–96, 104, 105, 107, 109, 113–115, 153, 188, 189, 191

G
General soil maps, 14–20, 26, 39
Gleization, 115, 121, 125, 127, 129, 132, 133, 137, 140, 167
Glossic horizon, 12, 59, 67, 74, 80, 94, 96, 113
Glossification, 115, 137, 139, 140
Great group, soil, 12, 25, 59, 67
Gulf of Mexico, 1

H
Hapludalfs, 24, 47, 51, 53, 59, 60, 72, 73, 76, 77, 97, 113, 115, 137, 152, 153, 167, 187–193
Hapluderts, 52, 53, 59, 97, 106, 107, 125, 126, 138, 188–193
Hapludults, 24, 25, 39, 59, 60, 70, 71, 74, 75, 97, 109, 110, 153, 158, 167, 187–193
Hilgard EW, 7, 8, 17, 167
Histic epipedon, 12, 133, 134, 167
Histosols, 12–16, 56, 59, 65, 133, 134, 139, 140, 167, 188–191
Holocene, 22, 32, 48, 50, 51, 54
Humification, 133, 137, 139, 140
Hydric soils, xv
Hyperthermic soil temperature regime, 56

I
Inceptisols, 12–16, 39, 47, 49, 51, 52, 59, 65, 121, 123, 125, 137–140, 167, 187–193
Intergrade, subgroups, 61
Intragrade, subgroups, 61

J
Jackson Group, 29, 51
Jenny, Hans, 7, 21

K
Kandic horizon, 12, 109
Köppen climate system, 21, 32

L
Land Capability Classes, 151, 152, 165, 168, 199
Lithic contact, 61
Lithic subgroup, 61
Loess, 1, 5, 7, 14, 17, 22, 24, 29, 31, 33, 35, 39, 40, 47, 48, 52, 56, 67, 71–73, 79, 83, 84, 88–90, 94, 96, 105, 109, 113–115, 121, 137–139, 148, 167, 173–178
Loess Hills, 1, 5, 17, 24, 29, 33, 40, 47, 109, 113
Logan, WN, 7, 14, 17

M
Mabry, TO, 7, 17, 48, 167
Major Land Resource Areas (MLRAs), 39, 40, 47, 49, 51–54, 56, 107, 109, 113, 125, 129, 131, 137, 141, 167, 187
Marine deposits, sediments, 33
Miocene, 42
Mississippi Delta, 1, 5, 7, 16, 18, 22, 29, 33, 47, 143, 144, 147, 148
Mississippi State University, 4, 7, 15, 27, 47
Mollic epipedon, 12, 13, 59, 97, 107, 125, 126, 133, 134
Mollisols, 12–14, 16, 59, 65, 133, 134, 139, 140, 167, 187, 188, 191

N
Natchez soil series, Mississippi's state soil, 19, 90
Natchez Trace, 4, 27
National Park Service (NPS), 27
Natric horizon, 12, 13, 114, 115, 139

O
Ochric epipedon, 12, 59, 67, 69–76, 78–80, 82–91, 93, 94, 96, 97, 103, 105, 109, 110, 113, 115, 121, 123, 125, 126, 129, 131, 133, 134, 167
Orchards, 147, 199, 201, 204
Order, soil, 5, 7, 12, 13, 15, 17, 39, 47, 49, 52, 54, 56, 59, 60, 65, 67, 110, 121
Orthents, 188

P
Paleaquults, 9, 25, 32, 56, 59, 91–93, 102, 103, 107, 109, 110, 187, 188, 190–192
Paleudalfs, 24, 52, 53, 59, 60, 79, 80, 84, 85, 97, 113, 115, 153, 158, 187–192
Paleudults, 9, 24, 25, 32, 40, 53, 54, 56, 59, 60, 67, 69, 72, 73, 80, 97, 109, 110, 152, 153, 158, 167, 187–193
Paludization, 133, 137, 139
Paralithic contact, 61, 91, 129
Pasture, 17, 110, 115, 132, 143, 146, 148, 151, 152, 165, 199–205
Pine, loblolly, 12, 17, 23, 27, 33, 39, 47, 52, 83, 85, 90, 92, 143, 144, 153, 158, 173–179, 200–205
Pine, longleaf, 14, 15, 25, 33, 39, 133, 143, 153, 158, 173–179, 199, 202, 203, 205
Pine, shortleaf, 14, 15, 23, 27, 33, 39, 47, 52, 143, 148, 153, 158, 167, 173–179, 199, 201–205
Pine, slash, 25, 27, 33, 39, 54, 143, 148, 153, 158, 167, 173–179, 201–203, 205
Pleistocene, 22, 40, 48, 50, 51, 53, 56, 109, 114, 115, 175
Plinthite, 12, 13, 61, 67, 69, 74, 109, 110, 137, 139, 181
Pliocene, 7, 8, 32
Podzolization, 139
Pontotoc Ridge, 1, 5, 14, 15, 26, 29, 33, 39, 167
Prime farmland, 32, 152, 153, 165
Psamments, 187, 188, 190, 191

Q
Quaternary, 29, 33

R
Rare soil, 141
Residuum, 31, 33, 35, 52, 53, 71–73, 75, 87, 90, 97, 114, 115, 125, 127, 129, 130, 133, 167, 173–179

S
Salt domes, 31, 32, 36, 37
Selma Group, 29
Seventh Approximation, 11
Shallow soil, 130
Site index, 151, 158, 165
Soil and water conservation districts, vi
Soil, definition, 7
Soil health, xv
Soil map, 8, 9, 12, 14–20, 26, 39, 67, 151, 152
Soil moisture regime, 21, 24, 39, 47, 50, 52, 54, 56, 63, 67, 69, 70, 73, 74, 76, 78, 80, 82, 84, 86, 88–91, 93, 96, 97, 110, 114, 115, 121, 123, 129, 132, 133, 141

Soil resource areas, 39, 41
Soil series, 7–12, 16, 17, 19–21, 31, 36, 39, 47, 49, 52, 54, 56, 57, 59–65, 67, 69–93, 95–97, 99, 101, 103, 105, 107, 109, 110, 113–115, 121, 125, 126, 129–131, 133, 134, 138–141, 143–146, 152, 153, 158, 165, 167, 168, 173, 181, 185, 187, 197, 199
Soil survey, 7–11, 17, 21, 39, 44, 67, 151, 165, 167, 169
Soil taxonomy, 7, 11, 12, 59, 65, 141
Soil temperature regime, 21, 39, 47, 49, 52, 54, 56, 63, 67, 69, 70, 73, 74, 76, 78, 80, 82, 84, 86, 88–91, 93, 96, 97, 110, 114, 115, 121, 123, 133
Spodic horizon, 12, 13, 59, 133, 134
Spodosols, 12, 13, 16, 59, 65, 133, 134, 167, 190
SSURGO database, 67
Subgroup, soil, 12, 60
Suborder, soil, 12, 141
Subsidence, 32, 148
Surveys, soil, 7–11, 17, 21, 39, 44, 67, 151, 165, 167, 169
Sweetgum, 24, 39, 49, 52, 78, 143, 158, 199–205
Swelling clays, 12

T
Thermic soil temperature regime, 21, 39, 47, 52, 54, 67, 69, 70, 73, 74, 78, 80, 82, 84, 86, 88, 90, 91, 93, 96, 97, 110, 114, 115, 121, 123, 133
Tupelo, black, 24, 203
Tupelo, water, 24, 49, 200, 201

U
Udalfs, 49, 52, 59, 113, 115, 167, 187–193
Udepts, 12, 59, 121, 123, 167, 187–193
Uderts, 125, 126, 188–193
Udic soil moisture regime, 47, 54, 63, 67, 69, 70, 73, 74, 80, 82, 90, 96, 97, 121, 123, 133
Udifluvents, 24, 25, 32, 47, 50, 59, 88, 89, 94, 95, 107, 129, 131, 152, 158, 188–193
Udults, 54, 59, 109, 110, 144, 167, 187–193
Ultisols, 12–16, 32, 39, 47, 53, 54, 59, 65, 109, 110, 121, 137, 139, 143, 167, 187–193
Umbric epipedon, 12, 59
US Forest Service (USFS), 7, 17, 23, 49

V
Vertisols, 12–14, 16, 49, 51–53, 59, 61, 65, 125–127, 137, 138, 140, 167, 187–193
Vertization, 125, 127, 137, 138, 140, 167

W
Web Soil Survey (WSS), 10, 169
Woodall Hill, Mountain, 39, 44

Y
Yellow poplar, 24, 158, 201, 202, 204, 205

GPSR Compliance

The European Union's (EU) General Product Safety Regulation (GPSR) is a set of rules that requires consumer products to be safe and our obligations to ensure this.

If you have any concerns about our products, you can contact us on ProductSafety@springernature.com

In case Publisher is established outside the EU, the EU authorized representative is:

Springer Nature Customer Service Center GmbH
Europaplatz 3
69115 Heidelberg, Germany

Batch number: 10731528

Printed by Printforce, the Netherlands